Lecture Notes in Physics

The Editorial Policy for Proceedings

The series Lecture Notes in Physics reports new developments in physical research and teaching – quickly, informally, and at a high level. The proceedings to be considered for publication in this series should be limited to only a few areas of research, and these should be closely related to each other. The contributions should be of a high standard and should avoid lengthy redraftings of papers already published or about to be published elsewhere. As a whole, the proceedings should aim for a balanced presentation of the theme of the conference including a description of the techniques used and enough motivation for a broad readership. It should not be assumed that the published proceedings must reflect the conference in its entirety. (A listing or abstracts of papers presented at the meeting but not included in the proceedings could be added as an appendix.)

When applying for publication in the series Lecture Notes in Physics the volume's editor(s) should submit sufficient material to enable the series editors and their referees to make a fairly accurate evaluation (e.g. a complete list of speakers and titles of papers to be presented and abstracts). If, based on this information, the proceedings are (tentatively) accepted, the volume's editor(s), whose name(s) will appear on the title pages, should select the papers suitable for publication and have them refereed (as for a journal) when appropriate. As a rule discussions will not be accepted. The series editors and Springer-Verlag will normally not interfere with the detailed editing except in fairly obvious cases or on technical matters.

Final acceptance is expressed by the series editor in charge, in consultation with Springer-Verlag only after receiving the complete manuscript. It might help to send a copy of the authors' manuscripts in advance to the editor in charge to discuss possible revisions with him. As a general rule, the series editor will confirm his tentative acceptance if the final manuscript corresponds to the original concept discussed, if the quality of the contribution meets the requirements of the series, and if the final size of the manuscript does not greatly exceed the number of pages originally agreed upon.

The manuscript should be forwarded to Springer-Verlag shortly after the meeting. In cases of extreme delay (more than six months after the conference) the series editors will check once more the timeliness of the papers. Therefore, the volume's editor(s) should establish strict deadlines, or collect the articles during the conference and have them revised on the spot. If a delay is unavoidable, one should encourage the authors to update their contributions if appropriate. The editors of proceedings are strongly advised to inform contributors about these points at an early stage.

The final manuscript should contain a table of contents and an informative introduction accessible also to readers not particularly familiar with the topic of the conference. The contributions should be in English. The volume's editor(s) should check the contributions for the correct use of language. At Springer-Verlag only the prefaces will be checked by a copy-editor for language and style. Grave linguistic or technical shortcomings may lead to the rejection of contributions by the series editors.

A conference report should not exceed a total of 500 pages. Keeping the size within this bound should be achieved by a stricter selection of articles and not by imposing an upper limit to the length of the individual papers.

Editors receive jointly 30 complimentary copies of their book. They are entitled to purchase further copies of their book at a reduced rate. As a rule no reprints of individual contributions can be supplied. No royalty is paid on Lecture Notes in Physics volumes. Commitment to publish is made by letter of interest rather than by signing a formal contract. Springer-Verlag secures the copyright for each volume.

The Production Process

The books are hardbound, and quality paper appropriate to the needs of the authors is used. Publication time is about ten weeks. More than twenty years of experience guarantee authors the best possible service. To reach the goal of rapid publication at a low price the technique of photographic reproduction from a camera-ready manuscript was chosen. This process shifts the main responsibility for the technical quality considerably from the publisher to the authors. We therefore urge all authors and editors of proceedings to observe very carefully the essentials for the preparation of camera-ready manuscripts, which we will supply on request. This applies especially to the quality of figures and halftones submitted for publication. In addition, it might be useful to look at some of the volumes already published.

As a special service, we offer free of charge LATEX and TEX macro packages to format the text according to Springer-Verlag's quality requirements. We strongly recommend that you make use of this offer, since the result will be a book of considerably improved technical quality.

To avoid mistakes and time-consuming correspondence during the production period the conference editors should request special instructions from the publisher well before the beginning of the conference. Manuscripts not meeting the technical standard of the series will have to be returned for improvement.

For further information please contact Springer-Verlag, Physics Editorial Department V, Tiergartenstrasse 17, W-6900 Heidelberg, FRG

V. J. Martínez   M. Portilla   D. Sáez   (Eds.)

New Insights into the Universe

Proceedings of a Summer School
Held in València, Spain
23-27 September 1991

Springer-Verlag
Berlin Heidelberg GmbH

Editors

Vicent J. Martínez
Departament de Matemàtica Aplicada i Astronomia
Universitat de València
46100 Burjassot, València, Spain

Miquel Portilla
Diego Sàez
Departament de Física Teòrica
Universitat de València
46100 Burjassot, València, Spain

ISBN 978-3-662-13908-0   ISBN 978-3-540-47296-4 (eBook)
DOI 10.1007/978-3-540-47296-4

Originally published by Springer-Verlag Berlin Heidelberg New York in 1992
Softcover reprint of the hardcover 1st edition 1992

Typesetting: Camera ready by author/editor using the TEX macro package
from Springer-Verlag Berlin Heidelberg GmbH
2158/3140-543210 - Printed on acid-free paper

Preface

This volume contains the lectures presented at the summer school "New Insights into the Universe" held in Valencia, 23 – 27 September 1991, at the Universidad Internacional Menéndez Pelayo.

The seminar was organized with the aim of facilitating interaction of Ph.D. students interested in modern aspects of cosmology with the community of experienced cosmologists. The contact was achieved successfully and we hope that the publication of this volume will extend it to a larger number of people interested in cosmological research.

The main topics overviewed are: the spatial distribution of galaxies and peculiar velocity fields, the lens effect and its cosmological observational consequences, and the microwave background radiation: spectrum and anisotropies.

Each of these subjects has been developed in various lectures. We asked the authors for a common structure in the presentation of the topics, giving firstly a general theoretical framework, and then an updated account of the observational status of the field. We thank them for the care they have had following our suggestion.

Bruce Patridge presented a nice basic lecture about the microwave background radiation, covering both theoretical and observational aspects, but as much of it was presented at a previous meeting held in Rio Janeiro, his lecture is not included in this volume. It will be published elsewhere by Kluwer. The lectures on the anisotropies of the cosmic background radiation by Nicola Vittorio had to be omitted, since it was not possible for Prof. Vittorio to prepare a manuscript at present.

We take here the opportunity to thank the Universidad Internacional Menéndez Pelayo and the Universitat de València for financial support and administrative assistance. We also thank both lecturers and students for their interest in the lectures and discussions, and Prof. R. Kippenhahn, who refereed the volume for publication by Springer Verlag.

València
May 13, 1992

V. J. Martínez
M. Portilla
D. Sáez

Contents

List of Participants

Aguirre García, Alberto	Madrid
Alba Villegas, Salvador	Valencia
Alberola Bufante, Amparo	Valencia
Alberola Vercher, José L.	Valencia
Atrio Barandela, Fernando	Salamanca
Ballesteros Roselló, Fernando J.	Valencia
Bardelli, Sandro	Bologna
Bartual Sanfeliu, M. José	Valencia
Baugh, Carlton	Oxford
Bernard Mañez, Carme	Valencia
Bernardeau, Francis	Paris
Bertoni, Carlo	Bologna
Blanco Solsona, Antonio	Valencia
Boronat Zarzeño, M. Soledad	Valencia
Brauer, Uwe	München
Buchert, Thomas	München
Campos Sancho, Beatriz	Valencia
Canelli, Carlo	Italy
Captyn, Dorte	Copenhagen
Carbo Aguilar, Santiago	Valencia
Carrasco Cordero, Maria José	Salamanca
Casasus Lacoma, Elena	Valencia
Castillo Fraile, Manuel	Barcelona
Catelan, Paolo	Trieste
Cayon Trueba, Laura	Santander
Cerdá Jordá, Salvador	Valencia
Cervera García, Oscar	Valencia
Clement León, Rosa	Valencia
Chaveli Gascón, José A.	Valencia
Chodorowski, Michal	Warsaw
Domínguz Tenreiro, Rosa	Madrid
Davies Andrew, G.	London
Domenech Carbo, Antonio	Valencia
Domenech Carbo, M. Teresa	Valencia
Estellés Villanova, Ramón	Valencia
Ferrando Bargues, Joan F.	Valencia
Fullana Alfonso, Marius Josep	Valencia

Garcia Dauder, César	Madrid
García García, F. Javier	Sevilla
Gasque Garcia, Eusebio	Valencia
Girardi, Marisa	Trieste
Gómez Berzosa, Ana P.	Valencia
Gómez Flechoso, M. Angeles	Madrid
González Casado, Guillermo	Barcelona
González Fernández, Antonio	Sevilla
González Sánchez, Alejandro	Brighton
Gutierrez de la Cruz, Carlos M.	La Laguna
Guzzo, Luigi	Padova
Ebeling, Harald	München
Hernaiz Guijarro, Moises	Valencia
Hernández Pastora, José L.	Salamanca
Hjorth, Jens	Aarhus
Ibañez Cabanell, José Maria	Valencia
Iglesias Paramo, Jorge	Madrid
Jorge Lázaro, Alejandro	Valencia
Lapiedra Civera, Ramon	Valencia
Leroy Faris, Peter	Valencia
Liern Carrión, Vicente	Valencia
López Pastor, Juan J.	Valencia
Lorrimer, Stephen J.	Durham
Mann, Robert G.	Edinburgh
Manrique Oliva, Alberto	Barcelona
Martínez, Vicent J.	Valencia
Martínez Carrasco, Pedro	Sevilla
Miralles Canals, Juan José	Valencia
Miyaji Takamitsu	Maryland
Molinari, Emilio	Milano
Mollerach, Silvia	Trieste
Monzó González, M. Carmen	Valencia
Moreno Mendez, José F.	Valencia
Murante, Giuseppe	Torino
Muriel, Herman	Toronto
Paredes Hernández, Silvestre	Valencia
Pedersen, Kristian	Copenhagen
Piccinelli Bocchi, Gabriella	México
Pons Bordería, M. Jesús	Valencia
Portilla Moll, Miquel	Valencia
Quilis Quilis, Vicent	Valencia
Reig Torres, Pablo	Valencia
Rey Ponce de León, Rosa	Valencia
Riccabone, Giuliano	Torino

Rodrigo Tortola, M. José	Valencia
Rodriguez Salvador, Laura M.	Valencia
Romero Bauset, José Vicente	Valencia
Sáez Milán, Diego	Valencia
Saiz Martínez, Andrés	Valencia
Sánchez González, Edilberto	Salamanca
Sánchez Ochando, Ana	Valencia
Schiller, Peter	München
Serna Ballester, Arturo	Madrid
Serra Ricart, Miquel	Barcelona
Sheth, Ravi	Cambridge
Signes Signes, Teresa M.	Valencia
Simo Melendez, Raquel	Valencia
Solis Garcia del Pozo, Alicia	Valencia
Stengler, Erik	Cambridge
Stompor , Radoslaw	Warsaw
Stoop Ritsaert	Leiden
Toffolatti, Luigi	Santander
Tormen, Giuseppe	Padova
Totosa Grau, Leandro	Valencia
Van den Bosch, Frank	Leiden
Vilchez Gómez, Rosendo	Barcelona
Xiang-Ping Wu	Paris
Xiang, Shouping	Dublin
Zamora Mestre, Javier	Valencia
Zucca, Elena	Bologna

Statistics of Cosmological Density Fields

J. A. Peacock

Royal Observatory, Blackford Hill, Edinburgh EH9 3HJ.

Abstract: These lectures deal with the statistical description of density perturbations in cosmology, concentrating on methods for relating present-day observations to linear initial conditions. The main topics covered are (i) correlations and power spectra; (ii) Gaussian density fields and density maxima; (iii) the Press-Schechter mass function and extensions; (iv) arguments for and against biased galaxy formation; (v) low-dimensional surveys (skewers and slices); (vi) recent 3D clustering results; (vii) topology and cellular models; (viii) constraints on galaxy-formation models from high-redshift galaxies.

1 Introduction

The idea of these lectures is to cover the background material which is useful in calculations involving cosmological density perturbation fields and their observational consequences. Useful references for more of the fundamentals may be found in Peebles (1980) and Efstathiou (1990). We shall aim to illustrate these results in action in a variety of topics of current research interest.

There are two main categories of model for galaxy formation: **gravitational instability** and **seeds**. The former is essentially noise amplification: small perturbations at some early time grow to nonlinearity under their own self-gravitation; most attention in the cosmological literature has been devoted to this class of model. The alternative is to have structure form in a way which is 'born nonlinear', and one can distinguish two ways of achieving this: active or passive seeds. The latter encompasses objects that disappear after achieving their initial effect, leaving a pure gravitational/hydrodynamical problem (textures, explosions). The former remain on the scene, complicating the picture (strings, point masses). A vast range of possibilities for the initial conditions has been investigated, and we cannot hope to cover them all. The following remarks cover those topics which currently seem more promising.

1.1 Inflationary models

This is the most popular class of models. Perturbations are generated during the epoch of vacuum energy domination, when the Universe is in an exponentially expanding de Sitter phase. The fact that such a spacetime contains a true event horizon (as opposed to the particle horizon familiar in Friedmann models) leads to the generation of Hawking radiation – the process which generates emission from black holes. There are thus thermal fluctuations which manifest themselves as **curvature fluctuations** which have a constant amplitude on the horizon scale as long as the inflationary phase persists. Thus, a natural prediction of the inflationary model is that the Universe should contain primordial fluctuations of scale-invariant form. Because these were generated by fluctuations in spatial curvature, they are **adiabatic** in nature – affecting the number densities of matter and radiation equally (*i.e.* $\delta_r = 4\delta_m/3$). The great defect of the inflationary model is that the amplitude of the fluctuations is not predicted, and has to be supplied by adjusting the parameters of the scalar field which drives inflation, leading to the conclusion that the field must be very weakly coupled (see *e.g.* Brandenberger 1990 for more details). One can also tinker with the details of the potential to remove the simple prediction of adiabatic scale-invariance (*e.g.* Salopek *et al.* 1989), but this leaves very little predictive power.

1.2 Isocurvature models

The mode orthogonal to curvature fluctuations is termed **isocurvature**. This results from isothermal initial conditions, in which the matter density is perturbed, but not the radiation density. Since matter is a negligible constituent of the Universe at very early times, this generates no perturbation in the total density, only in the number of photons per particle, leading to the alternative name of **entropy perturbations**. Older discussions of this mode concentrated on the case of 'pure' isothermal fluctuations, but these do not exist: an isocurvature mode generates radiation perturbations as it evolves. It is less natural to produce isocurvature fluctuations than adiabatic ones: any GUT-based model in which the baryon asymmetry of the Universe is generated via baryon number violation will produce a constant entropy per baryon. Models have been suggested in which later phase transitions (*e.g.* the quark-hadron transition at which the axion may acquire a mass) can generate entropy fluctuations, but scale-invariance is a less automatic outcome. See *e.g.* Efstathiou (1990) for more details.

1.3 Topological seeds

These models generate structure during some high-temperature phase transition. This may be associated with inflation, but an inflationary phase is not required. The phase transition will yield a direction for the scalar field on its internal N-dimensional space which is initially random; minimisation of gradient energy will then attempt to align the field over distances of order the horizon length. However, this may not be possible if there are closed loops over which the scalar field makes a complete rotation, since these cannot be removed by local field transformations. The resulting defects depend on the internal dimensionality: domain walls for $N = 1$; strings for $N = 2$; monopoles for $N = 3$; and textures for $N = 4$. Most attention has been given to the cosmic string picture, but recently the texture picture has been seen as particularly promising. The especially attractive feature of these models is that they have less freedom than inflationary models. Most results depend just on the energy scale of symmetry breaking, ϕ_0. In particular, the texture picture predicts that the amplitude of horizon-scale fluctuations is

$$\delta_{\mathrm{H}} \simeq \left(\frac{\phi_0}{E_{\mathrm{planck}}} \right)^2. \tag{1}$$

The fact that this number is of order 10^{-4} may then be accounted for in a satisfying way, if the GUT energy turns out to be somewhat higher than the figures of 10^{15} GeV usually discussed. The difficulty with these models is that they are much harder to calculate with, being intrinsically non-linear. A collection of recent papers on the cosmic string model may be found in Gibbons *et al.* 1990. See *e.g.* Turok (1991) for more details of the Texture model.

The task of modern cosmology is to confront the predictions of these models with observations. Ideally we will find the one model which explains what we see; if all current models fail, at least we will understand better the requirements for the ultimate theory.

2 Basic tools

In this section, we cover some of the main concepts relating to cosmological density fields, plus some of the main statistical tools used in their analysis.

The subject is in a sense the study of a special kind of noise, analogous to static on a radio broadcast, or waves on water. The important property of this noise field is, following the cosmological principle, that it must be isotropic: the statistical description of the field should contain no preferred axes. Since perfect isotropy implies homogeneity, we are interested in models of the Universe which are homogeneous on average. What this means is that widely-separated parts of space have density fields which are independent, even though statistical properties such as the variance in density will be the same in each region. Homogeneity in an inhomogeneous Universe means that the expectation properties of the field are the same in all volumes. The actual field found in a given volume is a **realization** of the statistical process, and will inevitably have properties which differ from the global average. These global properties can be recovered in two ways: either by averaging over many realizations, or by averaging over one large volume. Fields which satisfy this property of volume average → ensemble average are termed **ergodic**. Giving a formal proof of ergodicity for a random process is not always easy; in cosmology it is perhaps best regarded as a common-sense axiom.

2.1 Perturbation growth

For convenience, we summarise some of the main results on the linear growth of density perturbations. See Peebles (1980) for derivations. For adiabatic modes in a flat Universe, growth is proportional to the square of the conformal time in both radiation- and matter-dominated regimes:

$$\delta \propto \eta^2; \qquad \eta \equiv \int \frac{dt}{a(t)}. \tag{2}$$

This law applies to modes of wavelength above the Jeans' length $\lambda_J = c_s\sqrt{\pi/G\rho}$; this is of order the horizon length prior to recombination and ~ 10 kpc thereafter. For models with $\Omega \neq 1$, the linear growing mode is

$$\delta \propto \frac{3}{x} - k - \frac{3\sqrt{1-kx}}{x^{3/2}} S_k^{-1}(x^{1/2}), \tag{3}$$

where

$$x \equiv |\Omega(z)^{-1} - 1| = \frac{k(\Omega - 1)}{\Omega(1 + z)} \tag{4}$$

and matter domination is assumed. The decaying mode is proportional to $\sqrt{1-kx}/x^{3/2}$. A reasonable approximation for this evolution is given in terms of the ratio of the density field to that for $\Omega = 1$, for the same current perturbation value

$$\delta(z) = \delta(z \mid \Omega = 1) \left(\frac{\Omega(z)}{\Omega}\right)^{0.65}. \tag{5}$$

Thus, a Universe with $\Omega = 0.2$ implies fluctuations at recombination (when $\Omega(z) = 1$) which are roughly a factor 3 higher than in a flat Universe.

For isocurvature modes, the matter perturbation declines instead of growing while the mode is outside the horizon. For $\Omega = 1$, the behaviour can be expressed most neatly in terms of $y \equiv \rho_m/\rho_r$:

$$\delta_m/\delta_i = \frac{4}{y} - \frac{8}{y^2}(\sqrt{1+y}-1) \simeq 1 - y/2 + \cdots, \tag{6}$$

where δ_i is the value at $t = 0$. Once the Universe is matter-dominated, these modes match onto adiabatic growing modes after entering the horizon. The net result for the matter distribution today is similar to that produced from adiabatic fluctuations, but with different relative amplitudes of modes as a function of wavelength.

The peculiar velocities produced by this gravitational evolution are (for the growing mode)

$$\delta \mathbf{v} = \frac{-iaH}{k}g(\Omega)\,\hat{\mathbf{k}}\,\delta_k, \tag{7}$$

where $g = d\ln\delta/d\ln a$, and $g \simeq \Omega^{0.6}$ is an excellent approximation.

Non-linear evolution may be modelled using the case of a spherical over-density, for which the radius-time relation is a cycloid

$$\begin{aligned} r &= A(1 - \cos\theta) \\ t &= B(\theta - \sin\theta) \end{aligned} \tag{8}$$

and the mass is fixed through $A^3 = GMB^2$. With this model, it is easy to show that collapse to a point of infinite density occurs at a time when (for $\Omega = 1$) the linear density contrast is $3(12\pi)^{2/3}/20 \simeq 1.686$. This provides some justification for the often exploited rule of thumb whereby one follows linear evolution until $\delta \simeq 1$ and then decrees a bound object to have formed.

2.2 Fourier transforms

Having studied the growth of one mode, we now move to the construction of a general field by the superposition of many modes. For a flat comoving geometry, the natural way of achieving this is via Fourier analysis. For other models, plane waves are not a complete set and one should use instead the eigenfunctions of the wave equation in a curved space. Normally this complication is neglected: even in an open Universe, the difference only matters on scales of order the present-day horizon.

How do we make a Fourier expansion of an infinite density field? If the field was periodic within some box of side L, then we would just have a sum over wave modes:

$$F(\mathbf{x}) = \sum F_{\mathbf{k}}e^{-i\mathbf{k}\cdot\mathbf{x}}. \tag{9}$$

Now, if we let the box become arbitrarily large, then the sum will go over to an integral which incorporates the density of states in k-space – exactly as in statistical mechanics: The Fourier relations in n dimensions are thus

$$\boxed{\begin{aligned} F(x) &= \left(\frac{L}{2\pi}\right)^n \int F_k(k) \exp -i k \cdot \mathbf{x} \, d^n k \\ F_k(k) &= \left(\frac{1}{L}\right)^n \int F(x) \exp i k \cdot \mathbf{x} \, d^n x. \end{aligned}}$$

$$(10)$$

One advantage of this particular Fourier convention is that the definition of convolution is just a simple volume average, with no gratuitous factors of $(2\pi)^{-1/2}$:

$$f * g \equiv \frac{1}{L^n} \int f(\mathbf{x} - \mathbf{y}) g(\mathbf{y}) d^n y. \tag{11}$$

Although one can make all manipulations on density fields which follow using either the integral or sum formulations, it is usually easier to use the sum. This saves having to introduce δ-functions in k-space. For example, if we have $f = \sum f_k \exp(-ikx)$, the obvious way to extract f_k is via $f_k = (1/L) \int f \exp(ikx) \, dx$: because of the harmonic boundary conditions, all oscillatory terms in the sum integrate to zero, leaving only f_k to be integrated from 0 to L. There is less chance of committing errors of factors of 2π in this way than considering $f = (L/2\pi) \int f_k \exp(-ikx) \, dk$ and then using $\int \exp[i(k - K)x] \, dx = 2\pi \delta_{\mathrm{D}}(k - K)$.

2.3 Correlation functions

For example, consider the important quantity

$$\xi(\mathbf{r}) \equiv \langle \delta(\mathbf{x})\delta(\mathbf{x} + \mathbf{r}) \rangle, \tag{12}$$

which is the autocorrelation function of the density field – usually referred to simply as the **correlation function**. The angle brackets indicate an averaging over the normalization volume V. If we express δ as a sum, and note that reality means we can replace one of the two δ's by its complex conjugate, then we obtain

$$\xi = \left\langle \sum_k \sum_{k'} \delta_k \delta_{k'}^* e^{i(k'-k)\cdot \mathbf{x}} e^{-i k \cdot \mathbf{r}} \right\rangle. \tag{13}$$

An alternative way of obtaining this is to use the relation between modes with opposite wavevectors which holds for any real field: $\delta_k(-\mathbf{k}) = \delta_k^*(\mathbf{k})$. By the periodic boundary conditions, however, all the cross terms with $\mathbf{k}' \neq \mathbf{k}$ average to zero. Expressing the remaining sum as an integral, we have

$$\boxed{\xi(\mathbf{r}) = \frac{V}{(2\pi)^3} \int |\delta_k|^2 e^{-i k \cdot \mathbf{r}} d^3 k.}$$

$$(14)$$

In short, the correlation function is the Fourier transform of the **power spectrum**. We shall hereafter often use the alternative notation $P(k) \equiv |\delta_k|^2$.

Now, in an isotropic universe, the density perturbation spectrum should contain no preferred direction, and so we must have an **isotropic power spectrum**: $|\delta_k|^2(\mathbf{k}) = |\delta_k|^2(k)$. We can therefore perform the angular integral: introduce spherical polars with the polar axis along $\mathbf{k}$, and use the reality of ξ so that $e^{-i\mathbf{k}\cdot\mathbf{x}} \to \cos(kr\cos\theta)$. In three dimensions, this yields

$$\xi(r) = \frac{V}{(2\pi)^3} \int |\delta_k|^2 \frac{\sin kr}{kr} 4\pi k^2 \, dk. \tag{15}$$

The 2D analogue of this formula is

$$\xi(r) = \frac{A}{(2\pi)^2} \int |\delta_k|^2 J_0(kr) 2\pi k \, dk. \tag{16}$$

2.4 Power spectra

We shall usually express the power spectrum in dimensionless form, as the variance per $\ln k$ ($\Delta^2 = d\sigma^2/d\ln k \propto k^3 P[k]$):

$$\Delta^2(k) \equiv \frac{V}{(2\pi)^3} 4\pi k^3 \, P(k) = \frac{2}{\pi} k^3 \int_0^\infty \xi(r) \frac{\sin kr}{kr} r^2 \, dr. \tag{17}$$

With a pure power-law correlation function; $\xi(r) = (r/r_0)^{-\gamma}$, the corresponding 3D power spectrum is

$$\Delta^2(k) = \frac{2}{\pi} (kr_0)^\gamma \, \Gamma(2-\gamma) \sin \frac{(2-\gamma)\pi}{2} \equiv \beta(kr_0)^\gamma \tag{18}$$

($= 0.903(kr_0)^{1.8}$ if $\gamma = 1.8$). The power-law spectrum is conventionally expressed in terms of an index n via

$$P(k) \propto k^n \;\Rightarrow\; \Delta^2(k) \propto k^{3+n}. \tag{19}$$

Asymptotic homogeneity clearly requires $n > -3$. An upper limit on n comes from an argument due to Zeldovich. Suppose we begin with a totally uniform matter distribution and then group it into discrete chunks as uniformly as possible; the Fourier transform of the resulting perturbations looks like

$$\delta_k = \frac{1}{V} \int \delta(\mathbf{x}) \exp i\mathbf{k}\cdot\mathbf{x} \, d^3x. \tag{20}$$

Now, the Taylor expansion of $\exp i\mathbf{k}\cdot\mathbf{x}$ is $1 + i\mathbf{k}\cdot\mathbf{x} - (\mathbf{k}\cdot\mathbf{x})^2/2 + \cdots$; the integral over the first two terms vanishes by conservation of mass and momentum, but the third survives. This tells us that, if we try to create a power spectrum which goes to zero at small wavelengths more rapidly than $\delta_k \propto k^2$, we will fail. Thus, discreteness of matter produces the **minimal spectrum**: $n = 4$.

More plausible alternatives lie between these extremes. The value $n = 0$ corresponds to **white noise**: the same power at all wavelengths. This is also known as the **Poissonian** power spectrum, because it corresponds to fluctuations between different cells which scale as $1/\sqrt{M_{\text{cell}}}$ (see below). A density field created by throwing down a large number of point masses at random would therefore consist of white noise. Most important of all is the **scale-invariant** spectrum (also termed **Harrison-Zeldovich**). This corresponds to the value $n = 1$, *i.e.* $\Delta^2 \propto k^4$. To see where the name arises, consider perturbations in gravitational potential:

$$\nabla^2 \delta \Phi = 4\pi G \rho_0 \delta. \tag{21}$$

The two powers of k pulled down by ∇^2 mean that, if $\Delta^2 \propto k^4$ for matter, then δ_Φ^2 is a constant. Since potential perturbations govern the flatness of spacetime, this says that the scale-invariant spectrum corresponds to a metric which is fractal-like: it has the same degree of 'wrinkliness' on each resolution scale. The total curvature fluctuations diverge, but only logarithmically at either extreme of wavelength.

2.5 Filtering and moments

A common concept in the manipulation of cosmological density fields is that of **filtering**: convolution of the density field with some **window function**: $\delta \rightarrow \delta * f$. Many observable results can be expressed in this form. Some common 3D filter functions are

$$\text{Gaussian}: f = \frac{V}{(2\pi)^{3/2} R_G^3} e^{-r^2/2R_G^2} \Rightarrow f_k = e^{-k^2 R_G^2/2}$$

$$\text{Top-hat}: f = \frac{3V}{4\pi R_T^3} \quad (r < R_T) \Rightarrow f_k = \frac{3}{y^3}[\sin y - y \cos y] \quad (y \equiv k R_T)$$

$$\tag{22}$$

Note the factor of V in the definition of f; this is needed to cancel the $1/V$ in the definition of convolution. For some power spectra, the difference in these filter functions at large k is unimportant, and we can relate them by equating the expansions near $k = 0$, where $1 - |f_k|^2 \propto k^2$. This equality requires

$$R_T = \sqrt{5}\, R_G. \tag{23}$$

We are often interested not in the convolved field itself, but in its variance, for use as a statistic (*e.g.* to measure the rms fluctuations in the number of objects in a cell). By the convolution theorem, this means we are interested in a **moment** of the power spectrum times the squared filter transform. We shall generally use the following notation:

$$\sigma_n^2 \equiv \frac{V}{(2\pi)^3} \int P(k) |f_k|^2 k^{2n} d^3k;$$

(24)

the filtered variance is thus σ_0^2 (which we shall often denote by just σ^2). Moments may also be expressed in terms of the correlation function over the sample volume:

$$\sigma^2 = \int\int \xi(|x - x'|) f(x) f(x') d^3x \, d^3x'.$$

(25)

To prove this, it is easiest to start from the definition of σ^2 as an integral over the power spectrum times $|f_k|^2$, write out the Fourier representations of P and f_k, and use $\int \exp ik \cdot (x - x' + r) \, d^3k = (2\pi)^3 \delta_D^{(3)}(x - x' + r)$. Finally, it is also sometimes convenient to express things in terms of derivatives of the correlation function at zero lag. Odd derivatives vanish, but even derivatives give

$$\xi^{(2n)}(0) = (-1)^n \frac{\sigma_n^2}{2n+1}.$$

(26)

2.6 Transfer functions

Real power spectra result from modifications of any primordial power by a variety of processes: growth under self-gravitation, effects of pressure and dissipative processes. In general, modes of short wavelength have their amplitudes reduced relative to those of long wavelength in this way. The overall effect is encapsulated in the **transfer function**, which gives the ratio of the late-time amplitude of a mode to its initial value. The detailed result can be hard to calculate, mainly because we have a mixture of matter (both collisionless dark particles and baryonic plasma) and relativistic particles (collisionless neutrinos and collisional photons) which does not behave as a simple fluid. Particular problems are caused by the change in the photon component from being a fluid tightly coupled to the baryons by Thomson scattering, to being collisionless after recombination. Accurate results require a solution of the Boltzmann equation to follow the evolution in detail. The transfer function is thus a by-product of elaborate numerical calculations of microwave background fluctuations. Nevertheless, once we possess the transfer function, it is a most valuable tool. The evolution of linear perturbations back to last scattering obeys the simple relations summarised above, and it is easy to see how structure in the Universe will have changed during the matter-dominated epoch.

It is thus invaluable in practice to have some accurate analytic formulae which fit the numerical results for transfer functions. We give below results for some common models in the form of the transfer function needed to produce a scale-invariant power spectrum at large wavelength, $\Delta^2 \propto k^4 T_k^2$. For adiabatic models, T_k is the true transfer function; for isocurvature models, this is not the case and T_k is proportional to $1/k^2$ times the true transfer function. We assume $\Omega_B \ll \Omega$, so that all lengths scale with the horizon size at matter-radiation equality, leading to the definition $q \equiv k/(\Omega h^2 \mathrm{Mpc}^{-1})$. We consider the cases of (A) Adiabatic CDM; (B) Adiabatic massive neutrinos (1 massive, 2 massless); (C) Isocurvature CDM; and (D) Textures + CDM. The first three expressions come from Bardeen et al. (1986; BBKS); the last is a fit to the graph in Turok (1991).

(A) $T_k = \dfrac{\ln(1 + 2.34q)}{2.34q}[1 + 3.89q + (16.1q)^2 + (5.46q)^3 + (6.71q)^4]^{-1/4}$

(B) $T_k = \exp(-3.9q - 2.1q^2)$

(C) $T_k = (1 + [15.0q + (0.9q)^{3/2} + (5.6q)^2]^{1.24})^{-1/1.24}$

(D) $T_k = (1 + [19.4q + (6.6q)^{3/2} + (3.0q)^2]^{1.20})^{-1/1.20}$

$$(27)$$

Some plots of these transfer functions are given in figure 1.

2.7 Normalization

For scale-invariant spectra, a natural amplitude measure is the (constant) gravitational potential variance per unit $\ln k$:

$$\epsilon^2 \equiv \frac{V}{(2\pi)^3} \, 4\pi k^3 |\Phi_k|^2 / c^4 = \frac{9}{4}\left(\frac{ck}{H_0}\right)^{-4} \Delta^2(k). \qquad (28)$$

Two more commonly encountered measures relate to the clustering field around 10 Mpc. One is σ_8: the rms density variation when smoothed in spheres of radius $8h^{-1}$ Mpc; this is observed to be very close to unity. The other is an integral over the correlation function:

$$J_3 \equiv \int_0^r \xi(y)\, y^2\, dy = \int \Delta^2(k) W(k)\, \frac{dk}{k}, \qquad (29)$$

where $W(k) = (\sin kr - kr \cos kr)/k^3$. The observed value of this is $J_3(10h^{-1}\mathrm{Mpc}) = 277h^{-3}$ Mpc (from the CfA survey: Davis & Peebles 1983). Taking out a factor $r^3/3$ to make $W(0) = 1$, we see that this is very close in content to $\sigma_8 = 1$. However, it has been claimed that J_3 is rather insensitive to non-linear evolution of the density field on small scales, and so this is the measure which is often used to normalize cosmological theories.

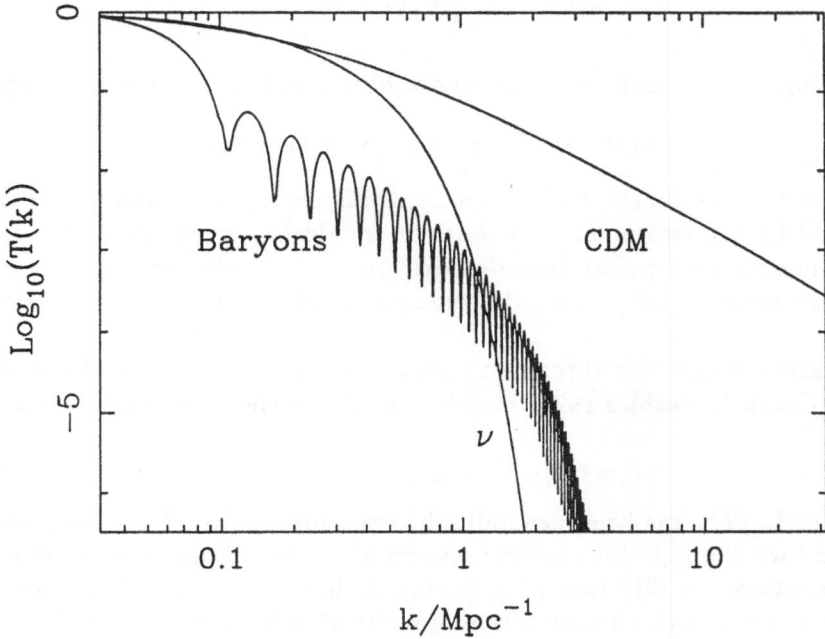

Fig. 1. A plot of transfer functions for various adiabatic models. The transfer functions for isocurvature CDM and CDM + textures have been omitted for clarity, as they are relatively similar in appearance to the adiabatic case.. As well as the smooth curves corresponding to dark matter models, we also show the case of an adiabatic pure baryon Universe. The plot assumes $\Omega h^2 = 1$; for the dark matter models, the wavenumber scales proportional to Ωh^2. The scaling for baryonic models is more complicated; empirically, it approximately obeys $k \propto (\Omega h^2)^{1/2}$.

2.8 *N*-point correlations

An alternative definition of the autocorrelation function is as the **two-point correlation function**, which gives the excess probability for finding a neighbour a distance r from a given galaxy. By regarding this as the probability of finding a pair with one object in each of the volume elements dV_1 and dV_2,

$$dP = \rho_0^2 [1 + \xi(r)] \, dV_1 \, dV_2, \tag{30}$$

this is easily seen to be equivalent to the autocorrelation definition of ξ. There is a straightforward extension to larger numbers of points: the probability of finding an n-tuplet of galaxies in n specified volumes is

$$dP = \rho_0^n [1 + \xi^{(n)}] \, dV_1 \ldots dV_n. \tag{31}$$

As with the two-point function, the probability is proportional to the product of the density field at the n points, and so

$$1 + \xi^{(n)} = \left\langle \prod_i (1 + \delta_i) \right\rangle . \tag{32}$$

Expanding the product gives a sequence of terms. For $n = 3$, for example,

$$\xi^{(n)} = \xi(r_{12}) + \xi(r_{23}) + \xi(r_{31}) + \zeta(\mathbf{r}_1, \mathbf{r}_2, \mathbf{r}_3). \tag{33}$$

The term ζ $(= \langle \delta_1 \delta_2 \delta_3 \rangle)$, which represents any excess correlation over that described by the two-point contributions we already know about, is called the **reduced three-point correlation function**. Similar expressions exist for larger numbers of points; the reduced n-point correlation function is $\langle \prod \delta_i \rangle$.

Observationally, the three-point function is non-zero. It has been suggested (Groth & Peebles 1977) that the results fit the so-called hierarchical form

$$\zeta = Q(\xi_{12}\xi_{23} + \xi_{23}\xi_{31} + \xi_{31}\xi_{12}), \tag{34}$$

with $Q \simeq 1$. This has been generalised to the **hierarchical ansatz**, which assumes that the reduced correlations can all be expressed in terms of two-point functions in this way (*e.g.* Balian & Schaeffer 1989). However, no convincing derivation of such a situation starting from linear initial conditions has ever been given.

The reduced functions are sometimes called the **connected correlation functions** by analogy with Greens functions in particle physics. The analogy allows diagrammatic techniques borrowed from that field to be used to aid calculation (Grinstein & Wise 1986). To describe all statistical properties of the density field requires the whole hierarchy of correlation functions (White 1979), which is a problem since only the first few have been measured. However, this is not a problem for Gaussian fields (see below), where all reduced correlations above the two-point level either vanish (if odd), or are expressible in terms of two-point functions (if even). It is because of this relation to the two-point function, and hence to the power spectrum, that the latter is often described as the 'holy grail' of cosmology: it tells us all that there is to know about the statistical properties of the density field.

2.9 Evaluation of clustering statistics

The correlation function is usually measured starting from the definition as the excess probability of finding a pair

$$dP = \rho_0^2 \left[1 + \xi(r) \right] dV_1 dV_2. \tag{35}$$

To implement this definition, we need to calculate the expected numbers of pairs in the absence of clustering, taking into account the sample limits. In practice, this number is usually estimated by creating a random catalogue much larger in number than the sample under study, and by counting pairs either within the two catalogues or between catalogues, giving two estimators for $1 + \xi$ as a ratio of pair counts, usually symbolized by

$$1 + \xi_1 = \langle DD \rangle / \langle RR \rangle$$
$$1 + \xi_2 = \langle DD \rangle / \langle DR \rangle. \tag{36}$$

In both cases one is effectively just measuring the expected number of pairs by monte-carlo integration. For a large volume, both methods should be equivalent; however, for a small sample the latter is more robust as it is less sensitive to whether there is a rich cluster close to the sample boundary.

Having made an estimate of $\xi(r)$, we now need to consider error bars. In the absence of clustering, $\langle \xi \rangle = 0$ and $\langle \xi^2 \rangle = 1/N_p$, where N_p is the number of independent pairs in a given bin of radius (Peebles 1980). For non-zero ξ, this suggests the usual 'Poisson error bar'

$$\frac{\Delta \xi}{1 + \xi} = \frac{1}{\sqrt{N_p}}. \tag{37}$$

This will usually be a lower limit to the uncertainty in ξ; Peebles (1973) shows that the right-hand side should be increased by a factor of approximately $1 + 4\pi n J_3$, $4\pi J_3$ being the volume integral of ξ out to the radius of interest, and n being the number density (see also Kaiser 1986). The problem with this expression is that J_3 may be hard to estimate. It has been suggested (Ling, Frenk & Barrow 1986) that the correct errors can be estimated via the bootstrap resampling method. However, even this provides an under-estimate: one is always concerned with sampling errors in the sense that large-scale variations in properties of the density field cause the area under study not to be a totally fair sample.

In an ideal sample, the correlation function can be transformed to give the power spectrum directly. However, in the presence of the noise inevitable with limited data, this is not always the best way of proceeding: it can be better to measure the power spectrum directly. Think about the Fourier relation between ξ and Δ^2: $\xi = \int \Delta^2 (\sin kr/kr) dk/k$. The result for ξ at large separations is a mixture of modes, with a large 'leakage' of high-k power because the window function $\sin kr/kr$ decines rather slowly. Since the small-scale power is often large, this can swamp any signal from low k. Worse, the *uncertainties* on ξ are dominated by this small-scale term, so it is hard to subtract the unwanted small-scale contribution with any accuracy. A further problem on large scales is that $\xi(r)$ is very sensitive to assumptions about the mean density (because $1 + \xi \propto \bar{n}^{-1}$ and $\xi \ll 1$); it is impossible to measure ξ once it falls below the uncertainty in $\bar{n}$. In contrast, uncertainty in the DC level of the density field just scales all power-spectrum coefficients by the same factor. So, suppose we do the simplest thing and take the discrete transform of the data. In the continuum case, we would like to evaluate the Fourier coefficients

$$\delta_k = \frac{1}{V} \int \delta(\mathbf{x}) \exp i\mathbf{k} \cdot \mathbf{x} \, d^3x; \qquad (38)$$

in practice, with a discrete set of N galaxies, we take the sum

$$\delta_k = \sum N^{-1} \exp i\mathbf{k} \cdot \mathbf{x}. \qquad (39)$$

In the absence of clustering, these coefficients execute a random walk on the Argand plane. The expectation value of the power is evaluated by splitting 3D space into a large number of cells with occupation numbers $n_i = 0$ or 1 (*cf.* Peebles 1980)

$$\langle |\delta_k|^2 \rangle = \sum n_i^2 / N^2 = 1/N. \qquad (40)$$

For $N \gg 1$ the central limit theorem yields the Rayleigh distribution for a single mode:

$$P(|\delta_k|^2 > X) = \exp -NX. \qquad (41)$$

This is the distribution of **Poisson** or **shot noise** which will overlay the true clustering signal that we are trying to measure. However, if we just subtract the shot noise from the power: $P \to P - 1/N$, this does not give the power spectrum of δ, since we do not have a uniform cubical survey. Suppose a varying background density is produced by selection of some varying fraction of objects, $n_b = f(\mathbf{x})\bar{n}$. The corrections this introduces are, in outline

(1) The Fourier coefficients obtained are proportional to the transform of $f(\mathbf{x})[1 + \delta(\mathbf{x})]$, so the transform of f needs to be subtracted.
(2) The selection function has a smaller effective volume than any cube in which it is embedded, which affects the amplitude of the power spectrum.

We now expand on these points a little (see Peacock & Nicholson 1991 for details): We first need to subtract the transform of the selection function, $f(\mathbf{x})$, which vanishes in the usual case of a uniform distribution:

$$\delta_k = \sum N^{-1} \exp i\mathbf{k} \cdot \mathbf{x}_i - \frac{1}{N} \int n_b(\mathbf{x}) \exp i\mathbf{k} \cdot \mathbf{x} \, d^3x. \qquad (42)$$

We can now subtract the Poisson contribution from the power; the quantity

$$P(k) \equiv \sum |\delta_k|^2 - \frac{m}{N} \qquad (43)$$

may therefore seem at first sight to be a good estimator for any true power in a region of k space containing m modes.

However, we have not yet properly accounted for the selection function. This has the effect of convolving our Fourier coefficients with the transform of the selection function. In the absence of phase correlations between clustering pattern and mask (fair sample hypothesis), we end up with just a convolution of power spectra:

$$P_{\text{obs}} = P_{\text{true}} * |f_k|^2.$$

(44)

Apart from smoothing the power spectrum, this also changes the normalization, as may be seen by considering the case of a uniform cubical survey of volume V embedded in a larger cube of volume V'. Using the density of states for the larger volume to count modes causes a multiple counting of modes, and therefore causes $P(k)$ to be too high by a factor V'/V. In general, we must therefore apply the correction

$$P(k) \rightarrow P(k)\frac{[\int f \, d^3x]^2}{\int f^2 \, d^3x \, \int d^3x}.$$

(45)

For a uniform survey in which Poisson noise dominates, Webster (1976) proved that different δ_k are uncorrelated and the result of summing over a band in k space containing m modes should be an approximately Gaussian variable with variance $2m/N^2$ (the factor 2 arises because only half the modes are independent owing to the reality of the density field). This procedure may be used to yield minimal Poisson error bars on the power, once the effective number of independent modes is scaled as above to allow for the selection function.

2.10 Integral constraints

We should now say a little more about the normalization effects which arise because the mean density is unknown and must be estimated from the sample to hand. This can clearly cause a systematic under-estimate of power in modes whose wavelengths approach the size of the sample volume. Similar problems exist in estimating the correlation function within a finite volume, but are much more severe in that case. The combination $1 + \xi$ is inversely proportional to the mean density; for finite samples, this can be significantly affected by modes larger than the sample size. This means that it is hard in practice to measure ξ when it drops below the fractional uncertainty in the mean density. The effect can however be modelled, as follows. Since we use the observed mean number density, $\bar{n}$, instead of the global $\langle n \rangle$, we have

$$(1 + \xi)_{\text{true}} = (1 + \xi)_{\text{obs}}(\bar{n}/\langle n \rangle)^2,$$

(46)

and hence,

$$\langle (1 + \xi)_{\text{true}} \rangle = (1 + \xi)_{\text{obs}}(1 + \sigma^2),$$

(47)

where σ^2 is the fractional rms density variation over volumes the size of the sample. We have here ignored any correlation between ξ_{obs} and fluctuations in $\bar{n}$; this will be correct to the extent that they are much smaller than correlations between $\bar{n}$ and ξ_{true} for a given field. The final effect is very nearly to subtract a constant from ξ. To apply this correction, however, we need to know the contribution to σ^2 from modes of wavelength larger than the sample, which is a serious drawback of the method.

With the power spectrum, the main effect of any offset in mean density is just to scale the amplitudes of all modes by some factor. However, the relative amplitude of large-wavelength modes is affected in a more subtle way, as follows. We shift the true D.C. level of δ by subtracting the mean value over our sample:

$$\delta' = \delta - \int \delta(x)f(x) \, d^3x. \tag{48}$$

In what follows, we will need to distinguish carefully between the power spectrum of the true density field, $P(k)$, and the power spectrum of the renormalized field (P'), together with their convolved counterparts $(P_*$ and $P'_*)$. Self-normalization forces $P'_*(0)$ to be zero, by subtracting a spike at the origin. Since we have established that the power spectrum we estimate is the true one convolved with $|f_k|^2$, the spike is also convolved over and we have $P'_*(k) = P_*(k) - P_*(0)|f_k|^2$, i.e.

$$P'_*(k) = P_*(k) \left(1 - |f_k|^2 \frac{P_*(0)}{P_*(k)} \right). \tag{49}$$

The effect that this has depends on the power spectrum; consider power-law spectra with $P \propto k^n$. Clearly, for an $n = 0$ spectrum, the 'normalization damping' factor is just $(1 - |f_k|^2)$; only modes with $|f_k|^2 \gtrsim 0.1$ will be affected at all significantly. Some numerical experiments at convolution of different power laws with Gaussian $|f_k|^2$ functions show that this statement holds reasonably well for all realistic values of n. For spectra with $P(0) = 0$ (i.e. $n > 0$), the convolution process produces spurious power at $k = 0$, which is removed by the above process. The net result can be that $P'_*(k)$ is actually closer to the true $P(k)$ than is $P_*(k)$.

2.11 Nonlinear evolution

Most of the above discussion has concentrated on linear density fields; how is the power spectrum altered by nonlinear evolution? As with most nonlinear questions, this cannot be answered analytically in general, but there is a useful scaling solution.

The density field under full nonlinear evolution may be thought of as consisting of a set of collapsed, virialized clusters. What is the density profile of one of these objects? The linear density contrast averaged out to radius

r will vary as $r^{-(n+3)/2}$; this also gives the scaling for $1 + z$ at collapse, if we regard $\delta \sim 1$ as marking the collapse epoch. The current density within this radius will be some multiple of that at collapse, scaling as $(1+z)^3$ – *i.e.* $\propto r^{-(9+3n)/2}$. Now convert to proper distance $\ell \propto r/(1+z) \propto r^{(5+n)/2}$; this yields a density contrast proportional to $\ell^{-(9+3n)/(5+n)}$. Thinking of the correlation function as the excess probability of finding a neighbour, this also gives us the nonlinear form of the correlation function. Since the linear form looks like $\xi \propto r^{-(3+n)}$, the effects of nonlinearity may be expressed in terms of an effective index:

$$n_{\text{eff}} = -\frac{6}{5+n}. \tag{50}$$

The power spectrum resulting from power-law initial conditions will evolve self-similarly with this index. Note the narrow range predicted: $-2 < n_{\text{eff}} < -1$ for $-2 < n < +1$, with an $n = -2$ spectrum having the same shape in both linear and nonlinear regimes.

How is this clustering expected to evolve with redshift? In the linear regime, we expect the usual growth law

$$\xi(z) \propto (1+z)^{-2} \quad \text{(linear)}. \tag{51}$$

This implies that the length-scale of nonlinearity must evolve as

$$r_0 \propto (1+z)^{-2/(n+3)} \tag{52}$$

Requiring this to match onto the nonlinear $\xi \propto r^{-(3+n_{\text{eff}})}$ implies a different rate of evolution in that regime:

$$\xi(z) \propto (1+z)^{-6/(5+n)} \quad \text{(non} - \text{linear)}. \tag{53}$$

A simpler way of seeing this result is to say that we have **stable clustering**: ξ is fixed in *proper* terms apart from a $(1 + z)^{-3}$ scaling owing to the changing mean density of unclustered galaxies which dilute the clustering at high redshift. Thus, with $\xi \propto r^{-\gamma}$, we obtain the comoving evolution $\xi \propto (1+z)^{\gamma-3}$, which is the above result. Whether this evolution has been seen or not is presently controversial. Efstathiou *et al.* (1991) have observed a very low amplitude for the angular clustering of galaxies at $B \simeq 26$, inferring that (if $\Omega = 1$) the clustering must evolve very rapidly – at about the linear-theory rate. However, their fields are very small and it is possible their small result is not representative. An indication that clustering may not decline this rapidly is given by the observed clustering of quasars at $z \simeq 1$; Shanks *et al.* (1987) find the relatively high value $r_0 = 7h^{-1}$ Mpc.

2.12 Redshift-space distortions

The distortions caused by working in redshift space are relatively simple to analyse if we assume we are dealing with a distant region of space which subtends a small angle, so that radial distortions can be considered as happening along one cartesian axis. In this case, the apparent amplitude of any linear density disturbance is just (Kaiser 1987)

$$\delta_{\text{obs}} = \delta \left(1 + \frac{\Omega^{0.6}\mu^2}{b}\right), \tag{54}$$

where μ is the cosine of the angle between the wavevector and the line of sight ($\mu = \hat{\mathbf{r}} \cdot \hat{\mathbf{k}}$). The parameter b allows for bias: the set of objects under study may be more clustered than the mass ($\delta = b\delta_{\text{mass}}$). This is discussed in more detail below. In azimuthal average, the effect is just to boost the power spectrum (and hence correlation function) by a constant factor (28/15 if the clustering is unbiased and $\Omega = 1$). On small scales, this linear correction does not apply; the clustering pattern is smeared out by virialized random velocity dispersions. Assuming these to be Gaussian and characterised by some velocity dispersion σ_v, we now have

$$\delta \to \delta \exp\left[-\tfrac{1}{2}\left(\frac{k\mu\sigma_v}{H}\right)^2\right]. \tag{55}$$

In azimuthal average, this now gives

$$P(k) \to P(k)\frac{\sqrt{\pi}}{2}\frac{\text{erf}(k\sigma_v/H)}{k\sigma_v/H}. \tag{56}$$

Modes at high k are thus only damped by one power of k, rather than exponentially.

When working with the correlation function, the effect of the small-scale random velocities may be dealt with by using the correlation function evaluated explicitly as a 2D function of transverse (r_p) and radial (π) separation. The projection along the redshift axis is then independent of the velocities

$$w(r_p) = \int_{-\infty}^{\infty} \xi(r_p, \pi)\, d\pi = 2\int_{r_p}^{\infty} \xi(r)\frac{r\, dr}{(r^2 - r_p^2)^{1/2}}, \tag{57}$$

and has the Abel integral inverse

$$\xi(r) = -\frac{1}{\pi}\int_r^{\infty} w'(y)\frac{dy}{(y^2 - r^2)^{1/2}}. \tag{58}$$

Looking at the function in the redshift direction allows the velocity dispersion to be estimated; it comes out at a *relative* dispersion of about 300 kms^{-1} for pairs of ~ 1 Mpc separation (Davis & Peebles 1983).

Sometimes these complications are neglected, and correlations are calculated in redshift space assuming isotropy. The result is a small increase in scale-length, as power on small scales is transferred to separations of order the velocity smearing. The result is a scale length around $7h^{-1}$ Mpc for the redshift-space $\xi(s)$ as opposed to the $5h^{-1}$ Mpc which applies for $\xi(r)$. This is in addition to the linear amplification discussed above. Clearly, such effects are unimportant for objects such as clusters which display clustering which is an amplified version of the true 3D pattern (see below).

3 Gaussian density fields

Apart from statistical isotropy of the fluctuation field, there is another reasonable assumption we might make: that the phases of the different Fourier modes δ_k are uncorrelated and random. This corresponds to treating the initial disturbances as some form of random noise, analogous to Johnson noise in electrical circuits. In fact, many mathematical tools which have become invaluable in cosmology were indeed first established with applications to communication circuits in mind. The random-phase approximation has a powerful consequence, which derives from the central limit theorem: loosely, the sum of a large number of independent random variables will tend to be normally distributed. This will be true, not just for the field δ, but all quantities which are derived from linear sums over waves (such as field derivatives) will be jointly Gaussian. The result is a **Gaussian random field**, whose properties are characterised entirely by its power spectrum.

3.1 Realizations

One immediate example of a Gaussian variable related to a Gaussian field arises when we consider a realization of such a field: a sample within a finite volume. The Fourier coefficients for the field within the box are linear sums over the field, and hence will themselves be (two-dimensional, complex) Gaussian variables. Call these a_k, where $\delta = \sum a_k \exp i k \cdot x$. If we call the realization volume V_R to distinguish it from the arbitrary normalization volume, the expectation value of the realization power follows from fixing the power in a given region of k-space in the continuum limit:

$$\langle |a_k|^2 \rangle = \frac{V}{V_R} |\delta_k|^2. \tag{59}$$

The power in a given mode will not take this expectation value, but will be Rayleigh-distributed about it (because the real and imaginary parts of a_k will be Gaussian):

$$P(|a_k|^2 > X) = \exp(-X^2/\langle |a_k|^2 \rangle). \tag{60}$$

We now have the prescription needed to set up the conditions in k-space for creating a Gaussian realization with a desired power spectrum; assign Fourier amplitudes randomly with the above distribution, and phases randomly between 0 and 2π. The real-space counterpart may then be found efficiently via the Fast Fourier Transform (FFT) algorithm. In fact, the central limit means that, even if the dispersion in amplitudes is neglected, the field will still be closely Gaussian if many modes are used. However, its statistical properties on scales approaching the box size (where there are few modes) may be suspect.

An alternative, simpler, method is to create a spatial array of white noise: Gaussian density variations in which each pixel is independent and has unit variance. This field can then be given the desired statistical properties by convolving it with a function f which is the 'square root' of the desired correlation function (*i.e.* $f * f = \xi(r)$). The squared Fourier transform of f is thus just the desired power spectrum; by the convolution theorem, this is also the power spectrum of the convolved white-noise field. From the computational point of view, this is not competitive with the FFT approach in terms of speed, although it may be simpler to code. However, it is often useful as a calculational tool to think of random fields as being generated in this two-step manner.

3.2 Density maxima

To see the power of the Gaussian-field approach, consider the question of **density peaks**: regions of local maxima in $\delta(\mathbf{x})$. These are the points which will go non-linear first as the perturbation field develops. If we adopt the 'collapse at $\delta = \delta_c$' prescription, then the distribution of collapse redshifts can be deduced once we know the distribution of overdensities for density maxima. This can be found as follows; we shall only give the 1D case in much detail, as larger numbers of dimensions become messier to handle. See Peacock & Heavens (1985) and Bardeen *et al.* (1986) for details of the 3D case. The field and its derivatives are jointly Gaussian

$$p(\delta, \delta', \delta'', \cdots) = \frac{|\mathbf{M}|^{1/2}}{(2\pi)^{m/2}} \exp(-\tfrac{1}{2}\tilde{\mathbf{V}} \cdot \mathbf{M} \cdot \mathbf{V}), \qquad (61)$$

for m variables in a vector $\mathbf{V}$, where $\mathbf{M} \equiv \mathbf{C}^{-1}$ is the inverse of the covariance matrix. The elements of the covariance matrix depend only on moments over the power spectrum (putting $[(\delta, \delta'', \delta']$ in a column vector):

$$\mathbf{C} = \begin{pmatrix} \sigma_0^2 & -\sigma_1^2 & \\ -\sigma_1^2 & \sigma_2^2 & \\ & & \sigma_1^2 \end{pmatrix} \qquad (62)$$

where

$$\sigma_m^2 \equiv \left(\frac{L}{2\pi}\right)^n \int |\delta_k|^2 k^{2m} d^n k. \tag{63}$$

We will be interested in points with $\delta' = 0$. At first sight, these may appear not to exist, since the probability of having δ' exactly zero vanishes. The way out of this paradox is to replace the 'volume element' $d\delta\, d\delta''\, d\delta'$ by using the Jacobian to replace δ' by position: $d\delta\, d\delta''\, |\delta''|\, d^n x$; in n dimensions, $|\delta''|$ denotes the Jacobian determinant of the Hessian matrix, $\partial^2 \delta/\partial x_i \partial x_j$. We then have only to integrate over the region of δ'' space corresponding to peaks, and we are done. To express the answers neatly, define some (mainly dimensionless) parameters as follows:

$$\boxed{\begin{aligned} \nu &\equiv \frac{\delta}{\sigma_0} \\[1mm] \gamma &\equiv \frac{\sigma_1^2}{\sigma_0 \sigma_2} \\[1mm] R_* &\equiv \sqrt{n}\,\frac{\sigma_1}{\sigma_2}. \end{aligned}} \tag{64}$$

The meaning of these variables is as follows: ν is the 'height' of the field, in units of the rms; R_* is a measure of the coherence scale in the field; γ is a measure of the width of the power spectrum: $\gamma = 1$ corresponds to a shell in k-space.

If we also put $x \equiv -\delta''/\sigma_2$, then the number density of stationary points is

$$dN = \frac{e^{-Q/2}}{(2\pi)^{3/2}(1-\gamma^2)^{1/2}\, R_*}\, |x|\, dx\, d\nu; \qquad Q = \frac{(\nu - \gamma x)^2}{1-\gamma^2} + x^2, \tag{65}$$

with peaks corresponding to $x > 0$. Doing the ν integral first, followed by that over x, gives the total peak density as

$$N_{\mathrm{pk}} = \frac{1}{2\pi R_*}. \tag{66}$$

Doing the x integral first is a tedious operation; however, the result can in fact be integrated again to yield the integral peak density (Cartwright & Longuet-Higgins 1956):

$$P(> \nu) = \frac{1}{2}\left[\mathrm{erfc}\left(\frac{\nu}{\sqrt{2(1-\gamma^2)}}\right) + \gamma e^{-\nu^2/2}\left\{1 + \mathrm{erf}\left(\frac{\gamma\nu}{\sqrt{2(1-\gamma^2)}}\right)\right\}\right] \tag{67}$$

To follow the same procedure in higher numbers of dimensions is quite involved; details are given by Bardeen *et al.* (1986) and Bond & Efstathiou (1987). The final result can be expressed as

$$\frac{dN_{\text{pk}}}{d\nu} = \frac{1}{(2\pi)^{(n+1)/2}} R_*^n e^{-\nu^2/2} G(\gamma, \gamma\nu), \tag{68}$$

where

$$G(\gamma, \gamma\nu) = \int_0^\infty F(x) \frac{\exp(-\frac{1}{2}\frac{(x-\gamma\nu)^2}{(1-\gamma^2)})}{[2\pi(1-\gamma^2)]^{1/2}} \, dx. \tag{69}$$

In one dimension, $F(x) = x$ (Rice 1954), whereas $F(x) = x^2 + \exp(-x^2) - 1$ in two dimensions (Bond & Efstathiou 1987). $F(x)$ is given for $n = 3$ by BBKS equations (A1.9) & (4.5). The integral densities in higher dimensions come out to be

$$N_{\text{pk}} = \frac{1}{4\pi\sqrt{3}R_*^2} \quad \text{(2D)}$$

$$N_{\text{pk}} = \frac{29 - 6\sqrt{6}}{8\pi^2 5^{3/2} R_*^3} \simeq 0.016 R_*^{-3} \quad \text{(3D)}. \tag{70}$$

In general, the high-peak probability density comes out as

$$\boxed{N_{\text{pk}}(> \nu) = \frac{1}{\sqrt{2\pi}} \left(\frac{\gamma}{\sqrt{2\pi}R_*}\right)^n \nu^{n-1} e^{-\nu^2/2}.}$$

$$\tag{71}$$

It is easy to see how this form arises: the correlation between δ and δ'' means that, for very high peaks, all principal second derivatives become equal and scale $\propto \nu$ – leading both to round peaks and to the above result.

The above results are illustrated in figure 2. We see that the height distribution for peaks is much narrower than the unconstrained positive tail of a Gaussian. Since we may use the spherical model to justify taking $\delta \simeq 1$ to mark the creation of a bound object, and since δ grows as $(1 + z)^{-1}$, we can convert a distribution of ν to a distribution of collapse redshifts. The relatively small fractional dispersion in ν for peaks means that it is possible to speak of a reasonably well-defined epoch of galaxy formation.

3.3 Clustering of peaks

Another important calculation which can be performed with density peaks is to estimate the clustering of cosmological objects. Peaks have some inbuilt clustering as a result of the statistics of the linear density field: they are 'born clustered'. For galaxies, this clustering amplitude is greatly altered by the subsequent dynamical evolution of the density field, but this is not true for clusters of galaxies, which are the largest nonlinear systems at the current epoch. The fact that we recognise clusters simply because they are the most spectacularly large galaxy systems to have undergone gravitational collapse has an important consequence, as first realised by Kaiser (1984).

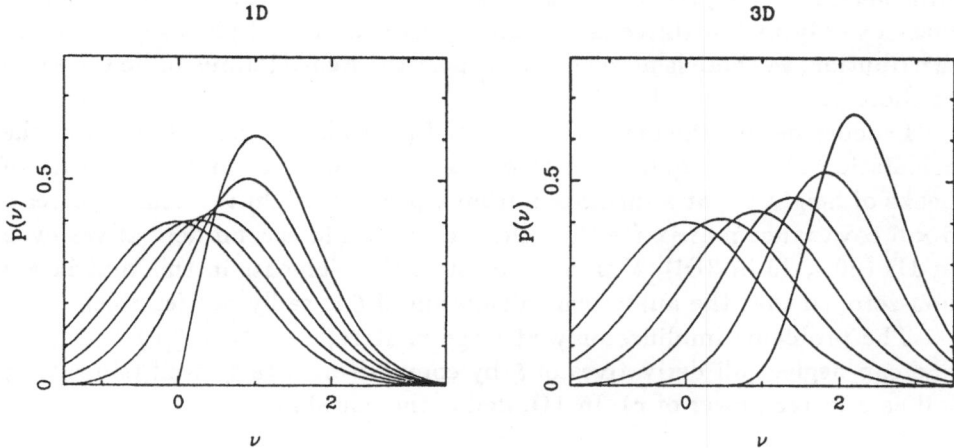

Fig. 2. The distribution of the 'height' (relative to the rms) of maxima in Gaussian density fields, in one and three dimensions, as a function of the spectral parameter γ ($\gamma = 0, 0.2, 0.4, 0.6, 0.8, 1$ is plotted). As $\gamma \to 0$, we recover the unconstrained Gaussian case. For γ closer to unity, most peaks are quite overdense ($\nu \simeq 2$) and the fractional dispersion in ν, which controls the dispersion in collapse times, is much smaller than in the unconstrained case.

The requirement that these systems have become non-linear by the present means that they must have been associated with particularly high peaks in the initial conditions. If we thus confine ourselves to peaks above some **threshold** in ν, the statistical correlations can be very strong – especially for the richer clusters corresponding to high peaks.

The main effect is easy to work out, using the **peak-background split**. Here, one conceptually decomposes the density field into short-wavelength terms (which generate the peaks) plus much longer wavelength terms (which modulate the peak number density). Consider the large-wavelength field as if it were some extra perturbation δ_+; if we select all peaks above threshold ν in the final field, this corresponds to taking all peaks above $\delta = \nu\sigma_0 - \delta_+$ in the initial field. This varying effective threshold will now produce more peaks in the regions of high δ_+ – leading to amplification of the clustering pattern. For high peaks, $P(> \nu) \propto \nu^2 e^{-\nu^2/2}$; the exponential is the most important term, leading to a perturbation $\delta P/P \simeq \nu(\delta_+/\sigma_0)$. Hence, we obtain the high-peak amplification factor for the correlation function:

$$\xi_{\rm pk}(r) \simeq \frac{\nu^2}{\sigma_0^2}\, \xi_{\rm mass}(r).$$

(72)

It is important to realise that the process as described is *nothing to do with biased galaxy formation*; it works perfectly well if galaxy light traces mass exactly in the Universe. Clusters occur at special places in the mass distribution, so there is no reason to expect their correlations to be the same as those of the mass field.

In more detail, the exact clustering of peaks is just an extension of the calculation of the number density of peaks. We want to find the density of peaks of height ν_2, at a distance r from a peak of height ν_1. This involves a 6×6 covariance matrix for the fields and first and second derivatives even in 1D (20×20 in 3D!). Moreover, most of the elements in this matrix are non-zero, so that the analytical calculation of ξ is sadly not feasible.

The problem simplifies only at large r. Here, $\psi \equiv \xi(r)/\xi(0) \ll 1$, and one can neglect all derivatives of ξ by comparison with ξ itself (since they fall as a faster power of r). In 1D, define the variable

$$\tilde{\nu} \equiv \frac{\nu - \gamma x}{1 - \gamma^2};\tag{73}$$

the large-separation limit of the correlation function may then be shown to be (Lumsden, Heavens & Peacock 1989)

$$\xi_{pk}(r) \to \frac{\langle \tilde{\nu} \rangle^2}{\sigma_0^2} \xi_{mass}(r).\tag{74}$$

A similar expression holds in 3D (see BBKS); there is an effective threshold which is usually rather less than ν for realistic values of ν (2 – 3, say). When we consider all peaks, *i.e.* let the threshold $\nu \to -\infty$, the effective threshold $\langle \tilde{\nu} \rangle \to 0$. The total number of peaks is determined by the small-scale structure in the power spectrum, not by any additional large-scale power.

Even if the peak clustering problem cannot be solved exactly, a related problem can be. This is to consider not the point process of peaks, but the correlations of thresholded *regions*. Assume that objects form with unit probability in all regions above the threshold, so that we need to deal with the correlation function of a modified density field which is constant above the threshold and zero elsewhere. This is

$$1 + \xi_{>\nu}(r) = \frac{1}{[P(> \nu)]^2} \int_\nu^\infty \int_\nu^\infty \frac{dx\, dy}{2\pi[1 - \psi^2(r)]^{1/2}} \times$$
$$\exp\left(-\frac{x^2 + y^2 - 2xy\psi(r)}{2[1 - \psi^2(r)]}\right)\tag{75}$$

Kaiser 1984). The complete solution of this equation is given by Jensen & Szalay (1986) (see also Kashlinsky 1991 for the extension to the cross-correlation of fields above different thresholds). A good approximation, which extends Kaiser's original result, is

$$1 + \xi_{>\nu} \simeq \exp\left(1 + \frac{\nu^2}{\sigma_0^2}\xi_{\text{mass}}\right). \tag{76}$$

There remains the question of the inclusion of dynamics into the above treatment. As the density field evolves, density peaks will move from their initial locations, and the clustering will alter. The general problem is rather nasty (see BBKS), but things are rather straightforward in the linear regime when the mass fluctuations are small. If statistical enhancement of correlations produces a fractional perturbation in the numbers of thresholded objects of $\delta_{\text{statistical}} = f\,\delta_{\text{mass}}$, then the effect of allowing weak dynamical evolution is just

$$\delta_{\text{obs}} = \delta_{\text{statistical}} + \delta_{\text{mass}}. \tag{77}$$

To see this, think of density perturbations arising in as in the Zeldovich approximation, via objects moving closer together. Density peaks will be convected with the flow and compressed in number density in the same way as any other particle. Thus, the effective enhancement ends up as $f \to f+1$. In an analytical *tour de force*, Bond & Couchman (1987) showed how the Zeldovich approximation may be used in this way to calculate the combined correlations exactly.

3.4 Application to Abell clusters

This is the class of object which forms the main application of the peak clustering method. For Abell richness $R \geq 1$ clusters, the comoving density is $6 \times 10^{-6}h^3$ Mpc^{-3}; once we choose a filter radius to select cluster-sized fluctuations, the threshold is specified mainly by the number density (although altering the power-spectrum model also has a slight influence through γ). For Gaussian filtering, the conventional choice of R_f is $5h^{-1}$ Mpc. For $h = 1/2$ CDM with $\gamma = 0.74$ on this scale, the required threshold is $\nu = 2.81$, although the effective threshold $\langle \bar{\nu} \rangle$ is only 2.15. These figures seem quite reasonable: Abell clusters are the rare high peaks of the mass distribution, and only collapsed recently. The only reason for setting a threshold at all is the requirement of gravitational collapse by the present, so it is inevitable that $\nu \sim 1$.

The observations of cluster-cluster clustering are somewhat controversial. The correlation function found by most workers is consistent with a scaled version of the galaxy function: $\xi = (r/r_0)^{-1.8}$, but values of r_0 vary. The original value found by Bahcall & Soneira (1983) was $25h^{-1}$ Mpc, but some more recent work favours values about half of this (Sutherland 1988; Sutherland & Efstathiou 1991; Dalton *et al.* 1991). The enhancement with respect to ξ for galaxies is thus between about 5 and 20. Since σ_0 is close to unity for this smoothing, the simple asymptotic scaling would imply a threshold between 2.2 and 4.5, which may seem promising. However, because we are interested in correlations on scales only ~ 5 times the smoothing radius, in practice the asymptotic scaling is not relevant: Lumsden *et al.* (1988) show that the $h = 1/2$ CDM model is inconsistent with r_0 values much greater than $10h^{-1}$ Mpc (see figure 3).

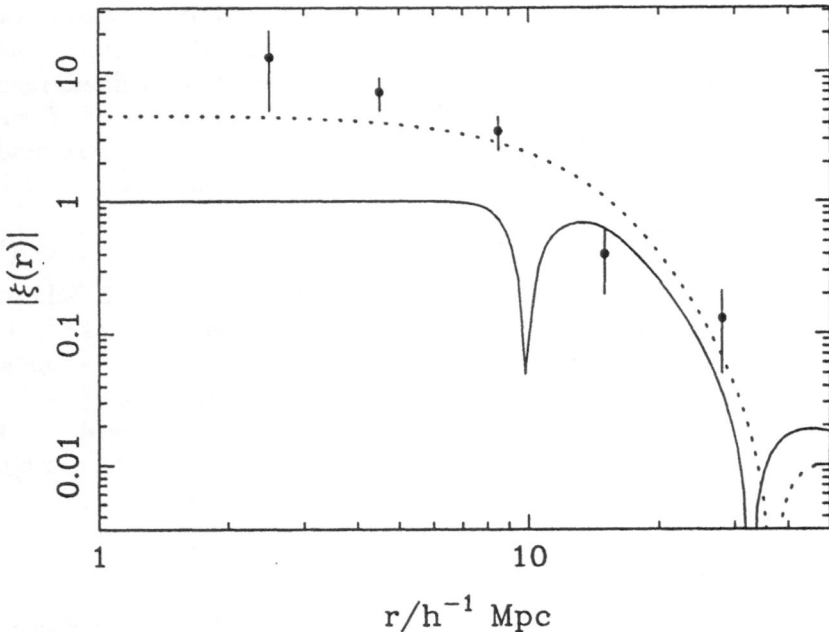

Fig. 3. The correlation function for Abell clusters in $h = 1/2$ CDM, as modelled via the peak-peak correlation function by Lumsden *et al.* (1988). This assumes that the relevant peak parameters are $R_f = 5h^{-1}$ Mpc filtering and a threshold of $\nu = 2.81$. The plot compares the 1D correlation function of such peaks (an excellent approximation to the full 3D result) with the result for thresholded regions, scaled via an effective threshold to agree with ξ_{pk-pk} at large r. Note the relatively small statistical clustering in this model, which is certainly inconsistent with $r_0 \simeq 20h^{-1}$ Mpc, as claimed by several workers. However, the data from the APM clusters of Dalton *et al.* (1991) do fit well at large r (the comparison is not really valid at small r close to the smoothing length).

The main weakness in this discussion is the functional form of the filter which is used to define clusters in the linear initial conditions. We have shown results obtained using a Gaussian filter; it would be nice if the results were insensitive to this choice, but this is not so. Lumsden *et al.* (1988) showed that filters which were more spatially extended could produce considerably stronger correlations than the Gaussian filter, even matching the filters so that R_* and ν were identical. Clearly, this is a degree of freedom which can only be constrained by comparison with N-body simulations. The question of what linear filter best predicts the location and mass of nonlinear clusters is something which can only be settled empirically. For the present, then, it should be regarded as an open question whether models such as CDM can account quantitatively for the enhanced correlations of galaxy clusters.

3.5 Non-Gaussian fields

This is a rather open area, since many new degrees of freedom open up once the restrictions of Gaussian behaviour are abandoned – in much the same way as 'non-dog' is a vaguer term than 'dog'. To specify a non-Gaussian field fully requires all the high-order correlation functions. In practice, therefore, attention has focused on cases which are simple modifications of Gaussian fields. Probably the most interesting of these is the lognormal field (Coles & Jones 1991):

$$1 + \delta \rightarrow \exp(\delta). \tag{78}$$

This should represent the weakly non-linear evolution of density perturbations (enforcing positivity and leading to positive skewness). Also, such a model is close to the 'peak-background split' behaviour used to analyse the amplification of cluster correlations and in some models for biased galaxy formation. The statistical properties for this model are especially simple; the two-point correlation is

$$1 + \xi_{\mathrm{LN}} = \exp(\xi_{\mathrm{G}}). \tag{79}$$

The three-point correlations obey the **Kirkwood scaling**

$$1 + \xi^{(n)} = (1 + \xi_{12})(1 + \xi_{23})(1 + \xi_{31}), \tag{80}$$

leading to a cubic term in ζ, whose existence is controversial (see Coles & Plionis 1991 and refs therein).

For a wider range of such modified non-Gaussian models and their properties, see Coles & Barrow (1987).

One question of great practical interest is whether the primordial fluctuations were Gaussian, and how to test for this. The problem with most of the models which have been studied is that they are markedly non-Gaussian: a simple inspection of the 1-point histogram of field values serves to demonstrate this. More subtle problems arise with fields which are Gaussian-distributed, but still non-Gaussian (*i.e.* having differing behaviour of the higher-order correlation functions). Figure 4 illustrates this with an example of two fields which differ from the 3-point level. We shall discuss ways in which any differences may be detected in section 8.

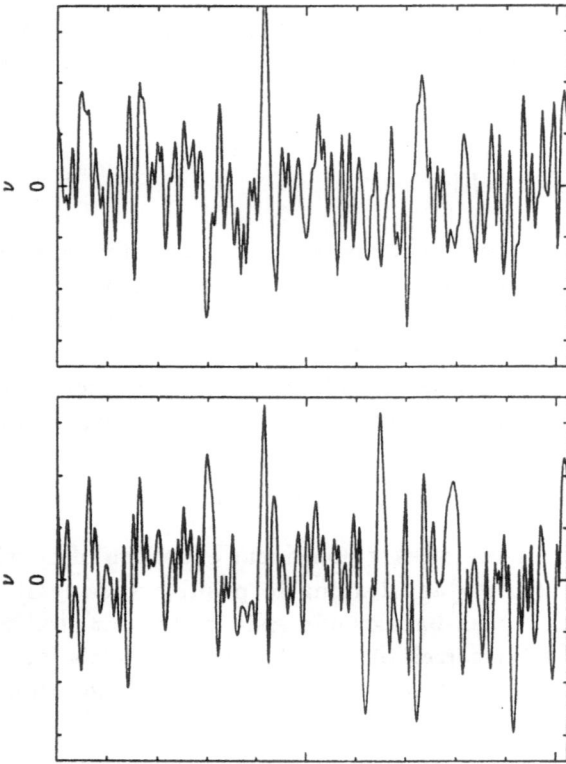

Fig. 4. Realizations of two 1D random fields. One is a pure Gaussian; the other is a product of two Gaussian realizations, with a vertical transformation to reimpose Gaussian behaviour. Both fields have the same distribution and power spectrum; the differences arise only at the 3-point function level and are not immediately obvious to the eye.

4 The Press-Schechter approach

In many of the cases discussed above, the density power spectrum contains no physical cutoff on small scales; the filtered density variance diverges as the filtering length is reduced to zero. Such fields (usually termed **hierarchical**) are always non-linear on small scales, and structure grows via merger of non-linear clumps. It may seem that such a situation cannot be analysed within the bounds of linear theory, but a way forward was identified by Press & Schechter (1974; PS); see Peacock (1990a) for a review of this method and its competitors. The critical assumption in the PS analysis is that, even if the field is non-linear, the amplitude of large-wavelength modes in the final field will be close to that predicted from linear theory. For this to be true requires the 'true' large-scale power to exceed that generated via non-linear coupling of small-scale modes, which turns out to require a spectral index $n > 1$ (Williams *et al.* 1991a). We now proceed by recognising that, for a massive clump to undergo gravitational collapse, the average overdensity in a volume containing that mass should (as usual) exceed some threshold, δ_c, of order unity. The location and properties of these bound objects can thus be estimated by an artificial smoothing (or filtering) of the initial linear density field. If the filter function has some characteristic length R_f, then the typical size of filtered fluctuations will be $\sim R_f$ and they can be assigned a mass $M \sim \rho_0 R_f^3$. The exact analytic form of the filter function is arbitrary and is often taken to be a Gaussian for analytic convenience.

The argument now proceeds in integral terms. For a given R_f, the probability that a given point lies in a region with $\delta > \delta_c$ (the critical overdensity for collapse) is

$$p(\delta > \delta_c \,|R_f) = \tfrac{1}{2}\left[1 - \mathrm{erf}\left(\frac{\delta_c}{\sqrt{2}\,\sigma(R_f)}\right)\right], \qquad (81)$$

where $\sigma(R_f)$ is the linear rms in the filtered version of δ. The PS argument now takes this to be proportional to the probability that a given point has ever been processed through a collapsed object of scale $> R_f$. This is really assuming that the only objects which exist at a given epoch are those which have just collapsed: if a point has $\delta > \delta_c$ for a given R_f, then it will have $\delta = \delta_c$ when filtered on some larger scale and will be counted as an object of the larger scale. The problem with this argument is that half the mass remains unaccounted for: this was amended by PS simply by multiplying the probability by a factor 2. Note that this procedure need not be confined to Gaussian fields; all we need is the functional form of $p(\delta > \delta_c \,|R_f)$. The factor of 2 problem will arise in any model for which this distribution is symmetric about $\delta = 0$.

This integral probability is related to the mass function $f(M)$ (defined such that $f(M)dM$ is the comoving number density of objects in the range dM) via

$$M f(M)/\rho_0 = |dp/dM|, \qquad (82)$$

where ρ_0 is the total comoving density. Thus,

$$\boxed{\frac{M^2 f(M)}{\rho_0} = \frac{2\delta_c}{\sqrt{2\pi}\,\sigma}\left|\frac{d\ln\sigma}{d\ln M}\right|\exp(-\tfrac{1}{2}\delta_c^2/\sigma^2).}$$

$$(83)$$

We have expressed the result in terms of the **multiplicity function**: $M^2 f(M)/\rho_0$ is the fraction of the mass which is carried by objects in a unit range of $\ln M$. For power-law spectra, this function takes a very simple form:

$$\frac{M^2 f(M)}{\rho_0} = \frac{n+3}{6} \sqrt{\frac{2}{\pi}} \, \nu \, e^{-\nu^2/2}, \tag{84}$$

Where ν is the threshold in units of the rms density fluctuation. The multiplicity function thus always has the same shape (a skew-negative hump around $\nu \simeq 1$); changing the spectral index only alters the mass scale via $\nu = (M/M_c)^{(n+3)/6}$.

4.1 Random walks and conditional mass functions

The factor of 2 'fudge' has long been recognized as the crucial weakness of the PS analysis. What one has in mind is that the mass from lower-density regions accretes onto collapsed objects, but it does not seem correct for this to cause a doubling of the total number of objects. Recent work has shed some light on the origin of this problem (Peacock & Heavens 1990; Bond et al. 1991). To see where the error crept in, consider the **random trajectory** taken by the filtered field at some fixed point as a function of filtering radius. This starts at $\delta = 0$ at $R = \infty$, and develops fluctuations of increasing amplitude as we move to smaller R. Thus, if $\delta < \delta_c$ at a given point, it is quite possible that it will exceed the threshold at some other point – indeed, if the field variance diverges as $R \to 0$, it is inevitable that the threshold will be exceeded. So, instead of ignoring all points below threshold at a given R, we should find the **first upcrossing** of the random trajectory: the largest value of R for which $\delta = \delta_c$.

This analysis is most easily performed for one particular choice of filter: sharp truncation in k-space. Decreasing R then corresponds to adding in new k-space shells, all of which are independent for a Gaussian field. The trajectory is then just a random walk, and the solution is very easy. Consider a point on the walk which has reached the threshold; its subsequent motion will be symmetric, so that it is equally likely to be found above the threshold as below at some smaller R. The probability of never having crossed the threshold (the **survival probability**) is then obtained by reflection of the Gaussian above threshold:

$$\frac{dP_s}{d\delta} = \frac{1}{\sqrt{2\pi}\,\sigma} \left[\exp\left(-\frac{\delta^2}{2\sigma^2}\right) - \exp\left(-\frac{(\delta - 2\delta_c)^2}{2\sigma^2}\right) \right]. \tag{85}$$

Integrating this up to get the probability of having crossed the threshold at least once gives

$$1 - P_s = 1 - \mathrm{erf}\left(\frac{\delta_c}{\sqrt{2}\,\sigma}\right), \tag{86}$$

which is just twice the unconstrained probability of lying above the threshold – thus supplying the missing factor of 2. Unfortunately, the above analysis is only valid for this special choice of filter: using k-space filters which are differentiable leads to just the original PS form (without the factor 2) at high mass, with the surplus probability being shifted to low masses, so that the shape of the function changes. Which filter is the best choice, we cannot say in advance; we are left with the empirical fact that the PS formula does fit N-body results quite well.

One useful extension of the random-walk model is that it allows a calculation of the **conditional multiplicity function**: given a particle in a system of mass M_0 at some epoch a_0, what was the distribution of masses where that particle resided at some earlier epoch a_1? This is now just the random walk with two absorbing barriers, at δ_c/a_0 and δ_c/a_1; we want the probability of not having crossed the second, subject to having crossed the first at M_0. The solution for the integral mass distribution is

$$P(> \mu) = 1 - \text{erf}\left[\frac{\nu(t^{-1} - 1)}{\sqrt{2(\mu^{-1} - 1)}}\right],\qquad(87)$$

where $\nu \equiv \delta_c/a_0\sigma_0$; $t \equiv a_1/a_0$; $\mu \equiv \sigma^2(M_0)/\sigma^2(M)$.

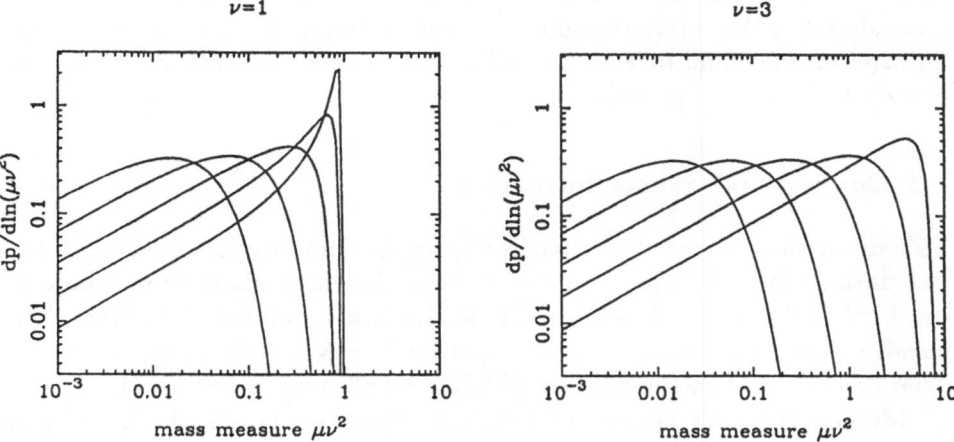

Fig. 5. The conditional multiplicity function. This figure plots the distribution of mass (as measured by reciprocal of filtered variance) for the early-time hosts of a particle in (a) a $\nu = 1$ and (b) a $\nu = 3$ fluctuation at the present. Redshifts 0.5, 1, 2, 4, 8 are shown. The histories of the particles are greatly divergent over most of the recent history of the Universe.

This function is illustrated in figure 5; this shows that the early histories of particles which end up in different mass systems are markedly different over a large range of expansion factor. This may provide a way of understanding many of the systematics of galaxy systems in terms of their merger histories; see Bower (1991).

4.2 Application to Abell clusters

The most important practical application of the PS formalism is to Abell clusters. As already discussed, these are the most massive non-linear systems in the current Universe, so a study of their properties should set constraints on the shape and normalization of the power spectrum on large scales. These issues are discussed by Henry & Arnaud (1991). They sidestep the issue of what mass to assign to a cluster by using the observed distribution of temperatures (see below for the relation between mass and virial temperature). Fitting their data with the PS form for top-hat filtering and $\delta_c = 1.69$ they deduce a *linear-theory* rms in $8h^{-1}$ Mpc spheres of 0.59 ± 0.02 and an effective power-law index of $n = -1.7^{+0.65}_{-0.35}$. The fact that σ_8 comes out close to the observed value of unity gives some encouragement for the gravitational collapse picture and/or argues against a very high degree of bias (*e.g.* the linear-theory $\sigma_8 = 1/2.5$ common in much work on the CDM model). The index limit is rather negative by comparison with CDM, but more in line with observations of clustering (see below). However, the analysis relies heavily on the accuracy of the PS formalism, which has yet to be tested in great detail in the relevant regime of mass. Refining this test of the fluctuation spectrum should be easily possible with the new generations of N-body models now becoming available.

4.3 Cooling and galaxy formation

This discussion of mass functions really applies only to 'haloes' of collisionless dark matter. Things are more complex for baryonic matter, where we must ask if the matter has been able to **dissipate** and turn into stars. This question was analyzed in a classic paper by Rees & Ostriker (1977), and has been reconsidered in the context of CDM by Blumenthal *et al.* (1984).

Mergers will heat gas up to the virial temperature via shocks; in order for the gas to form stars, it must be able to **cool** – to radiate away this thermal energy. Clearly, the redshift of collapse clearly needs to be sufficiently large that there is time for an object to cool between its formation at redshift z_{cool} (when $\delta\rho/\rho \simeq \delta_c$) and the present epoch. We shall show below that z_{cool} is a function of mass; it is therefore possible to put cooling into the Press-Schechter machinery simply by using the *mass-dependent* threshold $\nu(M) = \delta_c[1 + z_{\mathrm{cool}}(M)]/\sigma_0(M)$ in the mass function.

The cooling function for a plasma in thermal equilibrium has been calculated by Raymond, Cox & Smith (1976). For an H + He plasma with $Y = 0.25$ and some admixture of metals, their results for the cooling time $(t_{\mathrm{cool}} \equiv 3kT/2\Lambda(T)n)$ may be approximated as

$$t_{\mathrm{cool}}/\mathrm{years} = 1.8 \times 10^{24} \left(\frac{\rho_B}{M_\odot \mathrm{Mpc}^{-3}}\right)^{-1} \left(T_8^{-1/2} + 0.5 f_m T_8^{-3/2}\right)^{-1}, \quad (88)$$

where $T_8 \equiv T/10^8 K$. The $T^{-1/2}$ term represents bremsstrahlung cooling and the $T^{-3/2}$ term approximates the effects of recombination radiation. The parameter f_m governs the metal content: $f_m = 1$ for solar abundances; $f_m \simeq 0.03$ for no metals. In this model where so far dissipation has not been considered, the baryon density is proportional to the total density, the collapse of both resulting from purely gravitational processes. ρ_B is then a fraction Ω_B/Ω of the virialized total density. This is itself some multiple f_c of the background density at virialization (which we refer to as 'collapse'):

$$\rho_c = f_c \, \rho_0 \, (1 + z_c)^3. \tag{89}$$

The virialized potential energy for constant density is $3GM^2/(5r)$, where the radius satisfies $4\pi\rho_c r^3/3 = M$. This energy must equal $3MkT/(\mu m_p)$, where $\mu = 0.59$ for a plasma with 75% Hydrogen by mass. Hence, using $\rho_0 = 2.78 \times 10^{11}\Omega h^2 \, M_\odot \mathrm{Mpc}^{-3}$, we obtain

$$T_{\mathrm{virial}}/K = 10^{5.1}(M/10^{12} M_\odot)^{2/3} \, (f_c\Omega h^2)^{1/3} \, (1 + z_c). \tag{90}$$

So, for $\Omega = 1$, we must solve $f_t \, t_{\mathrm{cool}} = \frac{2}{3}H_0^{-1}[1 - (1 + z_c)^{-3/2}]$. If only recombination cooling was important, the solution to this would be

$$\boxed{\begin{aligned} (1 + z_c) &= (1 + M/M_{\mathrm{cool}})^{2/3} \\ M_{\mathrm{cool}}/M_\odot &= 10^{13.1} \, f_t^{-1} f_m f_c^{1/2}\Omega_B\Omega^{-1/2} \end{aligned}} \tag{91}$$

For high metallicity, where bremsstrahlung only dominates at $T \gtrsim 10^8 K$, this equation for z_c will be a reasonable approximation up to $z_c \simeq 10$, at which point Compton cooling will start to operate. Given that we expect at least some enrichment rather early in the progress of the hierarchy, we shall keep things simple by using just the above expression for z_c.

We see that cooling is rapid for low masses, where the luminous and dark mass functions are expected to coincide. Given that cooling of massive objects is ineffective, probability in the mass function must therefore accumulate at intermediate masses: the numbers of faint galaxies relative to bright are decreased. If $M_{\mathrm{cool}} \ll M_c$, then there is a power-law region between these two masses which differs from the PS slope: $M^2 f(M) \propto M^{(\frac{n+3}{6})+\frac{2}{3}}$; i.e. there is an effective change in n to $n+4$. This is illustrated for power-law spectra in figure 6.

We should not claim too much from the above analysis, as several potentially important points are neglected. Firstly, there is a danger that star formation in the first generation of the hierarchy may be so efficient that all gas immediately becomes locked up in low-mass objects. This seems implausible; the low binding energy should allow supernovæ to unbind the matter (Dekel & Silk 1986), leading to a very low overall efficiency of star formation. Indeed, the high fraction of baryonic material in galaxy clusters which is in

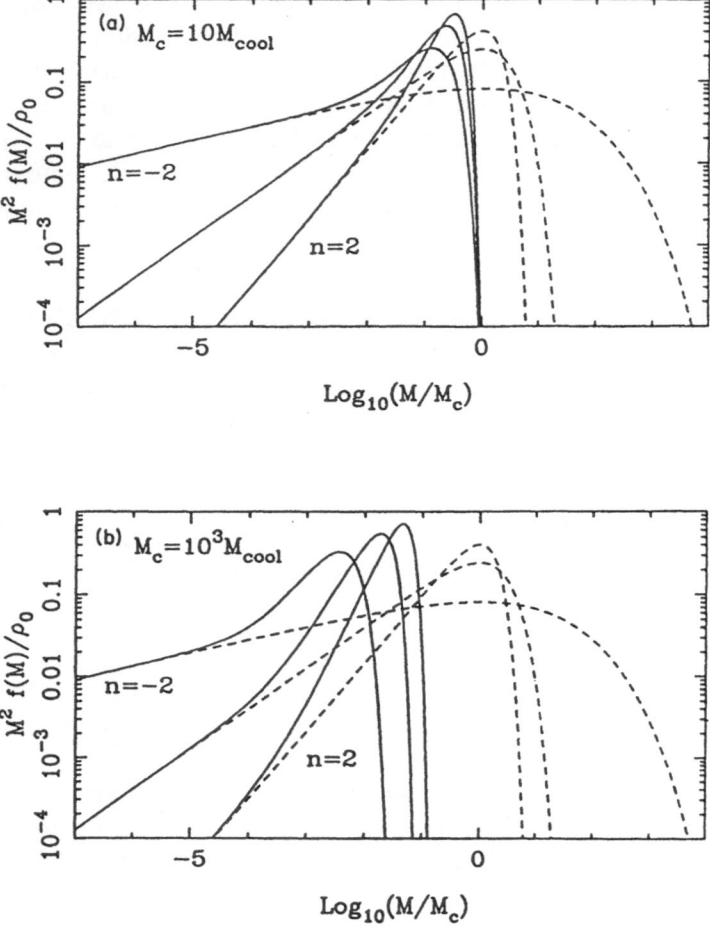

Fig. 6. The multiplicity functions resulting from the PS formalism with the adoption of the cooling threshold criterion $(\delta_c \rightarrow \delta_c[1 + M/M_{cool}]^{2/3})$. The solid lines show the mass functions with cooling, and the dashed lines show the mass functions without cooling; the two sets of curves coincide only at very low masses. M_c is the mass scale with $\sigma(M_c) = \delta_c$. Note (a) the relative insensitivity to M_{cool} and (b) that over several orders of magnitude in mass, the relative numbers of low-mass objects are greatly reduced by cooling.

the form of gas may be indicative of such a process having taken place. Nevertheless, erasure of baryonic sub-structure may be imperfect, which would lead us to underestimate the numbers of low-mass objects (White & Rees 1978). Second, the criterion of equating cooling time with look-back time will work only if an object is able to cool undisturbed over this time; if subsequent generations of the hierarchy collapse while the object is still cooling quasi-statically, then the gas will be reheated and collapse may never occur (see White & Rees for this point also). Objects are immune to this effect if the cooling time is shorter than the free-fall time, which turns out to be simply a criterion on mass (see *e.g.* Efstathiou & Silk 1983). If we relate collapse time ($t_{coll} \equiv [3\pi/32G\rho_c]^{1/2}$) to cooling time via $f_t \, t_{cool} = t_{coll}$, then the above figures for recombination cooling yield a mass

$$M_{coll}/M_\odot = 10^{13.5} f_t^{-1} f_m \Omega_B \Omega^{-1}. \qquad (92)$$

This is of the same order as M_{cool}, and so very massive structures may be truncated still more abruptly than we have assumed above. Detailed application of these arguments to the CDM model was made by Blumenthal *et al.* (1984), who conclude that cooling allows a very good understanding of the delineation between galaxies and clusters (see figure 7).

The remarkable conclusion of this section is that simple considerations of microphysics have allowed us to understand the characteristic masses of galaxies. The order of magnitude may be expressed in terms of fundamental constants as

$$M_G \sim \alpha^5 \left(\frac{\hbar c}{Gm_p^2} \right)^2 \left(\frac{m_p}{m_e} \right)^{1/2} m_p, \qquad (93)$$

where $\alpha \simeq 1/137$ is the fine-structure constant. By contrast, the typical mass of a star is $(\hbar c/[Gm_p^2])^{3/2} m_p$ (Efstathiou & Silk 1983).

4.4 Galaxy mergers

The PS method allows us to obtain an approximate expression for the merger rate in a gravitational hierarchy. For masses $\ll M^*$ (which is today an Abell cluster mass), The fraction, f, of the mass bound into objects larger than M is proportional to $\delta_c/\sigma_0(z)$ (if the fluctuations are symmetric about zero overdensity; this is more general than the above Gaussian analysis), so that dn/dz is a constant. If mergers are assumed to occur between two objects of equal mass, dn/dz is then also the number of merger events per dz. This epoch dependence is increased when we realise that mergers must be suppressed today. The PS theory governs only the behaviour of dark matter haloes; even if these merge, there is no guarantee that the stellar cores will do so also (merger of two clusters just gives a larger cluster). The key parameter is the relative velocity of two galaxies: only if this is below the relative escape velocity ($\simeq 200$ kms^{-1}) will a merger be probable. For collapsed clusters, the velocity dispersion is $\sigma_v \simeq GM/r \propto M^{2/3}\rho^{1/3} \propto M^{2/3}(1+z)$; since theories such as Cold Dark Matter have M^* changing quite rapidly with redshift, this was usually smaller in the past. This epoch dependence of merging may be important in understanding some of the observations of galaxy evolution (see below); for example, if we assume that AGN are

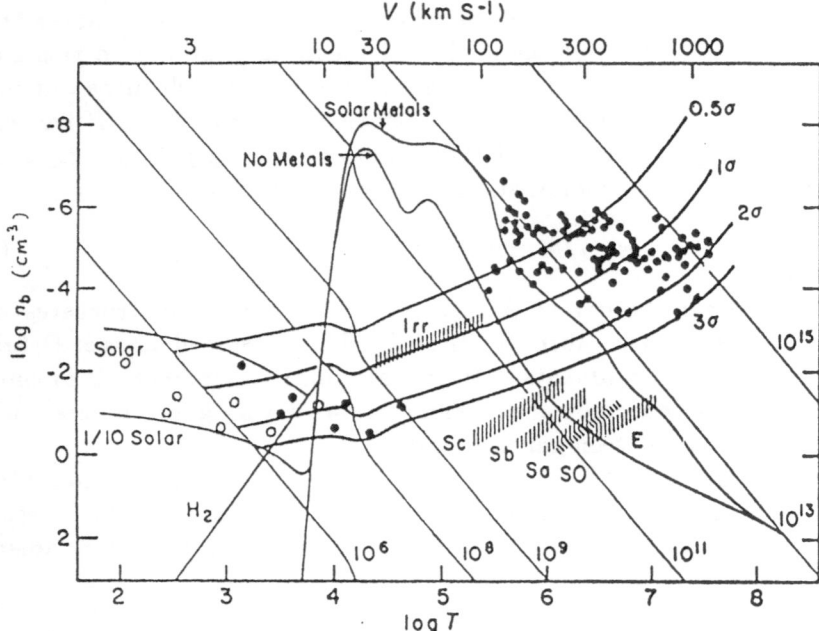

Fig. 7. The cooling diagram for CDM, taken from Blumenthal *et al.* (1984). The solid line shows the point at which the cooling time equals the collapse time. The diagram is a plot of baryonic density against temperature, but the conversion to velocity dispersion and mass is also shown (assuming 10% of the total mass to be baryonic). The horizontal lines indicate fluctuations of different height according to standard CDM. The locations of clusters (solid points) and galaxies (hatched areas) in this plot provide convincing evidence that it is dissipation which governs whether a collapsed object is described as a galaxy, or as a system of galaxies.

short-lived with a constant lifetime, then the comoving density of events should track the rate of mergers. Hence, the sort of epoch dependence of merging required to account for AGN density evolution ($[1 + z]^5$, say) can be produced (see Carlberg 1990).

5 Bias

One of the major advances of cosmology in the 1980s was the realisation that the distribution of galaxies need not trace the underlying density field. The main motivation for such a view may be traced to 1933 and Zwicky's measurement of the dark matter in the Coma cluster. A series of ever more detailed studies of cluster masses have confirmed his original numbers: if the Coma mass-to-light ratio was Universal, then the density parameter of the Universe was $\Omega = 0.1 - 0.2$. Those who argued that the value $\Omega = 1$ was more natural (a greatly increased camp after the invention of inflation by Guth) were therefore forced to postulate that the efficiency of galaxy formation was enhanced in dense environments: **biased galaxy formation**. This probably remains the best argument for the reality of bias, especially given that the measurements of large-scale peculiar velocity fields do favour values of Ω closer to unity.

A weaker argument surfaced at around the same time through the discovery of large voids in the galaxy distribution. There was a reluctance to believe that such large regions could be truly devoid of matter – although this was at a time before the discovery of large-scale velocity fields. This tendency was given further stimulus through the work of Davis, Efstathiou, Frenk & White (1985), who were the first to calculate N-body models with the CDM spectrum. Since the CDM spectrum curves slowly between effective indices of $n = -3$ and $n = 1$, the correlation function clearly steepens with time. There is therefore a unique epoch when ξ will have the observed slope of -1.8. Davis *et al.* identified this epoch as the present, and then noted that it implied a rather low *amplitude* of fluctuations: $r_0 = 1.3 h^{-2}$ Mpc. What seemed to be required was a galaxy correlation function which was an amplified version of that for mass. This was exactly the phenomenon analysed for Abell clusters by Kaiser (1984). Thus was born the idea of **high-peak bias**: bright galaxies form only at the sites of high peaks in the initial density field. This was developed in some analytical detail by Bardeen *et al.*, and was implemented in the simulations of Davis *et al.*, leading to the conclusion that the $\Omega = 1$ CDM model now gave a good match to observation.

As we have seen, the high-peak model produces a linear amplification of large-wavelength modes. This is likely to be a general feature of other models for bias, so it is useful to introduce the **linear bias parameter**:

$$\left. \frac{\delta\rho}{\rho} \right|_{\text{galaxies}} = b \left. \frac{\delta\rho}{\rho} \right|_{\text{mass}} .$$

$$(94)$$

This seems a reasonable assumption when $\delta\rho/\rho \ll 1$. Galaxy clustering on large scales therefore allows us to determine mass fluctuations only if we know the value of b. For example, the normalization in scale-invariant models may be specified by ϵ (the rms potential fluctuation per $\ln k$), but we can only measure the combination ϵb. However, the coinage of the bias parameter has been debased by its use in the mildly non-linear regime. It is conventional to quote the normalization of CDM models via the ratio of the observed optical-galaxy $J_3(r_c)$ $(= \int_0^{r_c} \xi r^2 \, dr)$ to the linear-theory value, even though $\xi \sim 1$ at the standard radius $r_c = 10 h^{-1}$ Mpc. We shall use explicitly the symbol b_J to denote the bias parameter defined in this way, since it is possible that $b_J \neq b$. Using the transfer function as given by Bardeen et al. (1986; BBKS), the following is a good approximation for CDM over the relevant range of h:

$$\epsilon b_J = \frac{9.5 \times 10^{-6}}{h - 0.03 - 0.15h^4}. \tag{95}$$

The favoured value of b_J has changed with time. The N-body work of Davis et al. (1985) advocated a value of $b_J = 2.5$, whereas BBKS preferred 1.7. However, the recent data on very large-scale structure (see below) increasingly seems to be pointing towards still lower bias values.

Why should the galaxy distribution be biased at all? In the context of the high-peak model, attempts were made to argue that the first generation of objects could propagate disruptive signals, causing neighbours in low-density regions to be 'still-born'. However, it turned out to be hard to make such mechanisms operate: the energetics and required scale of the phenomenon are very large (Rees 1985; Dekel & Rees 1987). A more promising idea, known as **natural bias**, was introduced by White et al. (1987). This relied on the idea that, in an overdense region, an object of a given mass will collapse sooner and thus have a higher density and circular velocity. Application of a circular-velocity threshold then yields a bias towards high-density regions. White et al. argued that such an effect was to be expected owing to the **Tully-Fisher effect**: a tight correlation between luminosity and circular velocity.

However, the problem with this model is that it still apparently predicts no bias if a strict selection by mass is performed. What is needed is some way in which star formation is biased (perhaps by epoch dependent efficiency) in order to produce more stars in the galaxies which collapse earlier. A general discussion of this problem was given by Cole & Kaiser (1989): suppose an object collapsing at redshift z generates a stellar luminosity

$$L \propto M^\alpha (1 + z)^\beta. \tag{96}$$

Cole & Kaiser show that a perfect Tully-Fisher relation then requires $\beta = 3\alpha/2$. This is easily proved from the usual expression for the virial velocity dispersion resulting from gravitational collapse: $V \propto M^{1/3}(1 + z_c)^{1/2}$. The above condition removes any redshift dependence and leaves $V \propto L^{1/3\alpha}$. The conventional Tully-Fisher slope of $1/4$ then implies $\alpha = 4/3$, $\beta = 2$. The natural bias mechanism implicitly depends on a strong epoch dependence of star-forming efficiency. In fact, Cole & Kaiser argue that a somewhat stronger epoch dependence ($\beta \gtrsim 3$) is required to achieve sufficient bias to understand cluster mass-to-light ratios. However, Peacock (1990b) showed that such a high value would predict a large scatter in the Faber-Jackson relation for ellipticals. The data are much more closely consistent with $\beta \simeq 1$.

In short, there appears to be little evidence for traditional bias schemes where one tinkers with the efficiency of star formation. There remains the alternative that galaxies were born unbiased but subsequently migrated into clusters more quickly than dark matter. These dynamical schemes are currently attracting the most attention (West & Richstone 1988; Carlberg *et al.* 1990; Carlberg 1991; Couchman & Carlberg 1991). The idea of bias being a phenomenon largely confined to clusters fits well with the emerging picture of large-scale structure: away from clusters, it seems that all types of galaxies follow the same density field, independent of luminosity or Hubble type (Thuan *et al.* 1987; Valls-Gabaud *et al.* 1989; Babul & Postman 1990). This is further evidence against earlier pictures in which the voids were filled with mass, but contained only dark 'failed' galaxies. Increasingly, it seems that the voids really have been emptied by gravity, as implied by the large-amplitude peculiar velocities on these scales (Yahil 1991).

6 Low-dimensional density fields

It is common to encounter datasets which are a projection of a three-dimensional density field, and there are various tools which are needed in order to extract the desired 3D information.

6.1 Skewers and slices

The simplest case is a skewer: data along a line in 3D space. If we choose this to correspond to the x_1 axis, then the 1D field is

$$\delta(x_1) = \frac{L^3}{(2\pi)^3} \int \delta_k \, \exp(ik_1 x_1) \, d^3 k, \tag{97}$$

and the correlation function (common to both 3D field and 1D skewer because of statistical isotropy) is

$$\xi(r) = \frac{L^3}{(2\pi)^3} \int |\delta_k|^2 \, \exp(ik_1 r) \, d^3 k. \tag{98}$$

Taking the 1D transform of this gives

$$P_{1D}(k) = \frac{L^2}{(2\pi)^2} \int P_{3D}(\sqrt{k^2 + k_2^2 + k_3^2}) \, dk_2 \, dk_3 = \frac{L^2}{2\pi} \int_{|k|}^{\infty} P_{3D}(y) \, y \, dy. \tag{99}$$

Note that what we have here is a projection: 1D modes of wavevector k receive contributions from all 3D modes with wavevectors $\leq k$. This just says that a long-wavelength disturbance along a skewer can be produced by a short-wavelength mode directed nearly perpendicular to the skewer. An interesting point is that P_{3D} is just proportional to the derivative of P_{1D};

the latter must therefore be a monotonically decreasing function. In terms of the variances per logarithmic interval of wave number $\Delta_{1D}^2 \equiv d\sigma_{1D}^2/d\ln k = kLP_{1D}/\pi$, the result may be re-expressed as

$$\Delta_{1D}^2(k) = k \int_k^\infty \Delta_{3D}^2(y)y^{-2}\,dy$$

(100)

(Note the factor 2 from $\pm k$ in one dimension). We can use this expression to derive the relation between moments over the 1D and 3D power spectra (Lumsden *et al.* 1988). In particular, we get $\gamma_{1D} = (\sqrt{5}/3)\gamma_{3D}$: the power spectrum of a skewer has a lower limit to its allowed width. Now, provided the 3D power spectrum has an effective index $n > -2$ as $k \to 0$, and assuming the integral converges at large k, the large-wavelength terms in the 1D power spectrum are dominated by power projected from short wavelengths in 3D, and are highly insensitive to the true 3D power on large scales for most spectra.

This projection of spurious large-scale power is an inevitable effect in lower-dimensional redshift surveys and will always weaken any conclusions which can be drawn concerning large-scale density perturbations. The effect for a two-dimensional slice is similar to that in one dimension:

$$\Delta_{2D}^2 = k^2 \int_k^\infty \Delta_{3D}^2(y)\frac{y^{-2}\,dy}{\sqrt{y^2 - k^2}}$$

(101)

Again, an effective $n = 0$ portion is produced at low k. Since Δ_{3D}^2 behaves very nearly as k^2, we see that $\Delta_{2D}^2/\Delta_{3D}^2 \simeq \cosh^{-1}(k_c/k)$, where the integral is cut off at k_c by the slice thickness; this power ratio will be a factor 3 – 5 in practice. Such enhancement of large-scale power needs to be kept in mind when interpreting surveys such as the CfA slice: the linear scale for unit variance in the projected areal density of galaxies will be rather higher than the corresponding three-dimensional length.

In the more realistic case where the skewer or slice have a non-zero thickness, the above formulae are modified as follows: $\Delta_{3D}^2(y) \to \Delta_{3D}^2(y)|\tilde{W}|^2(\sqrt{y^2 - k^2})$, where $\tilde{W}$ is the transform of the convolving function used to thicken the skewer or slice. For example, if a skewer is thickened to a cylinder of radius R, $\tilde{W}(x) = 2J_1(xR)/xR$; if a slice is thickened to a slab of depth $2L$, $\tilde{W}(x) = \sin(xL)/xL$.

As an illustration of these points, consider the redshift surveys compiled by Broadhurst *et al.* (1990; BEKS). They had a geometry which was roughly a tube of diameter $6h^{-1}$ Mpc, over a very large baseline. The redshift histogram appeared strikingly periodic, with a high spike in the 1D power spectrum at a wavelength of $128h^{-1}$ Mpc (see figure 8). This has caused

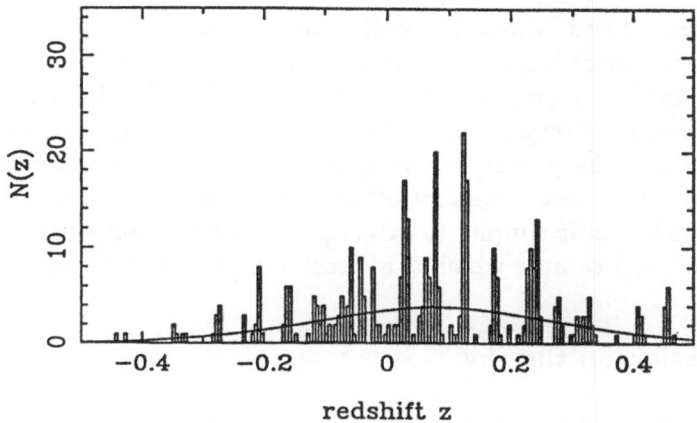

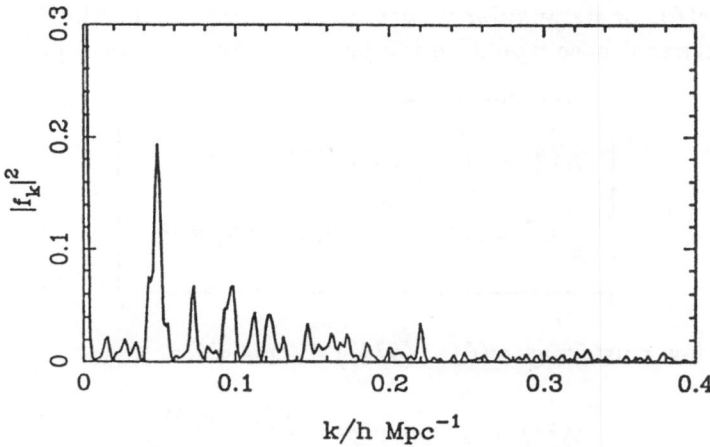

Fig. 8. (a) The redshift distribution and (b) the power spectrum of a set of synthetic data designed to mimic that of Broadhurst *et al.* (1990). The power spectrum units are such that $|f_k|^2$ is the contribution to the variance in $\delta n/n$ from a given mode. A random velocity with dispersion 800 kms^{-1} has been added to the model redshifts, greatly reducing the amplitude of the power spectrum for $k \gtrsim 0.1h$ Mpc^{-1}. It is thus dangerous to use the data at $k \gtrsim 0.1h$ Mpc^{-1} to estimate the power from small scale clustering: this can lead to a spurious degree of significance being assigned to the peak near the origin. See Kaiser & Peacock 1991 for more details.

some excitement, apparently challenging our fundamental concepts about galaxy clustering and homogeneity. However, from the formula for Δ^2_{1D}, we see that the power at large wavelengths will be dominated by the 3D power around the tube size (which cuts off the integral). This is of order unity, so the appearance of enormous large-scale power is readily understood. The lack of small-scale 1D power, which makes the spike all the more impressive, is to be attributed to a combination of redshift errors and random velocities. The conclusion is simple: to investigate clustering on some scale λ, we require a fully 3D sample which is at least λ in its shortest dimension.

6.2 Projection on the sky

A more common situation is where we lack any distance data; we then deal with a projection on the sky of a magnitude-limited set of galaxies at different depths. The statistic which is observable is the angular correlation function, $w(\theta)$, or its angular power spectrum Δ^2_θ. If the sky was flat, the relation between these would be the usual Hankel transform pair:

$$
\boxed{
\begin{aligned}
w(\theta) &= \int_0^\infty \Delta^2_\theta \, J_0(K\theta) \, dK/K \\
\Delta^2_\theta &= K^2 \int_0^\infty w(\theta) \, J_0(K\theta) \, \theta \, d\theta.
\end{aligned}
}
$$

$$(102)$$

For power-law clustering, $w(\theta) = (\theta/\theta_0)^{-\epsilon}$, this gives

$$
\Delta^2_\theta(K) = (K\theta_0)^\epsilon \, 2^{1-\epsilon} \, \frac{\Gamma(1-\epsilon/2)}{\Gamma(\epsilon/2)}, \tag{103}
$$

$(= 0.77[K\theta_0]^\epsilon$ for $\epsilon = 0.8)$. At large angles, these relations are not quite correct. We should really expand the sky distribution in spherical harmonics

$$
\delta(\hat{q}) = \sum a_\ell^m Y_{\ell m}(\hat{q}), \tag{104}
$$

where $\hat{q}$ is a unit vector which specifies direction on the sky. Since the spherical harmonics satisfy the orthonormality relation $\int Y_{\ell m} Y^*_{\ell'm'} \, d^2q = \delta_{\ell\ell'}\delta_{mm'}$, the inverse relation is

$$
a_\ell^m = \int \delta Y^*_{\ell m} \, d^2q. \tag{105}
$$

The analogues of the Fourier relations for the correlation function and power spectrum are

$$w(\theta) = \frac{1}{4\pi} \sum_{\ell} \sum_{m=-\ell}^{m=+\ell} |a_\ell^m|^2 \, P_\ell(\cos\theta)$$

$$|a_\ell^m|^2 = 2\pi \int_{-1}^{1} w(\theta) \, P_\ell(\cos\theta) \, d\cos\theta$$

(106)

For small θ and large ℓ, these go over to a form which looks like a flat sky, as follows. Consider the asymptotic forms for the Legendre polynomials and J_0 Bessel function:

$$P_\ell(\cos\theta) \simeq \sqrt{\frac{2}{\pi\ell\sin\theta}} \, \cos[(\ell+1/2)\theta - \pi/4]$$

$$J_0(z) \simeq \sqrt{\frac{2}{\pi z}} \, \cos[z - \pi/4].$$

(107)

This shows that we can approximate the correlation function in the usual way in terms of an angular power spectrum Δ_θ^2 and angular wavenumber K:

$$w(\theta) = \int_0^\infty \Delta_\theta^2(K) \, J_0(K\theta) \, \frac{dK}{K}; \qquad \Delta_\theta^2(K = \ell + \tfrac{1}{2}) = \frac{2\ell+1}{8\pi} \sum_m |a_\ell^m|^2.$$

(108)

An important relation is that between the angular and spatial power spectra. In outline, this is derived as follows. The perturbation seen on the sky is

$$\delta(\hat{\mathbf{q}}) = \int_0^\infty \delta(\mathbf{y}) \, y^2 \phi(y) \, dy,$$

(109)

where $\phi(y)$ is the **selection function**, normalized such that $\int y^2\phi \, dy = 1$, and y is comoving distance. The form $\phi \propto y^{-1/2} \exp -(y/y^*)^2$ is often taken as a reasonable approximation to the Schechter function. A flat Universe ($\Omega = 1$) is assumed. Now write down the Fourier expansion of δ. The plane waves may be related to spherical harmonics via the expansion of a plane wave in Spherical Bessel functions j_ℓ

$$e^{ikr\cos\theta} = \sum_0^\infty (2\ell+1) \, i^\ell \, P_\ell(\cos\theta) \, j_\ell(kr),$$

(110)

plus the spherical harmonic addition theorem

$$P_\ell(\cos\theta) = \frac{4\pi}{2\ell+1} \sum_{m=-\ell}^{m=+\ell} Y_{\ell m}^*(\hat{\mathbf{q}}) Y_{\ell m}(\hat{\mathbf{q}}'),$$

(111)

where $\hat{\mathbf{q}} \cdot \hat{\mathbf{q}}' = \cos\theta$. These relations yield the desired result:

$$\langle |a_\ell^m|^2 \rangle = 4\pi \int \Delta^2(k) \, \frac{dk}{k} \left[\int y^2 \phi(y) \, j_\ell(ky) \, dy \right]^2 .$$

(112)

What is the analogue of this formula for small angles? Rather than manipulating large-ℓ Bessel functions, it is easier to start again from the correlation function. By writing as above the overdensity observed at a particular direction on the sky as a radial integral over the spatial overdensity, with a weighting of $y^2\phi(y)$, we see that the angular correlation function is

$$\langle \delta(\hat{\mathbf{q}}_1) \delta(\hat{\mathbf{q}}_2) \rangle = \int\int \langle \delta(\mathbf{y}_1) \delta(\mathbf{y}_2) \rangle \, y_1^2 y_2^2 \phi(y_1) \phi(y_2) \, dy_1 \, dy_2. \qquad (113)$$

We now change variables to the mean and difference of the radii, $y \equiv (y_1 + y_2)/2$; $x \equiv (y1 - y2)/2$. If the depth of the survey is larger than any correlation length, we only get signal when $y_1 \simeq y_2 \simeq y$. If the selection function is a slowly-varying function, so that the thickness of the shell being observed is also of order the depth, the integration range on x may be taken as being infinite. For small angles, we then obtain **Limber's equation**:

$$w(\theta) = \int_0^\infty y^4 \phi^2 \, dy \int_{-\infty}^\infty \xi(\sqrt{x^2 + y^2 \theta^2}) \, dx.$$

(114)

Theory usually supplies a prediction about the linear density field in the form of the power spectrum, and so it is convenient to recast Limber's equation:

$$w(\theta) = \int_0^\infty y^4 \phi^2 \, dy \int_0^\infty \pi \Delta^2(k) \, J_0(ky\theta) \, dk/k^2. \qquad (115)$$

For the form of selection function discussed above, this becomes

$$w(\theta) = \frac{\pi}{2\Gamma^2(5/4)} \int_0^\infty \Delta^2(k) \, e^{-[k\theta y^*]^2/8} \, (1 - [k\theta y^*]^2/8) \, \frac{dk}{k^2 y^*}. \qquad (116)$$

The power-spectrum version of Limber's equation is already in the form required for relation to the angular power spectrum ($w = \int \Delta_\theta^2 J_0(K\theta) dK/K$), and so we obtain the direct small-angle relation between spatial and angular power spectra:

$$\Delta_\theta^2 = \frac{\pi}{K} \int \Delta^2(K/y) \, y^5 \phi^2(y) \, dy.$$

(117)

This is just a convolution in log space, and is considerably simpler to evaluate and interpret than the $w - \xi$ version of Limber's equation.

Finally, note that it is easy to make allowance for spatial curvature in the above discussion. All that is needed is to replace the $\Omega = 1$ volume element $y^2 \, dy$ by its generalised counterpart, $y^2 \, dy/(1 - ky^2)^{1/2}$.

6.3 Scaling

One important consequence follows from the above formula for the angular power spectrum. Suppose we increase the depth of the survey by some factor D, but keeping the functional form of $\phi(y)$ the same. The angular clustering therefore scales as

$$\Delta_\theta'^2(K) = \frac{1}{D}\Delta_\theta^2(K/D);$$

(118)

i.e. the clustering pattern moves to smaller angles, reducing the observed clustering, and is also diluted by a further $1/D$ factor. This test is an excellent way of discriminating true spatial clustering from spurious effects caused by foreground extinction or magnitude errors. Prior to redshift surveys, it was the only means of establishing galaxy clustering as a real effect.

7 Measuring the clustering spectrum

The history of attempts to quantify galaxy clustering is long. The major early landmarks were the angular analysis of the Lick catalogue, described in Peebles (1980), and the analysis of the CfA redshift survey (Davis & Peebles 1983). It has taken some time to obtain data on samples which greatly exceed these in depth, but several pieces of work have appeared recently, notably the angular correlation function for the APM survey (Maddox *et al.* 1990), shown in figure 9.

The data on $w(\theta)$ display a break at $\theta \simeq 10°$, which must reflect a break in the power spectrum. Fixing the clustering at small scales, where things are well determined, suggests the functional form

$$\Delta^2 = \frac{(k/k_0)^{1.6}}{1 + (k/k_c)^{-2.4}}.$$

(119)

This imposes the theoretical prejudice of scale invariance. As is shown in figure 9, this provides a good fit to the APM $w(\theta)$ data, with $k_0 = 0.19h\,\mathrm{Mpc}^{-1}$, and k_c in the range $0.015h - 0.025h\,\mathrm{Mpc}^{-1}$ (we have assumed $y^* = 233h^{-1}\,\mathrm{Mpc}$ in the selection function, as required to fit the original Lick data: see Peacock 1991).

The value of k_0 is highly insensitive to different assumptions about the form of the selection function, or about whether $\xi(r)$ might evolve as some power of $(1 + z)$. The numerical value of the break can be raised or lowered

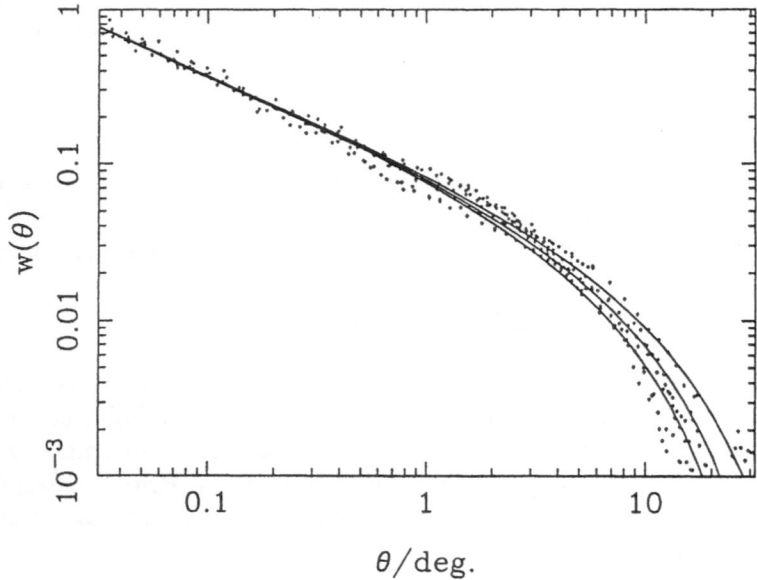

Fig. 9. The angular correlation function, $w(\theta)$, at Lick depth. The APM data are plotted as points. The solid lines show the result of applying Limber's equation to our fitting formula for the power spectrum, with $k_c = 0.015h$, $0.02h$ & $0.025h$ Mpc^{-1}, assuming a Schechter-type selection function (increasing k_c reduces w).

by about a factor 1.5 by adopting selection functions with varying widths, and allowing for realistic systematics in the data. Given that these latter are more likely to be in the sense of introducing some spurious large-scale clustering, we shall take the conservative approach of adopting the minimum level of large-scale fluctuations which is consistent with the APM data on the standard model: $k_c = 0.025h$ Mpc^{-1}. In what follows, we shall assume that the true value may lie anything up to a factor $\simeq 1.5$ either side of this.

We can attempt to compare this empirical fit with power-spectrum data derived from redshift surveys. Direct power-spectrum analysis has been performed for two samples: (a) the CfA survey (Baumgart & Fry 1991); (b) a survey of radio galaxies at $z < 0.1$. (Peacock & Nicholson 1991). Note that it is a much more robust procedure to determine the power spectrum directly than by transforming the correlation function: in the latter case, conclusions about large-scale clustering are very sensitive to the assumed mean density. Also, results from the IRAS QDOT survey are published in the form of the variance (σ^2) of δ in cubical cells of side ℓ (Efstathiou *et al.* 1990) and Gaussian spheres of radius R_G (Saunders *et al.* 1991). For a power-law spectrum ($\Delta^2 \propto k^{(n+3)}$), we have for the Gaussian sphere

$$\sigma^2 = \Delta^2 \left(k = \left[\tfrac{1}{2} \left(\tfrac{n+1}{2} \right)! \right]^{1/(n+3)} / R_{\mathrm{G}} \right). \qquad (120)$$

For $n \lesssim 0$, this formula also gives a good approximation to the case of cubical cells, with $R_{\mathrm{G}} \to \ell/\sqrt{12}$. The result is rather insensitive to assumptions about the power spectrum, and just says that the variance in a cell is mainly probing waves with $\lambda \simeq 2\ell$. Since we know the shape of Δ^2 reasonably well, we can get very accurate effective wavenumbers and plot the IRAS σ^2 values on the $\Delta^2 - k$ plane directly using these k_{eff} values. The IRAS points are on average well consistent with our formula for the APM $\Delta^2(k)$. This is a little surprising, as we do expect blue-selected galaxies to have slightly stronger correlations than IRAS galaxies, which avoid rich clusters. Cell-count analysis of a redshift survey from the APM galaxy catalogue gives variances approximately 1.5 times as high as the IRAS numbers (Maddox 1991). Uncertainties in the appropriate luminosity function to use in Limber's equation could yield a scale error of this order. We shall not apply any scaling, and so our results for Δ^2 refer to the redshift-space power spectrum of IRAS galaxies. This is to be preferred, as we have a better idea what degree of bias may apply to these objects.

The final compilation of power-spectrum results is shown in figure 10, after a vertical scaling to allow for the fact that (being luminous ellipticals) radio galaxies cluster more strongly than both optically selected and IRAS galaxies. There is an impressive degree of unanimity about these data, and it seems that a break in the spectrum at $\lambda \simeq 200h^{-1}$ Mpc has been detected. The fact that the break is at such a large scale shows why it has taken so long to obtain data of the necessary depth to detect it. It is worth noting that this conclusion is not dependent only on the APM results. The radio-galaxy data alone constrain the break quite well, setting a lower limit of $k_c \lesssim 0.05h$ Mpc^{-1}. Conversely, the Δ^2 estimate from the APM data would be inconsistent with the radio-galaxy upper limits if $k_c \lesssim 0.025h$ Mpc^{-1}. If future observations should prefer values of $w(\theta)$ which follow the upper bound of the APM data, this will be hard to reconcile with the radio-galaxy data.

Let us now compare these results with some theoretical models. figure 11 compares the observed $\Delta^2(k)$ with linear CDM power spectra, scaled to agree with observation at long wavelengths. This figure makes it clear that the problem with CDM is the *shape* of the power spectrum, rather than the absolute amount of power at large scales. The linear transfer function does not bend sharply enough at the break wavenumber. From the renormalized point of view, the problem with CDM is now that it produces too much *small-scale* power. If the model is to survive, the true galaxy power spectrum at small scales must be suppressed below the prediction of linear theory. This may not be so surprising: the discrepancy now occurs at a scale where $\Delta \sim 1$, and nonlinear effects could be significant. A much higher degree of caution is required in making comparisons between theory and observation

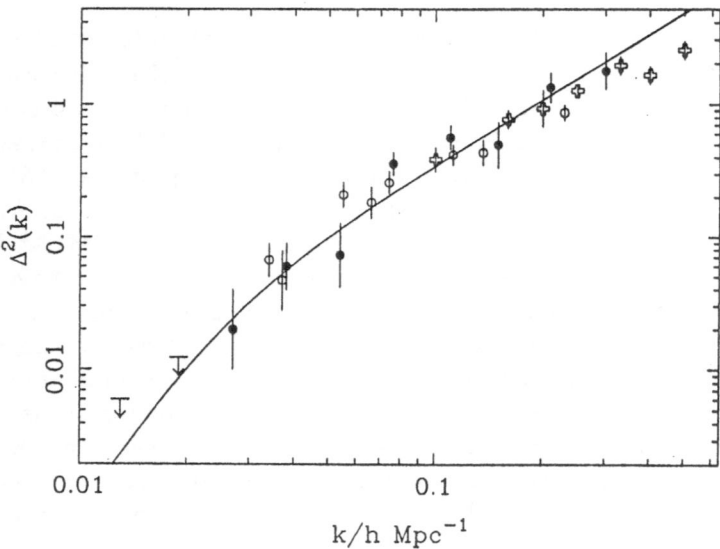

Fig. 10. The power spectrum in the form $\Delta^2 \equiv d\sigma^2/d\ln k$. The solid line shows a model fit to $w(\theta)$; filled points are radio galaxies (Peacock & Nicholson 1991) with Δ^2 reduced by a factor 3; open circles are IRAS (Efstathiou *et al.* 1990; Saunders *et al.* 1991) crosses are from the CfA (Baumgart & Fry 1991). As discussed in the text, the solid line is probably too low by a factor of 1.5 to represent the clustering of blue-selected galaxies; this plot therefore corresponds to the redshift-space power spectrum of IRAS galaxies. The fact that points lie below the line for $k \gtrsim 0.2$ is probably attributable to peculiar velocity smearing.

in the difficult non-linear regime than in the 'clean' linear world of > 100 Mpc perturbations.

Nevertheless, significant alterations to the CDM transfer function do appear to be required on disturbingly large scales, and it is worth looking for alternative models with linear power spectra that provide a better fit to the data. One could not do much better in this regard than isocurvature CDM with $h = 0.5$; this gives a rather sharper bend at the break scale, as the data seem to require. Sadly, this model does appear to conflict with CBR limits (Efstathiou & Bond 1986). Another way of altering the transfer function to reduce unwanted small-scale power is to consider damping resulting from the dark matter having non-zero rest mass. The general case is warm dark matter: $\Omega = 1$ is supplied by a particle somewhat less abundant than neutrinos. This class of dark matter has been unfashionable in recent years, but a new candidate has recently been suggested Rajagopal, Turner & Wilczek (1991): the axino (the superpartner of the axion). According to BBKS, the transfer function is modified in the warm dark matter case to $T_{\text{WDM}} = T_{\text{CDM}} \exp[-(kR_D)/2 - (kR_D)^2/2]$, where the damping length is $R_D = 0.2\,(\Omega h^2)^{1/3}\,[m/\text{keV}]^{-4/3}$ Mpc. However, this produces virtually negligible damping at the point in the power spectrum where it is required $(k \simeq 0.1h\,\text{Mpc}^{-1})$, unless the mass is low: of the order of 50 eV, for all h values. We thus end up with all the well-known problems of the hot dark matter picture for galaxy formation (*e.g.* Hut & White 1984).

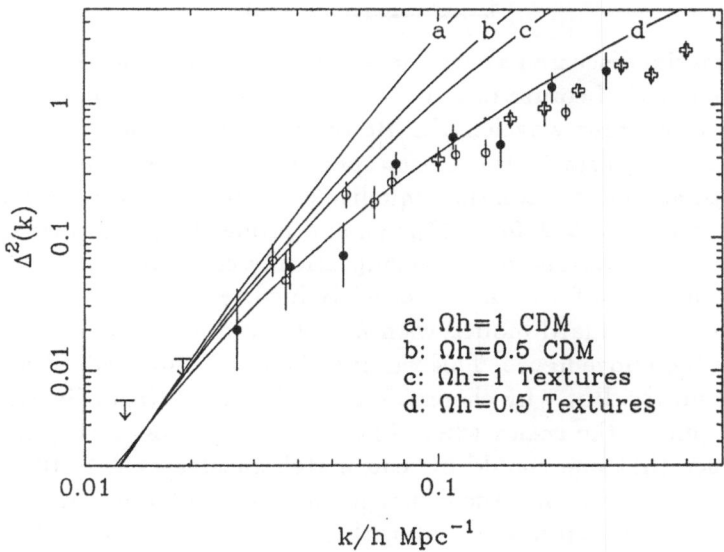

Fig. 11. A comparison of the 'observed' power spectrum as in Figure 10 with some theoretical linear power spectra. All curves are normalized to agree with the APM $\Delta^2(k)$ at $k = 0.015h$ Mpc^{-1}; for J_3 normalization, the CDM curves would cross the data at $k \gtrsim 0.1$ – producing a great apparent deficit of large-scale power.

A more radical alternative which looks promising is the case of cold dark matter with non-Gaussian fluctuations seeded by the topological defects known as 'textures' (Turok 1991). These produce a power spectrum similar in shape to standard CDM: scale invariant at large wavelengths, but breaking around the horizon length at matter-radiation equality. This model gives an excellent fit to the data for $h \simeq 0.5$.

We can use the power-spectrum data to obtain an estimate of the primordial normalization. First, there is a correction to make, because we have considered clustering in redshift space (Kaiser 1987). This boosts the amplitude of the large-scale power:

$$\Delta^2 \to \Delta^2(1 + 2\Omega^{0.6}/3b + \Omega^{1.2}/5b^2). \tag{121}$$

There is some evidence that IRAS galaxies are close to being unbiased in the linear sense, with values $1.0 \lesssim b \lesssim 1.6$ being inferred from comparisons of large-scale clustering and velocity fields (Kaiser & Lahav 1989; Yahil 1991); we may therefore try to make an estimate of ϵ. For our power-spectrum fitting formula, the redshift-space value of ϵb is 5.3×10^{-5}. For $\Omega = 1$, $b = 1.3$, the amplitude correction factor is 1.28, and our final estimate of ϵ becomes

$$\epsilon \simeq 3.2 \times 10^{-5}.$$

$$(122)$$

The systematics discussed above constrain ϵ to be within a factor of 2 of this figure. One could of course achieve any value for ϵ by altering the spectrum at unobservably large wavelengths. However, in most models one would not expect to see features in the power spectrum on scales much beyond the horizon size at matter-radiation equality ($15.7[\Omega h^2]^{-1}$ Mpc). Even for low Hubble constants, $\lambda > 200h^{-1}$ Mpc should probe the primordial spectrum. If the power spectrum is indeed asymptotically scale free, it seems implausible that our figure for ϵ can be seriously incorrect.

The CDM 'standard model' with $h = 0.5$ and $b_J = 2$ has ϵ a factor 3.1 lower than the above figure. This variant of CDM is thus indeed deficient in large-scale power. But why choose J_3 normalization? Normalization should be carried out at the scales where linear theory is most nearly applicable, which argues that one should instead match spectra at $\lambda > 100h^{-1}$ Mpc. Non-linear evolution may then change the shape of the power spectrum and allow CDM to survive, albeit with a higher amplitude of fluctuations (Carlberg & Couchman 1991).

8 Topology and cellular models

How are we to distinguish observationally whether the density field of the Universe is Gaussian? We need to look at more subtle statistics than just the power spectrum. In principle, one might use the higher-order n-point correlation functions, since these are directly related to the power spectrum for a Gaussian field; however, in practice these are rather noisy. An interesting alternative was suggested by Gott, Melott & Dickinson (1986): the topology of the density field. To visualise the main principles, it will help to think initially about a 2D field. Two extreme non-Gaussian fields would consist either of discrete 'hotspots' surrounded by uniform density, or the opposite: discrete 'coldspots'; the picture in either case is a set of polka dots. Both of these cases are clearly non-Gaussian just by symmetry: the contours of average density will be simply-connected circles containing regions which are all either above or below the mean density, but in a Gaussian field (or any symmetric case), the numbers of hotspots and coldspots must balance.

One might think that things would be much the same in 3D: the obvious alternatives are 'meatball' or 'Swiss cheese' models. However, there is a third topological possibility: that of the **sponge**. In our two previous examples, high- and low-density regions were distinguished by their **connectivity** (whether it is possible to move continuously between all points in a given set). In contrast, a sponge has both classes of region being connected: it is possible to swim to any point through the holes, or to burrow to any point within the body of the sponge; filling a sponge with cement and etching away the sponge produces a cement sponge. Again, just by the symmetry between overdensity and underdensity, a Gaussian field in 3D must have a sponge-like topology.

8.1 The genus

The above discussion has focused on the properties of contour surfaces. These properties can be studied quantitatively via the **genus**: the number of 'holes' in a surface (zero for a sphere, one for a doughnut *etc.*). This is related to the Gaussian curvature of the surface, $K = 1/(r_1 r_2)$ (where r_1 and r_2 are the two principal radii of curvature), via the Gauss-Bonnet theorem (see *e.g.* Dodson & Poston 1977)

$$C \equiv \int K \, dA = 4\pi(1 - G), \tag{123}$$

where G is the genus. Topological results are sometimes instead quoted in terms of the **Euler-Poincaré characteristic**, which is -2 times the genus.

For Gaussian fields, the expectation value of the genus per unit volume (denoted by g) is (see Hamilton, Gott & Weinberg 1986)

$$g = \frac{1}{4\pi^2} \left[\frac{-\xi''(0)}{\xi(0)} \right]^{3/2} (1 - \nu^2) e^{-\nu^2/2}. \tag{124}$$

For the median density contour ($\nu = 0$), the curvature is negative, implying that the surface has genus greater than unity. For $|\nu| > 1$, however, the curvature is positive – as expected if there are no holes. The contours become simply connected balls around either isolated peaks or voids.

It is interesting to note that the genus carries some information about the shape of the power spectrum, not in the behaviour with ν, but in the overall scaling. For Gaussian filtering,

$$g = \frac{1}{4\pi^2 R_G^3} \left(\frac{3 + n}{3} \right)^{3/2} (1 - \nu^2) e^{-\nu^2/2}, \tag{125}$$

and so the effective spectral index can be determined in this way.

A similar procedure can be carried out in 2D (see Melott *et al.* 1989; Coles & Plionis 1991). The Gauss-Bonnet theorem is now

$$C \equiv \int K \, dA = 2\pi(1 - G),$$ (126)

where the meaning of $1 - G$ is the number of isolated contours minus the number of contour loops within other loops (sometimes the 2D genus is defined with the opposite sign; our convention follows that in 3D and the signs below are consistent). The result for the 2D genus per unit area is

$$g = -\frac{1}{(2\pi)^{3/2}} \left[\frac{-\xi''(0)}{\xi(0)} \right] \nu \, e^{-\nu^2/2}.$$ (127)

We shall now indicate how this result is derived in 2D. The 3D case is analogous but, as always, a little more messy (see BBKS for details).

First, it helps to choose a special (x, y) coordinate system, in which a contour line is horizontal at the origin – *i.e.* $\delta'_1 = 0$. Let the equation of the contour line be $y(x)$; since $y' \equiv dy/dx = 0$ at the origin, the curvature is just y'' if $\delta'_2 > 0$, or $-y''$ if $\delta'_2 < 0$ (distinguishing whether high-density regions are 'inside' or 'outside' the contour). Making a Taylor expansion of δ about the origin and differentiating twice, we find for the curvature

$$K = -\delta''_{11}/|\delta'_2|$$ (128)

in all cases. The contribution to the curvature integral from the element $dx \, dy$ is therefore $K \, dx$; we just need the probability of crossing a contour in dy and we are done. As for peaks, this is achieved by saying we wish to consider the contour at $\delta = \delta_c$ and changing variables in the Gaussian probability distribution from $d\delta$ to dy, bringing in a factor $|\delta'_2|$ as the Jacobian. The contribution to the genus per unit area is then

$$\frac{dg}{dA} = \frac{-1}{2\pi}(-\delta''_{11}) \, P_G(\delta = \delta_c) \, d^3\delta'' \, d^2\delta',$$ (129)

where P_G is shorthand for the Gaussian probability density. Finally, we must allow for having assumed that the coordinate system is always oriented such that $\delta'_1 = 0$. This is achieved by replacing $d^2\delta'$ by $|\delta'_2| \, d\phi \, d|\delta'_2|$, where ϕ is a polar angle. Integrating ϕ over 0 to π gives the final answer, with no sign constraints on δ' or δ'':

$$\frac{dg}{dA} = -\tfrac{1}{2}(-\delta''_{11})|\delta'_2| \, P_G(\delta = \delta_c, \delta'_1 = 0) \, d^3\delta'' \, d\delta'_2.$$ (130)

Integration yields the above formula for g.

Applications of this method to real data (figure 12) naturally reveal departures from Gaussian behaviour – one wishes to test whether the initial

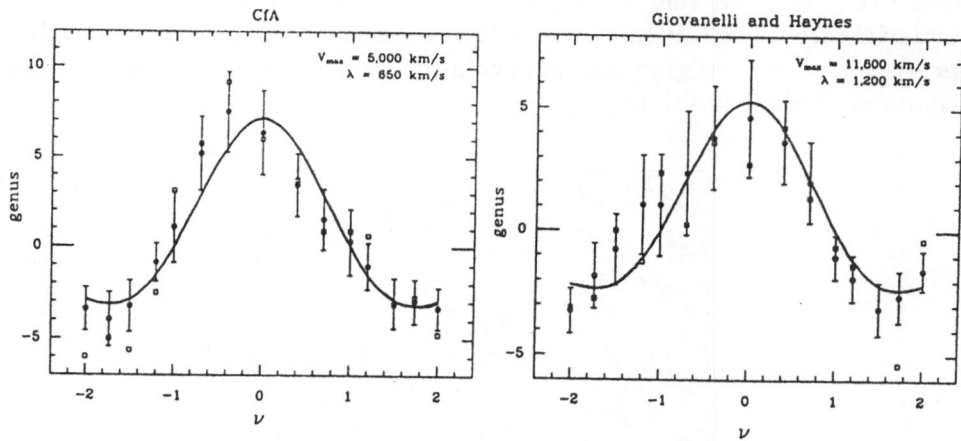

Fig. 12. Results from the Genus analysis applied to 3D redshift data, taken from Gott et al. (1989). The small 'meatball shift' seen here is argued by the authors to be consistent with non-linear evolution from Gaussian initial conditions. It is interesting that the behaviour becomes more nearly Gaussian as we move to deeper samples which allow larger filtering lengths. Such plots constitute the strongest evidence we have that cosmic structure did indeed form via gravitational instability from Gaussian primordial fluctuations.

conditions were Gaussian, realising that nonlinear evolution will cause the field to become non-Gaussian. This means that either N-body simulations have to be used to predict the degree of non-Gaussian behaviour (usually in the 'meatball' direction), or one is confined to smoothing the data heavily to probe only large linear/angular scales. These should still be Gaussian, but of course by smoothing over many small regions there is the danger that the central limit theorem will produce a Gaussian-like result in all cases.

8.2 Voronoi tesselations

The above indications of Gaussian behaviour for the median density contour notwithstanding, visual examination of redshift survey data has often suggested a cellular topology for the Universe. This need not be inconsistent with the idea of Gaussian initial conditions, as was first shown by Zeldovich (1970), who emphasised the tendency for gravitational collapse to yield flattened structures (pancakes). Subsequent extension of his work in the 'adhesion model' (Gurbatov, Saichev & Shandarin 1989) confirms this conclusion. They show that, at late times, matter will tend to congregate in a geometrical structure which consists of a network of pancakes, which meet in filaments, which in turn intersect at clusters. All this is saying is that the structure of the Universe tends to be dominated by the voids; because they are underdense, they expand faster than average, sweeping all material into a small volume where the expanding void walls meet. Such a process is generic to gravitational evolution, and will certainly result from Gaussian initial conditions.

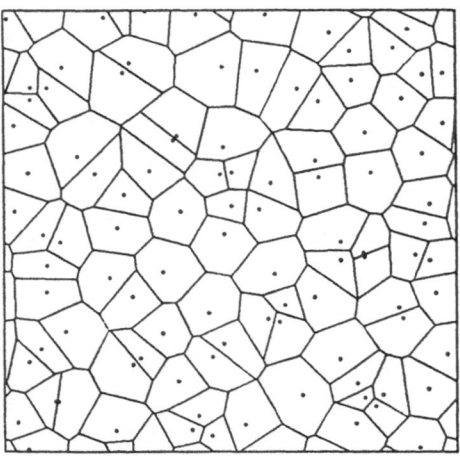

Fig. 13. Illustrating the construction of the Voronoi tesselation in 2D. The points mark the seeds – notionally peaks in the gravitational potential about which voids develop. If all voids overexpand at the same rate, matter accumulates in the cell walls which are equidistant from their nearest seeds.

A simple model of this process is provided by the **Voronoi tesselation**. Start with a number of randomly-distributed seeds; for each seed, construct the plane which bisects the line joining that seed to its neighbours. These planes define a cell around each seed such that all points in the cell are closer to that seed than to any other – the spatial analogue of the Wigner-Seitz k-space cells encountered in solid-state physics. In the adhesion model, the seeds would be identified with large-scale pits in the gravitational potential. The Voronoi model is inexact in that these pits are neither unclustered, nor are they all of the same depth. In practice, the influence of this imprecision is small, and the Voronoi model is useful both as a quantitative and heuristic tool for studying a cellular distribution of galaxies (van de Weygaert & Icke 1989; van de Weygaert 1991).

The predictions of the model are fixed by the density of seeds, n_s – or, rather, by the inter-seed separation $r_s \equiv n_s^{-1/3}$. There are three different sites for galaxies in the Voronoi network: nodes, filaments and walls, and they all have different clustering properties. The first two have approximately $\xi \propto r^{-2}$ with $r_0 = 0.28 r_s$ and $0.22 r_s$ respectively, whereas walls have $r_0 = 0.14r_s$ and the flatter slope $\xi \propto r^{-1}$ (Williams *et al.* 1991b). The flat slope for walls is easy to understand, and is a general prediction for any cellular model: the number of neighbours for a point on a sheet grows as r^2, whereas the expected number for uniform density grows as r^3. For example, the correlation function for a uniform cellular network composed of cubes of side λ is (Heavens 1985)

$$\xi(r) = \frac{\lambda}{6r} \left[1 + 2\operatorname{int}\left(\frac{r}{\lambda}\right) \right] - \frac{1}{3}. \tag{131}$$

Earlier work on the Voronoi model was motivated by the clustering of Abell clusters, identified as Voronoi nodes. This led to the adoption of the large scale-length $r_s \simeq 100h^{-1}$ Mpc (van de Weygaert 1991). However, the above results on the clustering of sheets clearly limit r_s to no more than about half this figure. Similar constraints arise with any cellular model: galaxy clustering limits the size of the cells to about $40h^{-1}$ Mpc. The fact that this is around the size of the largest voids shows that a large component of galaxy clustering is indeed induced by the geometrical structure which galaxies inhabit.

9 Galaxy evolution and galaxy formation

9.1 Evolution of normal galaxies

One area of great current interest in cosmology is the evolution of the population of galaxies, driven by technical developments which have allowed imaging and spectroscopy to be carried out to very faint levels (see *e.g.* Ellis 1990; Lilly *et al.* 1991). For many years it had been known that the numbers of faint galaxies seen in blue light exceeded that expected if there had been no evolution in the population. This was conventionally ascribed to stellar evolution: galaxies were expected to be more luminous in the past. However, recent spectral and infrared data have shown that what is going on is much more interesting.

To construct models for the evolution of galaxies requires a knowledge of the local luminosity function, plus the spectral shape of galaxies. This last is described via the K-correction which gives the difference between the observed dimming with redshift for a fixed observing waveband, and that expected on bolometric grounds:

$$m = M + 5\log_{10}\left(\frac{D_L}{10 \text{ pc}}\right) + K(z), \tag{132}$$

where D_L is luminosity distance. Thus, $K(z)$ is large and positive for a very red galaxy. For a $\nu^{-\alpha}$ spectrum,

$$K(z) = 2.5(\alpha - 1)\log_{10}(1 + z). \tag{133}$$

The effective spectral index of galaxies is a function of wavelength. Thinking of galaxies very crudely as black bodies, we expect the Rayleigh-Jeans $\alpha = -2$ in the far infrared, becoming positive at rest wavelengths $\gtrsim 1\mu$m, and returning to zero at very short wavelengths as we encounter the contribution of young blue O & B stars due to current star formation. In the B and K wavebands α values of about 4 and 0 respectively apply, but these are very rough figures. Accurate modelling requires the detailed spectral energy distribution of the galaxies.

Figure 14 shows the results of this modelling and reveals an apparent contradiction – we need strong evolution in the blue, but not in the infrared. Of course, the blue light is very sensitive to a small amount of star formation and so this need not be a problem – but what triggers the star formation? One natural explanation is galaxy mergers, but the non-evolving K data appear at first sight to rule this out. However, things may not be so simple. Glazebrook (1991; see also Broadhurst *et al.* 1991) has explored empirical merger models which conserve mass and evolve the luminosity function homologously:

$$\phi_* L_* = \text{constant}; \quad \phi_* \propto \exp(Q\tau), \tag{134}$$

where τ is the normalised look-back-time: $\tau \equiv Ht \simeq z$ for $z \ll 1$; a good approximation (exact for $\Omega = 1$ and 0) is

$$\tau \simeq [1 - (1+z)^{-\beta}]/\beta; \quad \beta = 1 + \Omega^{0.6}/2. \tag{135}$$

It turns out that large values of Q are not strongly excluded by the counts. Consider the limit of very large Q, and think about objects all with the same (evolving) luminosity L and density ρ. The number of objects above a given flux density, S, is just $\int \rho r^2 \, dr$. Since $S \propto L/r^2$, $L \propto 1/\rho$, and the limiting radius changes very slowly with S, this yields an integral count with $N \propto S^{-1}$ – rather close to the observed sub-Euclidean slope. The way to constrain such models is then through the redshift distributions; high Q models predict lower mean redshifts. Consider models which fit the blue counts by an empirical enhancement of blue light as expected in a merger-induced starburst:

$$L_B \rightarrow L_B \exp(b\tau). \tag{136}$$

The count data require $b \simeq 4$ but are again insensitive to Q. The redshift data for blue-selected samples (figure 15), however, strongly rule out models where only the blue luminosity changes; merger models give a much better fit. This in itself is not conclusive proof that mergers are occurring; tidal interactions without mergers might still generate starbursts (and perhaps this is a further way of generating bias in clusters: Lacey & Silk 1991). The final test of this hypothesis will be redshift data from deep infrared surveys. Here, we can study any evidence for mergers without the complications of understanding any associated starbursts. It will be amusing if it should

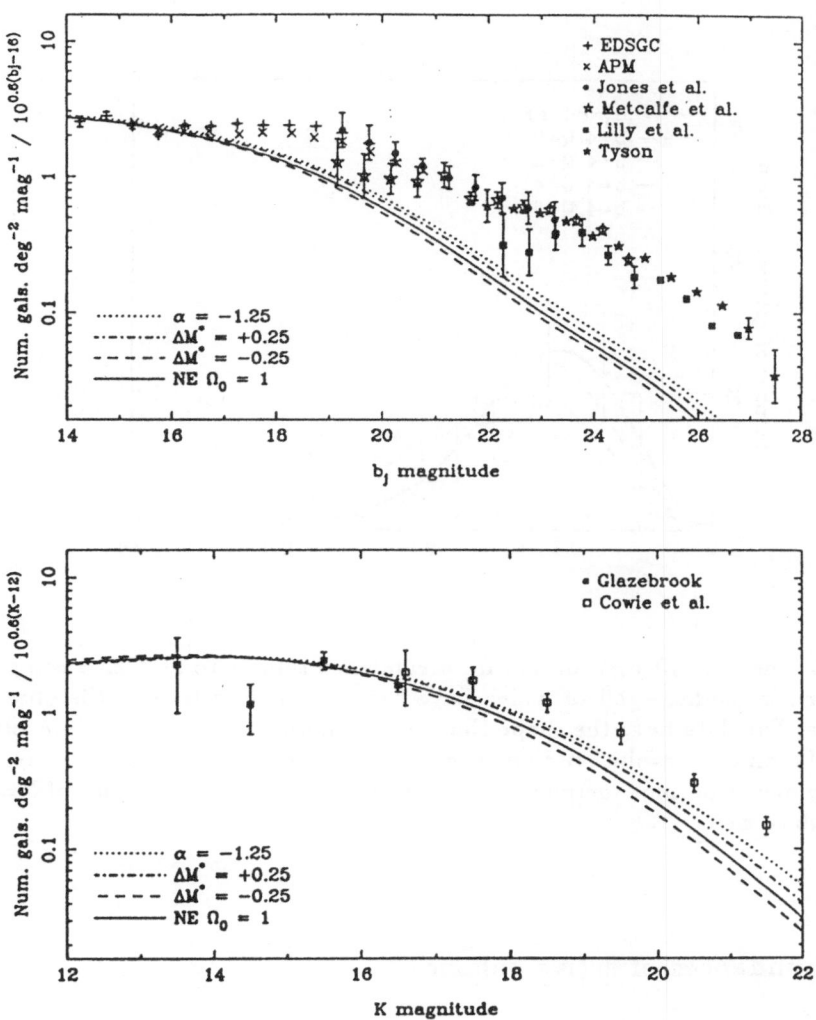

Fig. 14. Differential counts of galaxies in the K and B bands, normalised to the Euclidean prediction in order to reduce the dynamic range of the data (from Glazebrook 1991). In both cases, the numbers of faint galaxies decline more rapidly than the Euclidean form, as required if the sky brightness is not to diverge. The lines show some variants on no-evolution models, whereby the current galaxy population is transported unchanged to high redshifts. Although both sets of counts have very similar shapes, the different optical and infrared K-corrections mean that the blue counts were expected to cut off more rapidly, but this is not seen. The K counts are nearly consistent with the expectation for an unevolving galaxy population, whereas there is a great excess of faint blue galaxies.

turn out that a large amount of merging has been able to hide itself in this fashion.

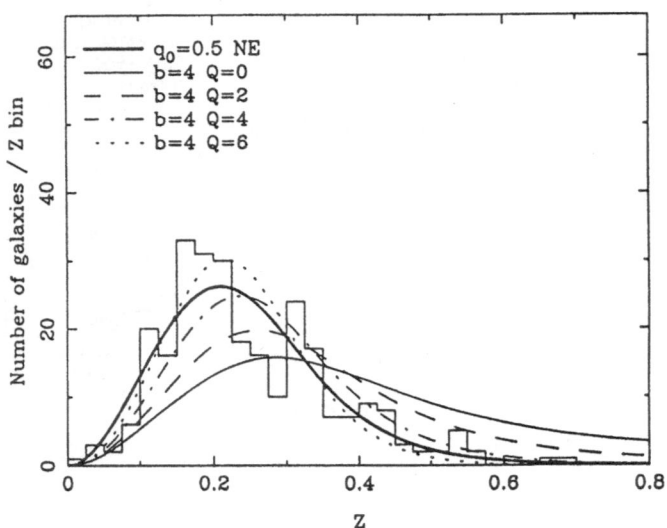

Fig. 15. The redshift distribution for a combined sample to $B \lesssim 22.5$ compared with various models – all of which have been normalised to give the observed numbers. The data have the shape (but not the normalisation) of a non-evolving model. Luminosity evolution alone predicts too many high-redshift objects. Luminosity evolution plus merging (equivalent to pure number evolution if $b = Q$) fits the data well.

9.2 Abundances of active galaxies

A more direct constraint on cosmological models arises from the abundance of objects at very high redshift; here, we are restricted to the study of active galaxies. For both radio-loud and radio-quiet AGN, it is a good approximation up to $z \simeq 2$ to describe the luminosity functions via Pure Luminosity Evolution: a conserved comoving density of objects with an evolving characteristic luminosity (L^*) which was a factor roughly 30 times larger at $z = 2$ than at the present (Boyle *et al.* 1987, 1988; Dunlop & Peacock 1990). The total comoving density is usually formally divergent at very low luminosities; restricting oneself to luminosities $L \gtrsim 0.1L^*$ gives characteristic densities of

$$\rho \simeq 10^{-6} h^3 \, \mathrm{Mpc}^{-3} \quad \text{radiogalaxies}$$
$$\rho \simeq 10^{-5} h^3 \, \mathrm{Mpc}^{-3} \quad \text{QSOs.}$$

(137)

These densities are so small because the host galaxies are highly luminous; the AGN densities are of the order of $\phi(L > 5L^*)$, and so it seems reasonable to use high-z AGN as a probe of the general population of *massive* galaxies.

At higher redshifts, data at radio and optical wavelengths suggest a deficit of AGN (Osmer 1982; Peacock 1985; Warren *et al.* 1988; Dunlop & Peacock 1990), although the detailed behaviour certainly remains controversial. Any 'redshift cutoff' is certainly rather gradual: perhaps only a factor 3 in comoving density between $z = 2$ and $z = 4$.

As first pointed out in an important paper by Efstathiou & Rees (1988), the above data on the abundance of massive galaxies may strongly constrain galaxy-formation models. Their argument assumes hierarchical growth of structure, as in the PS model. This merger picture can account in a reasonably natural way for the AGN redshift cutoff: at high z, M^* will eventually fall below the size of a massive galaxy, leading to an exponential decline in the number of available AGN sites. We could try to argue that we don't understand the reason that local AGN prefer massive galaxies, and that perhaps this changed at high z; however, the Eddington limit for quasars sets a lower limit to how far we can escape along this route. How long it takes M^* to evolve from the current Abell cluster value to that of a massive galaxy depends on the power spectrum. For Cold dark Matter, the critical redshift is probably in the region of 3 − 5 (Efstathiou & Rees 1988), and so we should not be surprised at a lack of very high-z quasars.

What mass is to be assigned to a given AGN? For QSOs, Efstathiou & Rees made various arguments involving timescales to constrain the black hole mass, which give rather similar answers to the Eddington limit. One then has the question of what fraction of a galaxy's total mass can be in the form of a central engine. For luminous quasars ($M_B < -27.6$, $h = 0.5$), Efstathiou & Rees concluded

$$M_{\rm BH} \gtrsim 10^9 M_\odot \Rightarrow M_{\rm tot} \gtrsim 10^{12} M_\odot. \qquad (138)$$

The line of argument here is clearly somewhat uncertain! However, we can short-circuit the difficulties by looking at the galaxy luminosities directly. We know that nearby radio galaxies are of giant elliptical to cD luminosity; examples such as M87 can have their total masses estimated, and answers in excess of $10^{12} M_\odot$ are obtained (*e.g.* White 1990). Furthermore, we know that the *stellar* mass of radio galaxies has not altered by more than a factor ~ 2 between $z = 2$ and the present (Lilly & Longair 1984; Lilly 1989). Add to this the fact that galaxy stellar luminosity (and thus presumably mass) correlates hardly at all with radio power, and we can conclude that the total density of radio galaxies may be used as the comoving density of objects above the Efstathiou-Rees limit:

$$\rho(M \gtrsim 10^{12} M_\odot) \simeq 10^{-6} h^3 \, {\rm Mpc}^{-3}. \qquad (139)$$

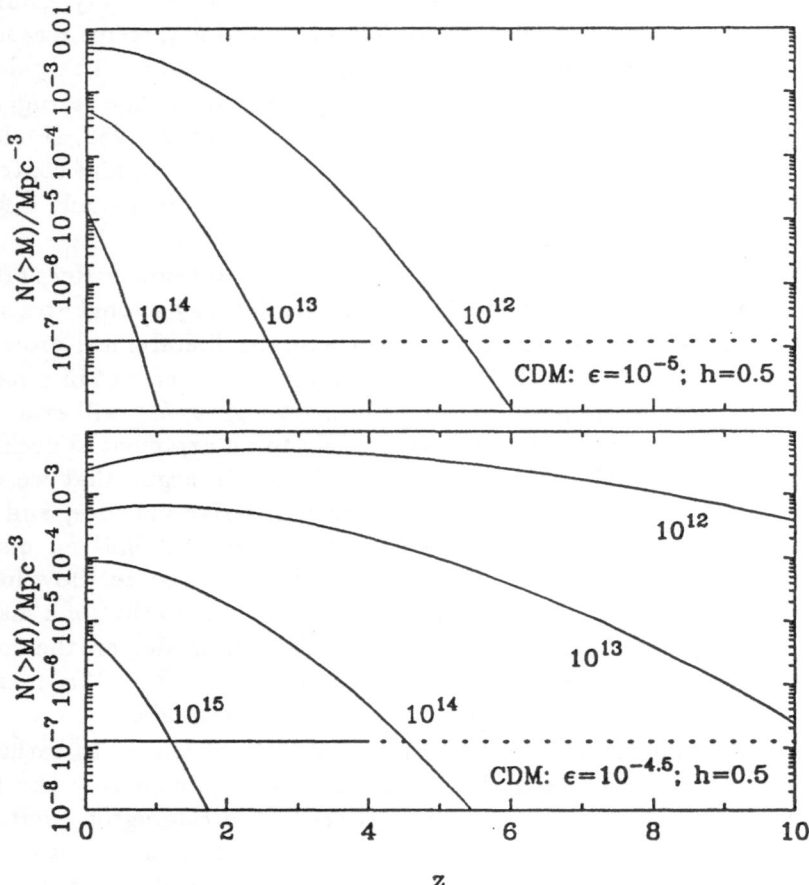

Fig. 16. The epoch dependence of the integral mass function in CDM, calculated as in Efstathiou & Rees (1988). Two normalizations in terms of the ϵ parameter are shown. These correspond to (a) 'conventional' $b = 2$ and (b) 'low-bias' $b = 0.8$ models. The horizontal line shows the observed comoving density for radio galaxies.

From figure 16, the constraint of achieving the radio-galaxy abundance at $z = 4$ would seem to require masses $\lesssim 10^{12.5}$ and $\lesssim 10^{14} M_\odot$ for the two models shown. At first sight, this agrees with Efstathiou & Rees, who concluded that high-bias CDM could stand the observed numbers of high-z QSOs. However, this conclusion is not warranted. The Press-Schechter formalism deals in fictitious total masses, being normalised to assign every particle in the Universe to the halo of one or other collapsed object. On this basis, a constant M/L ratio would assign $10^{13.2} h^{-1} M_\odot$ to an L^* galaxy (Efstathiou et al. 1988); for objects of radio-galaxy luminosity, the appropriate figure is around $10^{14}\,M_{\odot}$. Of course, most of this mass is in 'no-man's land' between galaxies and so is hardly observable, but this is the point: we are comparing Press-Schechter masses with those we *can* observe.

This suggests that even low-bias CDM is barely able to account for the observations. High-bias CDM seems to be in serious trouble from the abundance of massive high-z AGN (and, indeed, from several other directions). What about higher redshifts? Even though almost any hierarchical model (not just CDM) would have difficulty in accounting for an unaltered abundance of objects of radio-galaxy mass at $z > 10$, this does not mean that we should never expect to see AGN at these redshifts. The characteristic mass does not evolve so rapidly $(M^*(z = 10)/M^*(z = 4) \simeq 0.1$ for CDM), and so objects of correspondingly lower mass could be relatively abundant at these redshifts (in fact, the comoving densities would rise in proportion – by one or more orders of magnitude). Of course, the efficiency of black-hole formation at these redshifts is an open question, but radiation at the Eddington limit for proportionately smaller black holes could easily produce detectable numbers of AGN at $z > 10$.

10 Conclusions

These lectures have covered a rather restricted subset of the current vigorous activity in quantifying the cosmological density field. Progress is rapid, and it seems clear that the next few years will see most of the remaining problems solved – at least in terms of measuring the relation between mass and light on scales up to several hundred Mpc. The outstanding questions will then be whether we can link this observational progress to larger scales through the detection of microwave-background anisotropies – and whether we will have a theory to account for what is seen.

References

Babul, A. & Postman, M., 1990. *Astrophys. J.*, **359**, 280..

Bahcall, N.A. & Soneira, R.M., 1983. *Astrophys. J.*, **270**, 20.

Balian, R. & Schaeffer, R., 1989. *Astr. Astrophys.*, **220**, 1.

Bardeen, J.M., Bond, J.R., Kaiser, N. & Szalay, A.S., 1986. *Astrophys. J.*, **304**, 15. (BBKS)

Baumgart, D.J. & Fry, J.N., 1991. *Astrophys. J.*, **375**, 25.

Blumenthal, G.R., Faber, S.M., Primack, J.R. & Rees, M.J., 1984. *Nature*, **311**, 517.

Bond, J.R. & Efstathiou, G., 1987. *Mon. Not. R. astr. Soc.*, **226**, 655.

Bond, J.R. & Couchman, H.M.P., 1987. In: *proceeding of the Second Canadian Conference on General Relativity and Relativistic Astrophysics*, eds A. Coley & C. Dyer, World Scientific, Singapore.

Bond, J.R., Cole, S., Efstathiou, G. & Kaiser, N., 1991. *Astrophys. J.*, in press.

Bower, R.G., 1991. *Mon. Not. R. astr. Soc.*, **248**, 332.

Boyle, B.J., Fong, R., Shanks, T. & Peterson, B.A., 1987. *Mon. Not. R. astr. Soc.*, **227**, 717.

Boyle, B.J., Shanks, T. & Peterson, B.A., 1988. *Mon. Not. R. astr. Soc.*, **235**, 935.

Brandenberger, R.H., 1990. in *Physics of the Early Universe*, proc 36th Scottish Universities Summer School in Physics, eds Peacock, J.A., Heavens, A.F. & Davies, A.T. (Adam Hilger), p281.

Broadhurst, T.J., Ellis, R.S., Koo, D.C. & Szalay, A.S., 1990. *Nature*, **343**, 726. (BEKS)

Broadhurst, T.J., Ellis, R.S. & Glazebrook, K., 1991. *Nature*, in press.

Carlberg, R.G., Couchman, H.M.P. & Thomas, P.A., 1990. *Astrophys. J.*, **352**, L29.

Carlberg, R.G., 1990. *Astrophys. J.*, **350**, 505.

Carlberg, R.G., 1991. *Astrophys. J.*, **367**, 385.

Cartwright, D.E. & Longuet-Higgins, M.S., 1956. *Proc. R. Soc. London. A.*, **237**, 212.

Cole, S. & Kaiser, N., 1989. *Mon. Not. R. astr. Soc.*, **237**, 1127.

Coles, P. & Barrow, J.D., 1987. *Mon. Not. R. astr. Soc.*, **228**, 407.

Coles, P. & Plionis, M, 1991. *Mon. Not. R. astr. Soc.*, **250**, 75.

Coles, P. & Jones, B.J.T., 1991. *Mon. Not. R. astr. Soc.*, **248**, 1.

Couchman, H.M.P. & Carlberg, R.G., 1991. *Astrophys. J.*, in press.

Dalton, G.B., Efstathiou, G., Maddox, S.J. & Sutherland, 1991. *Astrophys. J.*, in press.

Davis, M., Efstathiou, G., Frenk, C.S. & White, S.D.M., 1985. *Astrophys. J.*, **292**, 371.

Davis, M. & Peebles, P.J.E., 1983. *Astrophys. J.*, **267**, 465.

Dekel, A. & Silk, J., 1986. *Astrophys. J.*, **303**, 39.

Dodson, C.T.J. & Poston, T., 1977. *Tensor Geometry* (London; Pitman).

Dunlop, J.S. & Peacock, J.A., 1990. *Mon. Not. R. astr. Soc.*, **247**, 19.

Ellis, R.S., 1990. in *proc. Evolution of the Universe of Galaxies*, ed R.G. Kron, Astr. Soc. Pacif. Conf. Ser., **10**, 248.

Efstathiou, G. & Silk, J., 1983. *Fund. Cosm. Phys.*, **9**, 1.

Efstathiou, G. & Rees, M.J., 1988. *Mon. Not. R. astr. Soc.*, **230**, 5P.

Efstathiou, G., Ellis, R.S. & Peterson, B.A., 1988. *Mon. Not. R. astr. Soc.*, **232**, 431.

Efstathiou, G. & Bond, J.R., 1986. *Mon. Not. R. astr. Soc.*, **218**,103.

Efstathiou, G., 1990. in *Physics of the Early Universe*, proc 36th Scottish Universities Summer School in Physics, eds Peacock, J.A., Heavens, A.F. & Davies, A.T. (Adam Hilger), p361.

Efstathiou, G., Kaiser, N., Saunders, W., Lawrence, A., Rowan-Robinson, M., Ellis, R.S. & Frenk, C.S., 1990. *Mon. Not. R. astr. Soc.*, **247**, 10P.

Efstathiou, G., Bernstein, G., Katz, N., Tyson, T. & Guhathakurta, P., 1991. *Astrophys. J.*, in press.

Gibbons, G.W., Hawking, S.W. & Vachaspati, T., 1990. *The formation and evolution of cosmic strings*, Cambridge University Press.

Glazebrook, K., 1991. *PhD thesis, University of Edinburgh.*

Gott, J.R. III, Melott, A.L. & Dickinson, M., 1986. *Astrophys. J.*, **306**, 341.

Gott, J.R. III, *et al.* 1989. *Astrophys. J.*, **340**, 625.

Grinstein, B. & Wise, M.B., 1986. *Astrophys. J.*, **310**, 19.

Groth, E.J. & Peebles, P.J.E., 1977. *Astrophys. J.*, **217**, 385.

Gurbatov, S.N., Saichev, A.I. & Shandarin, S.F., 1989. *Mon. Not. R. astr. Soc.*, **236**, 385.

Hamilton, A.J.S., Gott, J.R. III & Weinberg, D.H., 1986. *Astrophys. J.*, **309**, 1.

Heavens, A.F., 1985. *Mon. Not. R. astr. Soc.*, **213**, 143.

Henry, J.P. & Arnaud, K.A., 1991. *Astrophys. J.*, **372**, 410.

Hut, P. & White, S.D.M., 1984. *Nature*, **310**, 637.

Jensen, L.G. & Szalay, A.S., 1986. *Astrophys. J.*, **305**, L5.

Kaiser, N., 1984. *Astrophys. J.*, **284**, L9.

Kaiser, N., 1986. *Mon. Not. R. astr. Soc.*, **219**, 785.

Kaiser, N., 1987. *Mon. Not. R. astr. Soc.*, **227**, 1.

Kaiser, N. & Lahav, O., 1989. *Mon. Not. R. astr. Soc.*, **237**, 129.

Kaiser, N. & Peacock, J.A., 1991. *Astrophys. J.*, **379**, 482.

Kashlinsky, A., 1991. *Astrophys. J.*, **376**, L5.

Lacey, C. & Silk, J., 1991. *Astrophys. J.*, in press.

Lilly, S.J. & Longair, M.S., 1984. *Mon. Not. R. astr. Soc.*, **211**, 833.

Lilly, S.J., 1989. *Astrophys. J.*, **340**, 77.

Lilly, S.J., Cowie, L.L. & Gardner, J.P., 1991. *Astrophys. J.*, **369**, 79.

Ling, E.N., Frenk, C.S. & Barrow, J.D., 1986. *Mon. Not. R. astr. Soc.*, **223**, 21P.

Lumsden, S.L., Heavens, A.F. & Peacock, J.A., 1989. *Mon. Not. R. astr. Soc.*, **238**, 293.

Maddox, S.J., Efstathiou, G., Sutherland, W.J. & Loveday, J., 1990. *Mon. Not. R. astr. Soc.*, **242**, 43P.

Maddox, S.J., 1991. in proc. 2nd Rencontre de Blois, *Physical Cosmology*, eds A. Blanchard & M. Lachièze-Rey, Editions Frontières, Paris, in press.

Melott, A.L., Cohen, A.P., Hamilton, A.J.S., Gott, J.R. III & Weinberg, D.H., 1989. *Astrophys. J.*, **345**, 618.

Osmer, P.S., 1982. *Astrophys. J.*, **253**, 28.

Peacock, J.A. & Heavens, A.F., 1985. *Mon. Not. R. astr. Soc.*, **217**, 805.

Peacock, J.A. & Heavens, A.F., 1990, *Mon. Not. R. astr. Soc.*, **243**, 133

Peacock, J.A., 1985. *Mon. Not. R. astr. Soc.*, **217**, 601.

Peacock, J.A., 1990a. in proc. 10th Moriond astrophysics meeting, *Particle Astrophysics*, eds J.M. Alimi *et al.*, Editions Frontières, Paris, p375.

Peacock, J.A., 1990b. *Mon. Not. R. astr. Soc.*, **243**, 517.

Peacock, J.A. & Nicholson, D., 1991. *Mon. Not. R. astr. Soc.*, **253**, 307.

Peacock, J.A., 1991. *Mon. Not. R. astr. Soc.*, **253**, 1P.

Peebles, P.J.E., 1973. *Astrophys. J.*, **185**, 413.

Peebles, P.J.E., 1980. *The large-scale structure of the universe*, Princeton univ. press.

Press, W.H. & Schechter, P., 1974. *Astrophys. J.*, **187**, 425. (PS)

Raymond, J.C., Cox, D.P. & Smith, B.W., 1976. *Astrophys. J.*, **204**, 290.

Rees, M.J. & Ostriker, J.P., 1977. *Mon. Not. R. astr. Soc.*, **179**, 541.

Rees, M.J., 1985. *Mon. Not. R. astr. Soc.*, **213**, 75P.

Rice, S.O., 1954. *Selected papers on noise and stochastic processes*, p. 133, ed. Wax, N. Dover, New York.

Rajagopal, K., Turner, M.S. & Wilczek, F., 1991. Fermilab preprint 90/243-A, *Nucl. Phys. B.*, in press.

Salopek, D., Bond, J.R. & Bardeen, J.M., 1989. *Phys. Rev. D.*, **40**, 1753.

Saunders, W., Frenk, C., Rowan-Robinson, M., Efstathiou, G., Lawrence, A., Kaiser, N., Ellis, R., Crawford, J., Xia, X.-Y. & Parry, I., 1991. *Nature*, **349**, 32.

Shanks, T., Fong, R., Boyle, B.J. & Peterson, B.A., 1987. *Mon. Not. R. astr. Soc.*, **227**, 739.

Sutherland W.J., 1988. *Mon. Not. R. astr. Soc.*, **234**, 159.

Sutherland, W.J. & Efstathiou, G., 1991. *Mon. Not. R. astr. Soc.*, **248**, 159.

Thuan, T.X., Gott, J.R. III & Schneider, S.E., 1987. *Astrophys. J.*, **315**, L93.

Turok, N., 1991. in proc. Nobel Symp. no. 79., eds J.S. Nilsson, B. Gustafsson & B.-S. Skagerstam, *Physica Scripta*, **T36**, 135.

Valls-Gabaud, D., Alimi, J.-M., & Blanchard, A., 1989. *Nature*, **341**, 215.

van de Weygaert, R., 1991. *Mon. Not. R. astr. Soc.*, **249**, 159.

van de Weygaert, R. & Icke, V., 1989. *Astr. Astrophys.*, **213**, 1.

Warren, S.J., Hewett, P.C. & Osmer, P.S., 1988. in *Proceedings of a workshop on optical surveys for quasars*, eds Osmer, P.S., Porter, A.C., Green, R.F. & Foltz, C.B, A.S.P. conf. ser., **2**, 96.

Webster, A.S., 1976. *Mon. Not. R. astr. Soc.*, **175**, 61.

West, M.J. & Richstone, D.O., 1988. *Astrophys. J.*, **335**, 532.

White, S.D.M., 1979. *Mon. Not. R. astr. Soc.*, **186**, 145.

White, S.D.M. & Rees, M., 1978. *Mon. Not. R. astr. Soc.*, **183**, 341.

White, S.D.M., Davis, M., Efstathiou, G. & Frenk, C.S., 1987. *Nature*, **330**, 451.

White, S.D.M., 1990. in *Physics of the Early Universe*, proc 36th Scottish Universities Summer School in Physics, eds Peacock, J.A., Heavens, A.F. & Davies, A.T. (Adam Hilger), p1.

Williams, B.G., Heavens, A.F., Peacock, J.A. & Shandarin, S.F., 1991a. *Mon. Not. R. astr. Soc.*, **250**, 458.

Williams, B.G., Peacock, J.A. & Heavens, A.F., 1991b. *Mon. Not. R. astr. Soc.*, **252**, 43P.

Yahil, A., 1991. in proc. 10th Moriond astrophysics meeting, *Particle Astrophysics*, eds J.M. Alimi *et al.*, Editions Frontières, Paris, p483.

Zel'dovich, Ya. B., 1970, *Astron. Astrophys.*, **5**, 84.

Large-Scale Structures and Motions: Linear Theory and Statistics

Edmund Bertschinger

Department of Physics, MIT 6-207, Cambridge, MA 02139 USA

Abstract: These lecture notes focus on two aspects of large-scale structure theory: the linear theory of gravitational instability and the statistical description of random fields and point processes. Newtonian cosmological linear perturbation theory for density fluctuation growth is presented in both its Eulerian and Lagrangian formulations. The potential flow reconstruction algorithm is briefly described. A detailed presentation is given of the statistics of random fields and point processes, including N-point probability distributions and correlation functions, the cumulant expansion theorem, counts in cells, the void probability function and power spectra.

1 Introduction

As the attendance at this Summer School demonstrated, interest in cosmology and large-scale structure has skyrocketed in the last several years. The rise in the student interest parallels the rapid increase of new and often sensational observational discoveries and the influx of bold theoretical ideas, some from particle physics and some from astrophysics. As a result of this burst of inflation, there are no up-to-date textbooks giving a thorough presentation of the basic tools needed by a theorist working on large-scale structure.

Although my lectures covered cosmological N-body simulations in addition to linear theory and statistics, I have decided in these lecture notes to restrict myself to the latter topics. The reason for this is that, while there exist several introductory sources for N-body simulations (e.g., Bertschinger & Gelb 1991), there are few cogent pedagogical introductions to the analytical side of large-scale structure. There are a number of good broad reviews of large-scale structure, galaxy formation and related topics in astrophysical cosmology, including Peebles (1980), Efstathiou & Silk (1983), Zel'dovich & Novikov (1983), Blumenthal (1988), White (1990), Efstathiou (1990), and

Peacock (1992). However, none of these except Peebles' book, which is somewhat out of date, digs very deeply into the topics I have chosen to cover. The range of topics covered in these notes is fairly narrow, but with enough detail so that the student who masters this material will have little trouble diving into the research literature.

These lecture notes focus on two aspects of large-scale structure theory: the linear theory of gravitational instability and the statistical description of random fields and point processes. These subtopics of theoretical cosmology play a crucial role in the testing of theories of the early universe through astronomical observations. The ingredients for a theory of large-scale structure include: the background spacetime; specification of the types of matter present in the universe, the amount of each, and the equation of state of each; the initial fluctuations of density, entropy, and any other fields characterizing these fluctuations; and the laws governing the evolution of the fluctuations. We review these ingredients briefly in Sect. 2. Gravitational instability theory is discussed in Sect. 3, with the emphasis placed on Newtonian linear perturbation theory, in both its Eulerian and Lagrangian formulations. Section 4 is a brief summary of the potential flow reconstruction algorithm for inferring the three-dimensional distribution of mass from observations of galaxy redshifts and distances. A detailed presentation is given in Sect. 5 of the statistics of random fields and point processes, including N-point probability distributions and correlation functions, the cumulant expansion theorem, counts in cells, and power spectra.

2 Background cosmology

Let us recall the fundamental assumptions of physical cosmology. We make the usual assumption that the observable part of our universe may be approximated as part of a homogeneous and isotropic, expanding Friedmann-Robertson-Walker universe. (By attaching Friedmann's name to this model, we do not mean to imply that space is filled only with pressureless dust as in Friedmann's solution. There may be significant pressure from radiation or vacuum energy. Fans of vacuum energy, also known as a cosmological constant, may wish to add Lemaître's name to the list.) The assumption of large-scale homogeneity and isotropy is often called the *Cosmological Principle*. To this principle we add general relativity theory to describe the evolution and structure of spacetime and we recognize the obvious small-scale departures from homogeneity and isotropy. Our paradigm is that we live in a perturbed Friedmann-Robertson-Walker universe. By this we mean that metric fluctuations are small within our horizon, although density fluctuations are not necessarily small.

We account for the zeroth-order Hubble expansion using the cosmic scale factor $a(t)$, defined so that the proper separation between two objects varies with cosmic time, in the absence of perturbations, in proportion to $a(t)$. The scale factor obeys the Friedmann equation, derived in standard texts:

$$H^2(t) \equiv \left(\frac{1}{a}\frac{da}{dt}\right)^2 = \frac{8\pi}{3}G\bar{\rho}(t) - \frac{k}{a^2} \, . \tag{1}$$

Here $\bar{\rho}(t)$ is the total mean mass density (including relativistic and nonrelativistic components) and k is the curvature constant. The Hubble parameter $H(t)$ has present value $H_0 = 100\,h$ km s^{-1} Mpc^{-1}; h almost certainly lies between 0.5 and 1.0, with several recent determinations favoring $h \approx 0.8$. (This value is uncomfortably large for many specific theories of large-scale structure, a fact that will be ignored here.) Present values will be written with a subscript 0, except for the expansion factor, whose present value is defined to be $a(t_0) = 1$. With this choice, the curvature constant has units and therefore is not equal to the canonical ± 1 or 0. Instead, from (1), we see that

$$k = (\Omega_0 - 1)H_0^2 \, , \tag{2}$$

where Ω_0 is the present value of the density parameter, defined by

$$\Omega(t) \equiv \frac{8\pi G\bar{\rho}(t)}{3H^2(t)} \, . \tag{3}$$

(Many cosmologists drop the subscript on Ω_0 and refer to its present value simply as Ω. This can be confusing if $\Omega \neq 1$ because then Ω evolves with time.) Note that vacuum energy is included in $\bar{\rho}$ along with matter and radiation; if one prefers to separate out a vacuum density $\rho_{\text{vac}} = \Lambda/(8\pi G)$ and to exclude ρ_{vac} from the definition of Ω, then (2) becomes becomes $k = (\Omega_0 - 1)H_0^2 + \Lambda/3$. A positive cosmological constant Λ also has associated with it a negative pressure, $p_{\text{vac}} = -\rho_{\text{vac}}c^2$. We prefer not to single out Λ but rather to account for it in the background density and pressure.

In a universe dominated by nonrelativistic matter, in which the pressure is negligible compared with $\bar{\rho}c^2$, expansion dilutes the mass density inversely with volume: $\bar{\rho}(t) = \bar{\rho}_0 a^{-3}(t)$ where $\bar{\rho}_0 = 1.88 \times 10^{-29}\,\Omega_0 h^2$ g cm^{-3} is the present density. In this case, the solution to the Friedmann equation implies

$$\frac{1 - \Omega(t)}{\Omega(t)} = a(t)\frac{1 - \Omega_0}{\Omega_0} \, . \tag{4}$$

Note that $\Omega = 1$ is an unstable fixed point: as $a(t)$ increases, Ω deviates increasingly from $\Omega = 1$. The special case $k = 0$ and no pressure — a favored model among theorists — is called the Einstein-de Sitter universe. Inflation theories (Guth 1981) predict $k = 0$, implying the Einstein-de Sitter model if the universe is dominated by nonrelativistic matter.

Next we introduce briefly the metric characterizing the geometry of spacetime. In following the evolution of fluctuations we will adopt a Newtonian formulation, but it is instructive to see under what conditions this approximation arises from general relativity (which we assume to be applicable). Some familiarity with the basic concepts of general relativity is useful for this discussion, but the reader will not miss too much by skipping the next two paragraphs.

The Robertson-Walker line element for a homogeneous and isotropic spacetime is

$$ds^2 = -c^2 dt^2 + a^2(t) \left[\frac{dr^2}{1 - kr^2/c^2} + r^2 \left(d\theta^2 + \sin^2\theta \, d\phi^2 \right) \right] . \qquad (5)$$

We see that the spatial coordinate r is *comoving*, i.e., it must be multiplied by the scale factor $a(t)$ to give a proper distance. Note that the effect of curvature in the metric is small for distances much less than the Hubble length $cH_0^{-1} = 3000 \ h^{-1}$ Mpc (a variety of observations clearly favor $0 < \Omega_0 < 2$ and therefore $|k| < H_0^2$). We will restrict ourselves to structures much smaller than this, so that the curvature term may be neglected in the metric (although not necessarily in the Friedmann equation). Then, the Robertson-Walker line element may be written $ds^2 = -c^2 dt^2 + a^2(t) \left(dx^2 + dy^2 + dz^2 \right)$, where (x, y, z) are Cartesian comoving coordinates. In the following we will denote the comoving position by $\vec{x}$ and the corresponding proper position by $\vec{r} = a(t)\vec{x}$.

Now we must introduce the metric fluctuations arising because the universe is patently not homogeneous and isotropic, at least not on small scales. We will restrict ourselves to the post-recombination universe (i.e., after the temperature has fallen below 4000 K and hydrogen is almost fully unionized) and we assume that fluctuations in the radiation background and any other possible relativistic components are small compared with the density fluctuations of nonrelativistic matter, so that only the latter produce a significant gravitational effect. (This treatment allows for a uniform relativistic background energy density and pressure such as that provided by a cosmological constant. However, we neglect the gravitational effect of the momentum density, of fluctuations in pressure and in the relativistic component of energy density. This approximation is inappropriate in the early, radiation-dominated universe but is correct for our purposes — after recombination — if large-scale structure developed during a matter-dominated era.) Given this assumption, the metric fluctuations are determined by one scalar field, which we may take to be the Newtonian gravitational potential perturbation ϕ. [From this point onward, we will discard the spherical coordinates of (5) and we may use these symbols for other purposes.] One may then solve the Einstein field equations to first order in ϕ/c^2. The form of the results depends on the way that coordinates are assigned to the perturbed spacetime, i.e., on the choice of gauge. We will adopt the conformal Newtonian gauge, in which the line element becomes

$$ds^2 = a^2(\tau) \left[-\left(1 + 2\phi/c^2 \right) c^2 d\tau^2 + \left(1 - 2\phi/c^2 \right) \left(dx^2 + dy^2 + dz^2 \right) \right] . \qquad (6)$$

Equation (6) is a solution to the Einstein field equations for a perturbed Friedmann-Robertson-Walker universe in the weak-field, slow-motion approximation of general relativity (neglecting gravitational radiation and the weak gravitational effects of vorticity). The interested reader with some background in general relativity may find detailed discussions of the perturbed metric in many fine reviews of cosmological perturbation theory (Peebles & Yu 1970; Weinberg 1972; Peebles 1980; Zel'dovich & Novikov 1983; Kodama & Sasaki 1984; Efstathiou 1990; Mukhanov, Feldman & Brandenberger 1991).

Note that we have changed the time coordinate from proper time t to conformal time τ,

$$\tau \equiv \int_0^t dt'/a(t') \ . \tag{7}$$

The reason for making this change is that the zeroth-order metric is then manifestly conformally flat (i.e., it is just the Minkowski metric of special relativity, multiplied by an overall conformal factor a^2), which will lead to simpler forms for the equations of motion written in these coordinates. For example, the coordinate velocity of a particle is now just the peculiar velocity (the proper velocity relative to the expanding coordinate system, i.e., the proper velocity measured by a comoving observer minus the Hubble velocity):

$$\vec{v} \equiv \frac{d\vec{x}}{d\tau} = a\frac{d\vec{x}}{dt} = \frac{d(a\vec{x})}{dt} - H(a\vec{x}) = \frac{d\vec{r}}{dt} - H\vec{r} \ . \tag{8}$$

Another advantage of the use of conformal time is the fact that the solution to the Friedmann equation for a universe dominated by nonrelativistic matter then takes a particularly simple form. The problem is equivalent to the Kepler problem for radial orbits. For an open universe ($k < 0$ and $\Omega_0 < 1$) the solution is a hyperbolic function of conformal time:

$$a = \frac{\Omega_0}{1 - \Omega_0} \frac{\cosh\eta - 1}{2} \ , \quad \eta \equiv (-k)^{1/2}\tau = (1 - \Omega_0)^{1/2} H_0\tau \ . \tag{9}$$

Many workers use η to refer to the dimensionless conformal time as we have defined it here. For a closed universe ($k > 0$ and $\Omega_0 > 1$) the solution is obtained by analytic continuation. For $\eta^2 \ll 1$ both the open and closed universe solutions expand like the Einstein-de Sitter solution, $a = \Omega_0 H_0^2 \tau^2/4 = (9\Omega_0 H_0^2/4)^{1/3} t^{2/3}$.

Notice the gravitational potential perturbation in (6), appearing in both the time and space parts of the metric. The perturbation in the spacetime metric is responsible for the Newtonian gravitational force. The trajectory of a freely-falling particle is given by a geodesic in this metric. For a nonrelativistic particle of mass m and momentum $\vec{p} = m\vec{v}$, the geodesic equation obtained from (6) is

$$\frac{d(a\vec{p})}{d\tau} = -ma\vec{\nabla}\phi \; . \tag{10}$$

The spatial derivatives are with respect to the comoving coordinates: $\vec{\nabla} = \partial/\partial\vec{x}$. For $a = $ constant, we recover Newton's law. With expansion, (10) demonstrates the cosmological redshift: the momentum of a freely-falling particle, measured relative to the expanding coordinate system, decreases as a^{-1} because of the cosmic expansion. Similarly, the momentum, and hence energy, of photons decreases as a^{-1}. (As an exercise, the reader may wish to derive the geodesic equations for photons. In the absence of expansion they demonstrate the usual gravitational redshift and deflection of light.) Thus, absent perturbations, the redshift of the light observed today from a source that emitted the light at conformal time τ_e is $z = a^{-1}(\tau_e) - 1$. It is conventional to use redshift z as a measure of cosmic time through the relation $a(\tau) = (1 + z)^{-1}$.

The Einstein field equations, applied using the metric of (6), give the Friedmann equation to zeroth-order in ϕ/c^2. The first-order perturbations yield the Poisson field equation of Newtonian gravity:

$$\nabla^2\phi = 4\pi Ga^2\delta\rho \; , \tag{11}$$

where $\delta\rho \equiv \rho(\vec{x},\tau) - \bar{\rho}(\tau)$ is the fluctuation in the mass density. In order to obtain the Newtonian result, we neglect terms of order $(\lambda H/c)^2$, where λ is the proper wavelength of a Fourier component of ϕ. That is, we assume that perturbations are small on the Hubble scale. In addition, to interpret ϕ as the Newtonian potential we must make the weak-field, slow-motion approximation. That is, we assume ϕ/c^2 is small and that the gravitational source $\delta\rho$ does not involve relativistic motions with respect to our comoving coordinate system. Note that (11) does not assume that $\delta\rho/\bar{\rho}$ is small. For a fluctuation of proper wavelength λ, combining (1) and (11) we see that $\phi/c^2 \sim (H\lambda/c)^2\delta\rho/\bar{\rho}$. Thus, for fluctuations of size much smaller than the Hubble length, ϕ/c^2 may be small even if $\delta\rho/\bar{\rho}$ is large. This is why we can use a Newtonian treatment for large-scale structure. Our results will be consistent as long as the motions relative to the expanding background are nonrelativistic.

Let us summarize our assumptions. First, we assume that our observable universe is a perturbed Friedmann-Walker-Robertson spacetime and that it is nearly homogeneous and isotropic on the Hubble scale cH_0^{-1}. Second, we make the weak-field ($|\phi| \ll c^2$) and slow-motion ($|d\vec{x}/d\tau|^2 \ll c^2$) approximations in general relativity. Third, we restrict our discussion to scales much less than the Hubble length cH_0^{-1} and the curvature distance $|\Omega_0 - 1|^{-1}cH_0^{-1}$. Fourth, we assume that the density perturbations are dominated by nonrelativistic matter and we neglect the gravitational effect of stress (pressure) perturbations. This assumption is invalid in the early, radiation-dominated universe. It should be valid after recombination,

$z \ll 1000$. These assumptions lead to a cosmological Newtonian approximation. Rather than simply adopting a Newtonian approach without question, we have preferred to outline the approximations needed to justify this approach in the context of general relativistic cosmology.

Finally, we state the basic paradigm of structure formation theory, which we will call the *Cosmogonical Principle*: Galaxy formation and large-scale structure arose from the growth of initial small-amplitude density fluctuations, by gravitational instability. Without the last three words, we may call this the weak cosmogonical principle. Essentially all cosmologists would agree with the weak principle. The near-perfect isotropy of the cosmic microwave background radiation provides strong evidence that the universe was much more homogeneous at early times than it is today. Most, though not all, cosmologists would agree with the stronger statement that cosmological structure probably developed via gravitational instability. In the following, we will not consider alternatives to gravitational instability.

3 Equations of motion for the matter

We are interested in describing the motion of the nonrelativistic matter in the universe, a tracer of which we see in the form of galaxies. The matter is both responsible for, and responsive to, the metric perturbations given by the Newtonian gravitational potential perturbation ϕ. In the gravitational instability paradigm followed here, small-amplitude fluctuations generated somehow in the early (or possibly late) universe grow in amplitude when nonrelativistic matter begins to dominate the total energy density of the universe. Here we ignore the evolution of the fluctuations before recombination; for this, a relativistic treatment would be needed because of the coupling of the ordinary matter and the radiation and because of the gravitational importance of fluctuations in the latter. For a detailed treatment of relativistic cosmological perturbation theory, the interested reader is referred to the reviews mentioned above.

We assume that after recombination there exist density fluctuations $\delta\rho$ in some nonrelativistic component or components of matter, and that these fluctuations have small amplitude at recombination. We will make no assumptions about the amount of nonrelativistic matter or the equation of state of any relativistic components, except as noted below. The nonrelativistic matter includes the ordinary baryonic matter (it would better be called nucleonic matter because there are baryons other than the protons and neutrons present in ordinary matter) as well as the collisionless dark matter that clusters around galaxies. It may also include non-baryonic matter such as exotic hypothetical dark matter particles (axions, supersymmetric particles, massive neutrinos, etc.).

The total mean density of nonrelativistic matter is written $\rho_m(\tau)$. Expansion dilutes this as $\rho_m(\tau) \propto a^{-3}(\tau)$. We do not assume that this matter closes the universe; the fraction of closure density in nonrelativistic matter, $\Omega_m = 8\pi G \rho_m/3H^2$, is left unspecified. We denote the density contrast by $\delta(\vec{x}, \tau) \equiv \delta\rho_m(\vec{x}, \tau)/\rho_m(\tau)$. We assume that any relativistic matter (e.g., radiation or vacuum energy) makes a negligible contribution to the mass density *fluctuations* so that gravitational potential perturbations arise only from $\delta\rho_m$. (This is exactly true for a cosmological constant and should be an excellent approximation for other relativistic fields on scales much less than the Hubble length.) In the following we will drop the subscript m except when referring to the mean density.

We will present two alternative formulations of the equations of motion of the nonrelativistic matter, Eulerian and Lagrangian. For each, we will compare the equations in comoving coordinates with those in proper coordinates, and we will obtain solutions to the linearized equations for the growth of small-amplitude perturbations. Higher-order perturbation theory has been discussed in the Eulerian formulation by many authors (e.g., Peebles 1980, Sect. 18); the Lagrangian version has recently been developed by Giavalisco *et al.* (1991).

3.1 Eulerian formulation

The Eulerian description considers motion relative to some coordinate system unattached to the matter. The matter is treated as a fluid. This approach is valid both for collisional and collisionless matter, although it is usually associated with the former. The Eulerian fluid equations may be derived by taking velocity moments of the Boltzmann equation for the time evolution of the phase space distribution function. These velocity moments exist regardless of whether the fluid is collisional or collisionless; the only difference is that in the former case collisions drive the velocity distribution to be nearly equal to the Maxwell-Boltzmann distribution (for a perfect fluid, exactly equal to it) while in the latter case collisions are unimportant. In either case the fluid velocity is defined as the average momentum per unit mass of the particles in the vicinity of a given Eulerian position.

We present below the Eulerian fluid equations in both proper $(t, \vec{r})$ and comoving $(\tau, \vec{x})$ coordinates. Gradients are taken with respect to the appropriate spatial coordinates. Partial derivatives have their usual meaning. Thus, the partial derivative of time at fixed proper position $\vec{r}$ differs from the partial derivative at fixed comoving position $\vec{x}$, $a(\partial/\partial t) = \partial/\partial\tau - H\vec{r}\cdot\partial/\partial\vec{x}$. As an exercise, the reader may wish to derive the equations in comoving form from those in proper coordinates.

Eulerian Fluid Equations

Proper Coordinates	Comoving Coordinates

$$(t,\ \vec{r} = a\vec{x}),\quad \vec{u} \equiv \frac{d\vec{r}}{dt}\ ; \qquad (\tau,\ \vec{x}),\quad \vec{v} \equiv \frac{d\vec{x}}{d\tau} = \vec{u} - H\vec{r}\ . \quad (12)$$

$$\frac{\partial \rho}{\partial t} + \vec{\nabla} \cdot (\rho \vec{u}) = 0\ ; \qquad \frac{\partial \delta}{\partial \tau} + \vec{\nabla} \cdot [(1 + \delta)\vec{v}] = 0\ . \quad (13)$$

$$\frac{\partial \vec{u}}{\partial t} + \left(\vec{u} \cdot \vec{\nabla}\right)\vec{u} = -\frac{1}{\rho}\vec{\nabla}p - \vec{\nabla}\phi_{\mathrm{p}};\frac{\partial \vec{v}}{\partial \tau} + \left(\vec{v} \cdot \vec{\nabla}\right)\vec{v} = -\frac{\dot{a}}{a}\vec{v} - \frac{1}{\rho}\vec{\nabla}p - \vec{\nabla}\phi\ . (14)$$

$$\nabla^2 \phi_{\mathrm{p}} = 4\pi G\rho\ ; \qquad \nabla^2 \phi = 4\pi G a^2 \delta\rho\ . \quad (15)$$

Equations (12)–(15) show the nonrelativistic fluid equations in proper and comoving coordinates. Equation (13) is the continuity equation (expressing mass conservation) and (14) is the Euler equation (conservation of momentum). The comoving continuity equation is exact even for large δ; δ appears rather than ρ because the comoving density is $\rho(\vec{x}, \tau)/\rho_{\mathrm{m}}(\tau) = 1 + \delta$. In the Euler equation, the only difference in the comoving form is the extra "Hubble drag" term $-(\dot{a}/a)\vec{v}$ (here, $\dot{a} \equiv da/d\tau$). This term arises because comoving coordinates define an expanding, noninertial reference frame. (In a similar way, the use of rotating coordinates leads to centrifugal and Coriolis forces.) We already saw in (10) that the use of comoving coordinates leads to a decrease in the momentum. The similarity of (10) and (14) is clear given that the convective time derivative following a given fluid element is $d/d\tau = \partial/\partial\tau + \vec{v} \cdot \vec{\nabla}$.

The only significant approximation in our fluid equations is the neglect of anisotropic stresses. For a perfect fluid, the pressure force is given by the gradient of a scalar. For a collisionless (or imperfect collisional) fluid, p must be generalized to a stress tensor $\overleftrightarrow{\Pi}$: $\nabla p \to \nabla \cdot \overleftrightarrow{\Pi}$. One could obtain equations for the evolution of the stress tensor by taking higher moments of the Boltzmann equation. However, the problem arises of how to truncate the series of moment equations. If collisions are sufficiently rapid, the series converges quickly and one may approximate higher-order corrections by adding viscosity terms to the Euler equation. For a collisionless gas, however, it is better in general to use the Lagrangian description unless the pressure is small.

For a collisional gas in which the pressure is important, the fluid equations must be supplemented with an equation of state relating the pressure and density and other relevant thermodynamical quantities such as entropy. This generally entails adding a heat equation (or equations, if the gas is a mixture of components). For a perfect gas, $p = p(\rho, S)$ where S is the specific entropy (the entropy per baryon), and S is conserved for each fluid element:

$$\frac{\partial S}{\partial \tau} + \left(\vec{v} \cdot \vec{\nabla} \right) S = 0 \ . \tag{16}$$

The pressure gradient in this case is

$$\vec{\nabla} p = c_s^2 \, \vec{\nabla} \rho + \left(\frac{\partial p}{\partial S} \right)_\rho \vec{\nabla} S \ , \tag{17}$$

where $c_s^2 = (\partial p / \partial \rho)_S$ is the square of the adiabatic sound speed. The entropy gradient term is generally unimportant in the post-recombination universe but it can yield important new behavior in the early universe (though the relativistic fluid equations must be used in this case).

Equation (15) is the Poisson equation, with a distinction between ϕ_p and ϕ needed because of the difference in the source term. Actually, the proper Newtonian potential ϕ_p is ill-defined in cosmology because of the problem of boundary conditions. The usual integral solution of the Poisson equation diverges for an unbounded mass distribution. On the other hand, because the spatial average of $\delta\rho$ vanishes (by definition), the comoving potential ϕ generally is finite. One solution is to measure only local potential differences by defining $\phi_p = 0$ at the origin of coordinates. In the absence of perturbations, it is easy to show that $\phi_p = (2\pi/3) \, G\rho_m(\tau) \, r^2$. The two forms of Poisson equation are then equivalent in general if we define $\phi_p = \phi + (2\pi/3) \, G\rho_m(\tau) \, r^2$.

Cosmic fluid dynamics is simplest in comoving coordinates. This becomes clear when we consider the zeroth-order fluid motion, uniform Hubble flow. The solution to (13)–(15) in proper coordinates is $\vec{u} = H\vec{r}$, $\rho = \rho_m(\tau)$, $\vec{\nabla} S = 0$, $\phi_p = (2\pi/3) \, G\rho_m(\tau) \, r^2$, apparently depending on the origin of coordinates. (Actually there is no dependence; this solution has the same form for any origin.) Expressed in comoving coordinates, the uniform Hubble flow solution is much simpler: $\vec{v} = \delta = \vec{\nabla} S = \phi = 0$.

3.1.1 Linear solutions

Now we linearize the fluid equations in comoving coordinates so as to consider small-amplitude perturbations about the zeroth-order Hubble flow solution. There are two motivations for seeking linear perturbation solutions. The first is that these solutions will apply at early times when, according to our assumptions, $\delta^2 \ll 1$. The second is that these solutions will also be approximately correct for the long-wavelength components of the density and velocity fields (i.e., the spatially averaged fields) if these smoothed fields have small fluctuations, even if $\delta \gg 1$ on small scales (i.e., for the unsmoothed field). This second motivation may be treated as an assumption that will be justified later based on fully nonlinear calculations.

Linearizing (13)–(16), we get

$$\dot{\delta} + \vec{\nabla} \cdot \vec{v} \approx 0 \ , \quad \dot{\vec{v}} \approx -\frac{\dot{a}}{a}\vec{v} - c_{\rm s}^2 \vec{\nabla}\delta - \frac{1}{\rho_m}\left(\frac{\partial p}{\partial S}\right)_\rho \vec{\nabla} S - \vec{\nabla}\phi \ , \tag{18}$$

$$\dot{S} = 0 \ , \quad \nabla^2\phi = 4\pi G\rho_m a^2\delta \ ,$$

where a dot indicates a conformal time derivative and we have assumed a perfect gas. Equations (18) are a fifth-order system of linear equations for five fluid variables (density and entropy perturbations and three components of velocity). We therefore expect five linearly independent solutions. These solutions may be classified according to the nature of the velocity field. A general vector field may be decomposed into potential (longitudinal) and rotational (transverse) parts:

$$\vec{v}(\vec{x}, t) = \vec{v}_\parallel + \vec{v}_\perp \ , \quad \vec{\nabla}\times\vec{v}_\parallel = \vec{\nabla}\cdot\vec{v}_\perp = 0 \ . \tag{19}$$

As we will see, there are three independent longitudinal solutions and two independent transverse solutions.

3.1.2 Longitudinal isentropic solutions

The longitudinal solutions obey $\vec{\nabla}\times\vec{v} = 0$ so that $\vec{v} = \vec{v}_\parallel = -(1/a)\vec{\nabla}\Phi_v$, where Φ_v is the proper velocity potential. We first consider the isentropic case $\vec{\nabla}S = 0$. The solutions in this case are often called "adiabatic modes" although they should instead be called isentropic (i.e., $\vec{\nabla}S = 0$) because we assume that the motion of individual fluid elements is adiabatic in any case [i.e., $dS/d\tau = 0$, (16)].

Combining the time derivative of the linearized continuity equation with the divergence of the linearized Euler equation [(18)], we get the equation of motion for the density perturbation for longitudinal isentropic modes,

$$\ddot{\delta} - c_{\rm s}^2\nabla^2\delta = -\frac{\dot{a}}{a}\dot{\delta} + 4\pi G\rho_m a^2\delta \ . \tag{20}$$

Equation (20) is a linear acoustic wave equation with Hubble damping and gravitational driving terms. Because the coefficients of δ are spatially homogeneous, (20) may be solved by expanding $\delta(\vec{x}, \tau)$ in plane waves. For one plane wave, $\delta(\vec{x}, \tau) = e^{i\vec{k}\cdot\vec{x}}D_{\vec{k}}(\tau)$, we obtain a second-order ordinary differential equation,

$$\ddot{D}_{\vec{k}} + \frac{\dot{a}}{a}\dot{D}_{\vec{k}} = c_{\rm s}^2\left(k_{\rm J}^2 - k^2\right)D_{\vec{k}} \ , \tag{21}$$

where we have defined the comoving Jeans wavenumber $k_{\rm J}$ by

$$k_{\rm J} = a\left(\frac{4\pi G\rho_m}{c_{\rm s}^2}\right)^{1/2} = \frac{1}{8\,{\rm Mpc}}\left(\frac{c_{\rm s}}{300\,{\rm km\,s^{-1}}}\right)^{-1}\left(\frac{\Omega_m h^2}{0.1}\right)^{1/2} \ . \tag{22}$$

Note that the plane waves have constant comoving wavelength $2\pi/k$. [The comoving wavenumber k should not be confused with the curvature constant of (2).] Thus, the waves stretch with the expansion of the universe, with the proper wavelength of a density perturbation increasing as $a(\tau)$ just as for photons.

Exact solutions to (21) exist for a variety of cases (see, e.g., Weinberg 1972). The qualitative behavior of the solutions can be readily discerned from (21). For $k < k_J$ (long wavelengths), the right-hand side of (21) is positive and gravity defeats pressure so that gravitational instability may take place. In general there is one monotonically growing solution and one monotonically decaying solution. Because of the Hubble damping and because the density decreases with expansion, the growth is power-law rather than exponential in time. For $k > k_J$ (short wavelengths), the right-hand side of (21) is negative, pressure is more important than gravity, and the two solutions are acoustic waves (traveling along $\pm \vec{k}$). This is the situation prevailing prior to recombination for nearly all waves shorter than the Hubble length, because the effective sound speed of the coupled photon-baryon gas then approached the relativistic value $c_s = c/\sqrt{3}$. After recombination the sound speed of the uncoupled baryon gas dropped below 10 km s^{-1} and the Jeans mass $(\rho_m k_J^{-3})$ dropped to less than $10^6 \, M_\odot$. After that, the pressure of the baryon gas was generally unimportant for galaxy formation. The density fluctuations of the uncoupled relativistic components (photons, neutrinos and possibly other relativistic fields) become unimportant on scales much less than the Hubble length because these fluctuations do not grow for wavelengths less than the Jeans length $(2\pi/k_J)$. That is why we can generally neglect the gravitational effect of pressure perturbations even if the background cosmology has nonnegligible pressure.

In hot dark matter scenarios, the pressure of the gravitationally dominant dark matter component remains important for the evolution of density fluctuations. By perturbing the collisionless Boltzmann equation, one can show (Zel'dovich & Novikov 1983) that the criterion for collisionless ("free-streaming") damping is similar to the Jeans criterion $k > k_J$, where $k_J^2 = 4\pi G \rho_m a^2 \langle v^{-2} \rangle$ is the Jeans wavenumber for a collisionless gas (angle brackets denote a phase-space average). For a collisionless gas short-wavelength perturbations $(k \gg k_J)$ are strongly damped in a few oscillation periods unless they are regenerated by some non-decaying seed. For massive neutrinos, the prototypical hot dark matter, while the neutrinos were relativistic $v \sim c$ yielding $k_J \propto \tau^{-1}$; after the temperature of the neutrinos fell below their mass the neutrinos were nonrelativistic and $v \propto a^{-1}$ so that $k_J \propto a^{1/2}$. All primordial fluctuations whose wavelength is less than the maximum Jeans length (about 13 $\Omega_{v,0}^{-1} h^{-2}$ Mpc comoving) are strongly damped by free-streaming damping. For longer wavelengths, or for reseeded fluctuations after k_J increases above k, free-streaming is unimportant and hot dark matter fluctuations evolve the same way as cold collisionless matter.

In the long-wavelength limit ($k \ll k_J$) pressure is unimportant and all waves grow (or decay) at the same rate. This situation is relevant for baryon-dominated and cold dark matter-dominated models after recombination (and in the latter case at all times after the universe became dominated by nonrelativistic matter). The general solution of (20) is then

$$\delta(\vec{x}, \tau) = D_+(\tau)\,\delta_+(\vec{x}) + D_-(\tau)\,\delta_-(\vec{x}) , \tag{23}$$

where $D_\pm$ are the growing and decaying solutions to (21) for $k \ll k_J$ (Heath 1977). For a pressureless universe with $\Omega = \Omega_m < 1$, the solutions are

$$D_+(\eta) = \frac{3\sinh\eta\,(\sinh\eta - \eta)}{(\cosh\eta - 1)^2} - 2 , \quad D_-(\eta) = \frac{2\sinh\eta}{(\cosh\eta - 1)^2} , \tag{24}$$

where $\eta = \eta(\tau)$ is the dimensionless conformal time [(9)]. For $\Omega_m = 1$, $D_+ \propto a \propto \tau^2$ and $D_- \propto a^{-3/2} \propto \tau^{-3}$. In all practical cases, the growing mode quickly comes to dominate. The logarithmic growth rate, needed below for the velocity, depends on Ω_m and on the background cosmology, and is conventionally denoted f:

$$\frac{d\log D_+}{d\log a} = \frac{a\dot{D}_+}{\dot{a}D_+} \equiv f(\tau) . \tag{25}$$

Using the exact solution of (24) for a universe dominated by nonrelativistic matter, one finds that $f \approx \Omega_m^{0.6}$ is a good approximation (Peebles 1980, Sect. 14). For an Einstein-de Sitter universe ($\Omega_m = 1$), $f = 1$. For a given Ω_m, the growth rate depends very weakly on the cosmological constant (Peebles 1984).

Given a solution to the density perturbation field δ, the velocity, gravitational potential and gravity field follow simply. The gravity field is $\vec{g} \equiv -(1/a)\vec{\nabla}\phi$ where ϕ is the gravitational potential. For longitudinal modes, $\vec{v} = -(1/a)\vec{\nabla}\Phi_v$. Then, from (18), for $k \ll k_J$, we get [using (25) for the growth rate neglecting the decaying mode],

$$\Phi_v = Hfa^2\nabla^{-2}\delta , \quad \phi = \frac{3}{2}\Omega_m H^2 a^2 \nabla^{-2}\delta = \frac{3}{2}\frac{\Omega_m H}{f}\Phi_v , \tag{26}$$

$$\vec{v} = \vec{v}_\parallel = \frac{2}{3}\frac{f}{\Omega_m H}\vec{g} .$$

In Fourier space, $\vec{\nabla} = i\vec{k}$ and so $\vec{v}$ is parallel to the wavevector. This is the origin of the term "longitudinal." Notice the simple relation between peculiar velocity and gravity. For an Einstein-de Sitter universe, whose age is $t = 2/(3H)$, this relation becomes $\vec{v} = \vec{g}t$. This relation can also be derived from (7) and (10), noting that in an Einstein-de Sitter universe, ϕ is a constant for the growing mode [(26) with $\delta \propto a$ and $H^2 a^3 = H_0^2$]. More generally, ϕ is constant for the linear growing mode in a matter-dominated universe as long as $\Omega_m^{-1}|1 - \Omega_m| \ll 1$. In all cases, the peculiar velocity induced by gravity is in the direction of gravity. Why? In linear theory, the density fluctuation field grows in amplitude but its spatial dependence (in comoving coordinates) does not change. Thus, the gravity field does not change direction and, as long as their displacements are small, particles are accelerated in a fixed direction.

3.1.3 Longitudinal entropy solutions

Now that we have obtained the longitudinal solutions with $\vec{\nabla}S=0$, it is easy to add a nonzero entropy gradient. There is exactly one new mode, because there is only one evolution equation for S, $\dot{S} = 0$. The entropy per baryon is therefore constant in time but may have nonzero gradient. From (18), entropy gradients induce density and longitudinal velocity perturbations. Equation (21) acquires a source term:

$$\ddot{D}_{\vec{k}} + \frac{\dot{a}}{a}\,\dot{D}_{\vec{k}} + c_s^2\left(k^2 - k_J^2\right)D_{\vec{k}} = -\frac{1}{\rho_m}\left(\frac{\partial p}{\partial S}\right)_\rho k^2 S_{\vec{k}}\,, \tag{27}$$

where we have decomposed S into plane waves $e^{i\vec{k}\cdot\vec{x}}S_{\vec{k}}$ and $S_{\vec{k}}$ is constant. Given the homogeneous solutions $D_{\vec{k}}(\tau)$ from the previous section, it is straightforward to find a particular solution for nonzero $S_{\vec{k}}$ in terms of integrals over the homogeneous solutions. The solutions will not be given here. The velocity and gravitational potentials follow in a manner similar to the isentropic case analyzed above.

The most important application of entropy perturbations is to the coupled photon-baryon fluid before recombination. [Relativistic effects are important in this case and the left-hand side of (27) must be suitably modified.] The energy density of the radiation is $\rho_r \propto T^4$ and the entropy density is proportional to $T^3 \propto \rho_r^{4/3}$, where $T \propto a^{-1}$ is the radiation temperature. Consequently, the relative fluctuation in entropy per baryon for a photon-baryon gas is

$$\frac{\delta S}{S} = \frac{3}{4}\,\delta_r - \delta_b\,, \tag{28}$$

where $\delta_r \equiv \delta\rho_r/\rho_r$ and $\delta_b \equiv \delta\rho_b/\rho_b$ are the relative energy density fluctuations of radiation and baryons (actually, nucleons), respectively. It is possible to generate (in a phase transition, for example) variations in the entropy per baryon with zero net density fluctuation, i.e., $\delta\rho_r + \delta\rho_b = 0$ but $\delta S \neq 0$. The initial fluctuations of velocity and potential also vanish. The solution with nonzero entropy fluctuation but vanishing initial density, velocity and curvature (i.e., gravitational potential) perturbations is often called the "isocurvature mode." This mode has also been called "isothermal" because, in the radiation-dominated era (when $\rho_r \gg \rho_b$), the relative perturbation in the radiation energy density (and hence radiation temperature) is much smaller than that of the baryons: $\delta_r = -(\rho_b/\rho_r)\,\delta_b$.

Starting with zero net density fluctuation, entropy gradients act as a source term for the generation of velocity and density fluctuations. Entropy fluctuations may be viewed as spatial variations in the equation of state $p = p(\rho, S)$ that lead to density fluctuations because of the corresponding variations in pdV work done during expansion. The isocurvature mode evolves independently of the isentropic ("adiabatic") mode during the radiation-dominated era. Once the universe becomes dominated by cold non-relativistic matter, the entropy-induced pressure gradients become unimportant and all that is left is the zero-pressure growing longitudinal mode described above. For a detailed discussion of isocurvature perturbations using general relativity, see Kodama & Sasaki (1986).

3.1.4 Transverse solutions

Transverse modes are defined by the condition $\vec{\nabla} \cdot \vec{v} = 0$. ¿From (18) it follows that the associated perturbations in density, entropy and gravitational potential all vanish: $\delta = S = \phi = 0$. The curl of the perturbed Euler equation becomes simply

$$\frac{\partial}{\partial \tau}\left(a \vec{\nabla} \times \vec{v}\right) = 0 , \tag{29}$$

whose solution is

$$\vec{v}(\vec{x}, \tau) = \vec{v}_\perp(\vec{x}, \tau) = a^{-1} \vec{\nabla} \times \vec{A}(\vec{x}) . \tag{30}$$

This velocity field is rotational and has nonzero vorticity $\vec{\omega} \equiv (1/a) \vec{\nabla} \times \vec{v}$. Without loss of generality we may set $\vec{\nabla} \cdot \vec{A} = 0$. Thus, we see that there are two independent modes, corresponding to the two independent components of the transverse vector potential $\vec{A}(\vec{x})$. In Fourier transform space, there are two independent components of the velocity transverse to the wavevector ($\vec{k} \cdot \vec{v}_\perp = 0$).

Equation (30) has a very simple physical interpretation. From the conservation of angular momentum, the circulation around a closed loop comoving with the expanding fluid is constant:

$$\oint \vec{v} \cdot a d\vec{x} = \text{constant} . \tag{31}$$

Equation (31) is equivalent to (30) for the transverse velocity.

In fact, the conservation of circulation is applicable beyond linear theory. ¿From the curl of (14) we get the general equation of motion for vorticity,

$$\frac{\partial \vec{\omega}}{\partial \tau} = -2\frac{\dot{a}}{a}\vec{\omega} + \vec{\nabla} \times (\vec{v} \times \vec{\omega}) , \tag{32}$$

where we have assumed $\vec{\nabla}\rho \times \vec{\nabla}p = 0$. The Hubble drag term $-2(\dot{a}/a)\vec{\omega}$ is absent in proper coordinates. Integrating the vorticity through a surface element comoving with the fluid, (32) implies (31) in general. The conservation of circulation is known as the Kelvin circulation theorem. According to this theorem, an irrotational flow remains irrotational. The only way to generate vorticity is if $\vec{\nabla}\rho \times \vec{\nabla}p \neq 0$ (possible as a result of nonlinear dissipation) or if there are tangential stresses such as magnetic stress (shear viscosity alone is not enough) or if there are oblique shock waves in the flow. As a side note, (32) is identical to the magnetic induction equation (with $\vec{\omega}$ replaced by the magnetic field $\vec{B}$) in the magnetohydrodynamic approximation where the magnetic field is frozen into a highly conducting plasma.

Equation (32) does not hold for a collisionless gas because the condition $\vec{\nabla}\rho \times \vec{\nabla} \cdot \overset{\leftrightarrow}{\Pi} = 0$ can easily be violated by mixing of trajectories. The density and stress are defined by taking the phase space averages at a given Eulerian position. This averaging is a kind of dissipation and it can lead to vorticity (for the mean velocity) even if gravity is the only force. However, the circulation theorem is still valid outside of regions where trajectories intersect.

In cosmological perturbation theory, transverse modes (vorticity perturbations) are often called vector perturbations while longitudinal modes (density or entropy perturbations) are often called scalar perturbations. This classification is based on the symmetry of the solutions under three-dimensional rotations. The scalar modes are fully characterized by scalar functions of position (δ, Φ_v, ϕ, and S) while the vector modes are characterized by a vector field (the transverse velocity or vorticity). Just as scalar perturbations are accompanied by perturbations in the spacetime metric [(6)], vector perturbations also induce perturbations in the metric. These perturbations have been neglected in (6); $\phi = 0$ for vector perturbations but the spacetime is not Robertson-Walker. When vorticity is present, a transverse vector field $\vec{w}(\vec{x},\tau)$ should be added to the metric as a $\vec{w}\cdot d\vec{x}\,d\tau$ term in the line element. The transverse gravity modifies (29)–(32) but the corrections are of order $(\lambda H/c)^2$ and are negligible on scales small compared with the Hubble length.

3.2 Lagrangian formulation

The Lagrangian formulation follows the trajectories of individual particles or fluid elements. With a finite number of particles, this is the approach used in cosmological N-body simulations. A Lagrangian description is also valid for a continuous fluid, as long as particles are labeled by a continuous index that attaches a unique name to every infinitesimal mass element. In cosmology it is most convenient to label each particle by its initial comoving position, $\vec{q} \equiv \vec{x}(\tau = 0)$. The trajectories are then $\vec{r} = \vec{r}(\vec{q},\tau)$ in proper coordinates or $\vec{x} = \vec{x}(\vec{q},\tau)$ in comoving coordinates. In the Lagrangian view, motion is simply a mapping from $\vec{q}$ to $\vec{x}(\vec{q},\tau)$.

The Lagrangian approach may be used for both collisional and collisionless fluids. In the former case, exemplified by smooth-particle hydrodynamics, pressure forces must be computed and so each particle, representing a fluid element, must be endowed with thermal energy and other properties in addition to mass, position and velocity. For simplicity, we will restrict ourselves to the case of a collisionless gas with gravitational forces only. In this case, particle trajectories are simply geodesics of the perturbed Robertson-Walker metric [(6)]. The geodesic equations are simply the equations of motion. In proper coordinates these are the familiar Newton's laws,

$$\frac{d\vec{u}}{dt} = -\vec{\nabla}\phi_{\mathrm{P}} , \qquad \frac{d\vec{r}}{dt} = \vec{u} , \tag{33a}$$

becoming, in comoving coordinates [cf. (10)],

$$\frac{d\vec{v}}{d\tau} = -\frac{\dot{a}}{a}\vec{v} - \vec{\nabla}\phi , \qquad \frac{d\vec{x}}{d\tau} = \vec{v} . \tag{33b}$$

The time derivatives are all taken at fixed $\vec{q}$. We have assumed that the particles are nonrelativistic and have applied the weak-field and slow-motion approximations of general relativity. In the following we will use comoving coordinates.

Even with a Lagrangian treatment of the matter we must still solve the Poisson equation [(15)]. This first requires evaluating the Eulerian density field $\delta\rho(\vec{x},\tau)$. In the Lagrangian formulation, the density field is given by applying mass conservation. The mass in a given Eulerian volume is gotten by summing over the Lagrangian mass elements present in the volume:

$$\rho(\vec{x},\tau)\,d^3x = \sum \rho_{\mathrm{m}}(\tau)\,d^3q . \tag{34}$$

Here we have used the fact that the Lagrangian coordinates are simply the initial comoving positions (assumed to be unperturbed at $\tau = 0$) so that the mass of an element of Lagrangian volume d^3q is $a^3\rho_{\mathrm{m}}(\tau)\,d^3q$. The sum is needed only if the mapping $\vec{q} \to \vec{x}(\vec{q},\tau)$ is multi-valued, i.e., if trajectories intersect. By computing the Jacobian of this mapping, we may rewrite (34) as

$$1 + \delta(\vec{x},\tau) = \frac{\rho(\vec{x},\tau)}{\rho_{\mathrm{m}}(\tau)} = \sum \left\| \frac{\partial \vec{x}}{\partial \vec{q}} \right\|^{-1} . \tag{35}$$

The double vertical bars indicate the Jacobian determinant. The sum is over all $\vec{q}$ leading to the given $\vec{x}$ at time τ.

Our system of equations is now given by the Lagrangian equations of motion for the particles [i.e., (33b)], mass conservation [i.e., (35)] and the Poisson equation [i.e., (15)]. These equations are not difficult to solve numerically for a discretized mass distribution. Even so, it is useful to investigate analytic solutions valid in the limit of small density perturbation.

3.2.1 Linear solution

The solution of the Lagrangian equations for small density perturbations ($\delta^2 \ll 1$) was first given by Zel'dovich (1970). The solution method is to use perturbation theory for the trajectories, writing

$$\vec{x}(\vec{q},\tau) = \vec{q} + \vec{x}^{(1)}(\vec{q},\tau) + \vec{x}^{(2)}(\vec{q},\tau) + \cdots , \tag{36}$$

where successive terms are considered to diminish rapidly. (Specifically, the eigenvalues of $\partial(\vec{x} - \vec{q})/\partial\vec{q}$ are supposed to be small.) The density and gravitational potential may be similarly expanded. Evaluating the density [(35)] to first order gives

$$\delta^{(1)}(\vec{x},\tau) = -\vec{\nabla}_{\vec{q}} \cdot \vec{x}^{(1)}(\vec{q},\tau) , \tag{37}$$

where the gradient is with respect to $\vec{q}$ and it is assumed that the mapping is single-valued. The Poisson equation may now be inverted. First we must decompose $\vec{x}^{(1)}(\vec{q},\tau)$ into longitudinal and transverse parts [cf. (19)]:

$$\vec{x}^{(1)}(\vec{q},\tau) = \vec{x}^{(1)}_{\parallel} + \vec{x}^{(1)}_{\perp} , \quad \vec{\nabla}_{\vec{q}} \times \vec{x}^{(1)}_{\parallel} = \vec{\nabla}_{\vec{q}} \cdot \vec{x}^{(1)}_{\perp} = 0 . \tag{38}$$

Only the longitudinal displacement generates a density perturbation and hence a gravitational perturbation:

$$\vec{\nabla}\phi^{(1)} = -4\pi G \rho_{\mathrm{m}} a^2 \, \vec{x}^{(1)}_{\parallel} . \tag{39}$$

The equations of motion now give, in first order,

$$\frac{\partial^2 \vec{x}^{(1)}}{\partial\tau^2} + \frac{\dot{a}}{a}\frac{\partial\vec{x}^{(1)}}{\partial\tau} = 4\pi G \rho_{\mathrm{m}} a^2 \, \vec{x}^{(1)}_{\parallel}(\vec{q},\tau) , \tag{40}$$

where the time derivatives are now written as partial derivatives to make clear that (40) should be interpreted as a partial differential equation for $\vec{x}^{(1)}(\vec{q},\tau)$.

The transverse part of (40) gives us the familiar rotational perturbations, $\vec{v}^{(1)}_{\perp}(\vec{q},\tau) = a^{-1}\vec{\nabla}_{\vec{q}} \times \vec{A}(\vec{q})$ and $\vec{x}^{(1)}_{\perp}(\vec{q},\tau) = \vec{\nabla}_{\vec{q}} \times \vec{A}(\vec{q}) \int d\tau/a$. Because $\vec{A}$ is considered a first-order quantity, we may change the argument from $\vec{q}$ to $\vec{x}$, reproducing (30).

The longitudinal part of (40) is equivalent to (20) for the Eulerian density perturbation itself in the limit of zero pressure. Thus, we know that the general solution is analogous to (23), with the growing and decaying longitudinal parts each factoring into a function of time (the same growing and decaying functions $D_{\pm}$ introduced above) multiplying a function of $\vec{q}$. Retaining only the growing solution, we get, to first order,

$$\vec{x}(\vec{q},\tau) \approx \vec{q} + D_{+}(\tau)\vec{\psi}(\vec{q}) , \tag{41}$$

where $\vec{\psi}(\vec{q})$ is a longitudinal vector field called the linear growing-mode displacement field. To first order we may consider $\vec{\psi}$ to be a function of $\vec{x}$, allowing us to get the corresponding Eulerian density and velocity fields simply:

$$\delta^{(1)} = -D_+(\tau)\vec{\nabla}\cdot\vec{\psi} , \quad \vec{v}_{\parallel}^{(1)}(\vec{q},\tau) = \dot{D}_+\vec{\psi}(\vec{x}) = Hfa(\vec{x}-\vec{q}) . \quad (42)$$

Comparing this with (26), we see that the linear displacement field is proportional to the gradient of the velocity potential: $\vec{\psi} = -(a^2 HfD_+)^{-1}\vec{\nabla}\Phi_v$.

Zel'dovich's important contribution was not to rederive the results of Eulerian perturbation theory as we have done here. It was rather to obtain a qualitative correct quasi-nonlinear solution by extrapolating the linear trajectories of (41) beyond the range of strict applicability. The so-called Zel'dovich approximation consists of using (41) with the linear displacement field but then computing the density using (35) without expanding it in a perturbation series. Assuming that trajectories do not intersect, the quasi-nonlinear density is

$$1 + \delta(\vec{q},\tau) = [(1 + D_+\lambda_1)(1 + D_+\lambda_2)(1 + D_+\lambda_3)]^{-1} , \quad (43)$$

where $\lambda_i = \lambda_i(\vec{q})$ are the eigenvalues of the "deformation tensor" $\partial\vec{\psi}/\partial\vec{q}$. Note that because the displacement field is longitudinal the deformation tensor is symmetric and therefore has orthogonal eigenvectors. Equation (43) must be supplemented by the mapping of (41).

¿From (43), one sees that collapse to infinite density may occur under the Zel'dovich approximation, at those points in space where the deformation tensor has negative eigenvalues. Collapse first occurs along the direction given by the eigenvector corresponding to the most negative eigenvalue. In the general case ($\lambda_1 \neq \lambda_2 \neq \lambda_3$), this collapse is one-dimensional, leading to a two-dimensional surface of infinite density called a caustic (forming at those places where trajectories first intersect) and often referred to as a "Zel'dovich pancake." After the initial caustic formation, collisionless matter streams through the pancake (collisional matter would be halted by shock waves), which becomes bounded on either side by caustic surfaces in the collisionless matter and shock waves in the collisional component. After that, collapse next occurs in the direction corresponding to the second most negative eigenvalue (at those points in space where there exists a second negative eigenvalue). This collapse compresses matter in the plane of the pancake (recall that the eigenvectors of the deformation tensor are orthogonal) toward a one-dimensional curve often called a "filament." Finally, collapse occurs along the third direction, leading to a zero-dimensional "clump." This qualitative behavior is confirmed by numerical calculations of gravitational instability. An excellent discussion is given by Shandarin & Zel'dovich (1989).

A significant shortcoming of the Zel'dovich approximation is the fact that after pancake formation, matter continues to stream through the pancake without ever turning around to fall back in. The pancakes dissolve. A clever repair was made by Gurbatov, Saichev & Shandarin (1989 and references therein). With the addition of a tiny amount of viscosity to the Euler equation — converting it in the absence of pressure into a Burgers' equation (Whitham 1974) — the matter is halted at the pancakes. In this "adhesion approximation," even an arbitrarily small viscosity is enough to form a viscous boundary layer preventing the pancakes from dissolving. The adhesion approximation also benefits from the existence of an exact solution to the three-dimensional Burgers' equation with an intuitive geometrical construction. In practice the method is nontrivial to implement because it still requires a mapping from Lagrangian ($\vec{q}$) space to Eulerian ($\vec{x}$) space. Moreover it is still an approximation as the gravitational field is never computed after the perturbations become large. The adhesion approximation is useful for developing intuition but it is not a substitute for exact or numerical nonlinear solution.

4 Potential flow reconstruction

We have seen that gravitational instability produces a velocity field that is irrotational (i.e., a potential flow) in the linear regime [(26) and (42)] and that remains irrotational outside of regions where trajectories intersect. Moreover, in the linear regime the velocity field is proportional to the gravity field. With these facts in mind, Bertschinger & Dekel (1989) suggested a method for reconstructing the smoothed three-dimensional peculiar velocity, gravitational potential and mass density fields from observations of galaxy distances and redshifts. Their method, named POTENT, enables one to map the large-scale distribution of gravitating matter in the universe. It is based on Galileo's observation that all bodies fall the same way in a gravitational field, so that measurements of the motions of freely-falling galaxies may be used to trace out the gravity field even if those galaxies do not contain all of the mass. Through POTENT, such measurements may ultimately be used to determine whether galaxies trace the mass, to estimate Ω_m, to test specific models for the initial density fluctuations, and to test the strong Cosmogonical Principle itself. The method is discussed briefly here. See Bertschinger (1990) for a more complete review and Dekel, Bertschinger & Faber (1990), Bertschinger et al. (1990) and Nusser et al. (1991) for further details. Kaiser and Stebbins have developed a similar technique, described in the review of Kaiser (1990). Recent results obtained using the POTENT reconstruction procedure are summarized by Dekel (1992).

The goals mentioned above all require a measurement of the present peculiar velocity field $\vec{v}(\vec{x})$. Measurements of galaxy distances (actually, of the combination $H_0 r$) and redshifts provide the radial component $v_r = cz - H_0 r$ (using the linear Doppler formula for the proper velocity of a galaxy at small redshift). The transverse velocity components are measurable in principle from proper motions, but the angular velocities are too small to measure in practice beyond the Local Group. So, we are missing two-thirds of the velocity components for each galaxy. How can we proceed?

Fortunately, for a potential flow the radial component of the velocity field is sufficient to reconstruct the missing two components. One scalar field, the velocity potential Φ_v, fully specifies a potential flow: $\vec{v} = -\vec{\nabla}\Phi_v$ (we set $a = 1$ in referring to the present time). The velocity potential itself may be obtained from the radial component by performing line integrals in the radial direction:

$$\Phi_v(r,\theta,\phi) = -\int_0^r v_r(r',\theta,\phi)\,dr' \ . \tag{44}$$

(Note that here ϕ is the azimuthal spherical angle, not the gravitational potential.) If the radial velocity field can be determined along each radial path, the potential is fully determined (aside from an irrelevant constant that is set to zero at $r = 0$). The gradient of the potential in the tangential directions then restores the missing components. Skeptical readers may wish to add some other scalar field to Φ_v so computed and to take its gradient, bearing in mind that $\vec{v}$ is finite at the origin. However, the physics is as simple as electrostatics: the line integral of a potential field gives differences in potential. Gravity works the same way.

Bertschinger & Dekel (1989), Dekel, Bertschinger & Faber (1990) and Nusser et al. (1991) tested the potential flow reconstruction method using N-body simulations of nonlinear gravitational clustering. They found that when the simulation is smoothed to remove small-scale nonlinear structure (using a Gaussian convolution window of radius exceeding two mass correlation lengths), the reconstructed potential, velocity and density fields agree with the actual smoothed fields to an accuracy of a few percent, given perfect sampling and no noise. These tests show that the vorticity created by nonlinear mixing on small scales is negligible on large scales, and that linear theory is a good approximation on large scales (where the fluctuations have small amplitude) even when the unsmoothed density fluctuations are large. This result justifies the premise stated in Sect. 3.1.1, viz., that small-scale nonlinearity does not vitiate linear theory on large scales, at least for hierarchical clustering models with decreasing power for long wavelengths. The meaning of power as a function of wavelength will be made clear below when we discuss the power spectrum.

Equation (44) cannot be evaluated directly from the data because we do not have perfect radial velocity measurements at all points in space. In practice we have galaxy measurements sparsely spread through space (in some places, densely sampled in galaxy clusters) with distance measurements that contain large statistical uncertainties (and possibly systematic errors). There is no foolproof way to measure distances to galaxies, as the embarrassingly large range of values of the Hubble constant attests. (Fortunately, for measuring peculiar velocities, actual distances need not be measured but rather only the combination $H_0 r$. This quantity may be obtained with much greater precision than r itself, through measurements of distance ratios.) The statistical uncertainties in v_r arise from the lack of perfect correlation in the galaxy properties used to estimate distance (generally, some measure of internal velocity compared with luminosity or diameter) and usually not from errors of measurement. Thus, the uncertainties in the peculiar velocity field may not be reduced by averaging repeated measurements using the same technique (although they may be reduced by combining the results of disparate distance measurement methods). Rather, we must average together the measurements of different galaxies. This necessitates smoothing and interpolation. The smoothing has the added effect of filtering out small-scale nonlinear vorticity. Dekel, Bertschinger & Faber (1990) thoroughly discuss the pros and cons of various ways to smooth the data. In the end they perform a weighted convolution sum over the galaxy measurements:

$$v_r(\vec{r}) = \sum_i W_i(\vec{r}) v_{r,i} , \quad W_i(\vec{r}) \propto n_i^{-1} \sigma_i^{-2} \exp\left(-|\vec{r} - \vec{r}_i|^2 / 2R_s^2\right) , \quad (45)$$

where the sum is over objects (galaxies or galaxy clusters, the latter treated as single points), R_s is the Gaussian smoothing radius (generally set to 1200 km s^{-1}, i.e., 12 h^{-1} Mpc), σ_i is the estimated standard error of the radial peculiar velocity $v_{r,i}$ (around 20% of the distance to a galaxy, reduced for a cluster by the square root of the number of measured galaxies) and n_i is the local density of objects in the sample, estimated using the distance to the fourth object nearest to object i. While it may be possible to improve the smoothing, the very significant problems introduced by sparse sampling and distance errors suggest that it will be difficult to do much better. Just how significant the errors are becomes clear when one realizes that 20% of the distance equals the typical peculiar velocity for distances beyond 3000 km s^{-1} (i.e., 30 h^{-1} Mpc). In short, there is a terrible signal-to-noise problem beyond the Local Supercluster.

The POTENT procedure consists of the following steps. (1) Construct a smooth radial velocity field $v_r(r, \theta, \phi)$ [(45)]. (2) Integrate the radial velocity field along radial paths to get the velocity potential Φ_v [(44)]. (3) Take the tangential derivatives to get the tangential velocity components: $\vec{v} = -\vec{\nabla}\Phi_v$. (4) Use gravitational instability theory to get the gravitational potential and mass density fluctuation fields [(26)].

Concerning the last step, in linear theory we use the simple relation between the present density and velocity fields,

$$\delta(\vec{x}) \approx -(H_0 f)^{-1} \vec{\nabla} \cdot \vec{v} , \quad (46)$$

where f is the growing mode logarithmic growth rate [(25)] and H_0 may be absorbed into the distances. Dekel, Bertschinger & Faber (1990) show how to use the Zel'dovich approximation to obtain a more accurate estimate of the density field in the quasi-linear regime ($\delta \lesssim 4$). Their implementation requires an iterative solution of the mapping between Eulerian and Lagrangian space. Recently, Nusser et al. (1991) have shown that the Zel'dovich approximation may be applied without iteration directly in Eulerian space. Their idea is to use (35) but to invert the relation between $\vec{q}$ and $\vec{x}$ (assuming that the mapping is single-valued, which breaks down on small scales in regions where trajectories intersect). Using the linear relation between displacement and velocity [(42)], we may invert (41) using the measured Eulerian peculiar velocity field:

$$\vec{q} \approx \vec{x} - (H_0 f)^{-1} \, \vec{v}(\vec{x}) \ . \tag{47}$$

¿From (35) with a single-valued mapping we then obtain the general result

$$1 + \delta(\vec{x}, \tau) = \left\| \overset{\leftrightarrow}{1} - (aHf)^{-1} \frac{\partial \vec{v}}{\partial \vec{x}} \right\| \ . \tag{48}$$

Thus, the present Eulerian density perturbation field in the Zel'dovich approximation may be found without iteration. Notice that the reconstructed density depends on Ω_m through the growth rate f. Recall that in linear theory, $f \approx \Omega_m^{0.6}$.

A very important application of potential flow reconstruction is to dynamical estimation of Ω_m on large scales. If it can be assumed that galaxies trace mass on large scales, then from a complete galaxy redshift survey one can measure the fluctuations of the smoothed density, $\delta\rho/\rho = \delta(\vec{x})$. By comparing this density field with the peculiar velocity field, one may then infer f and hence Ω_m using (45) or (47). (We do not know a priori that galaxies trace mass, so this may instead give us a measure of the ratio of galaxy and mass fluctuations combined with f. The uncertainty of the relation between the galaxy and mass distributions is known as the biased galaxy formation problem.) This idea for measuring Ω_m dates back at least to Silk (1974) and Peebles (1976) but only recently has significant progress been made with large samples of redshifts and distances sampling a large volume. Yahil (1992) discusses the comparison of the POTENT density reconstruction with the IRAS galaxy redshift survey (Strauss et al. 1990) in this volume. See Bertschinger (1990), Yahil (1990) and Kaiser et al. (1991) for further discussion.

5 Statistical measures of structure

As we have learned from linear theory, in a wide class of theoretical models the initial conditions for structure formation are fully specified by one scalar field. For adiabatic (i.e., isentropic) perturbations, this is the perturbed gravitational potential $\phi(\vec{x})$ of the post-recombination linear growing mode, which is related in relativistic cosmological perturbation theory to the metric fluctuations generated much earlier. For entropy (isocurvature) perturbations it is enough to specify the initial specific entropy distribution $S(\vec{x})$ and the equation of state. For perturbations induced by seed objects like cosmic strings, domain walls or textures, the initial distribution of seeds and of matter must be specified plus the equations of motion of the seeds. For simplicity we will assume that one scalar field, denoted $\phi(\vec{x})$, is enough to fully specify the initial fluctuations. In principle, given $\phi(\vec{x})$ and a sufficiently large computer capable of simulating all relevant nonlinear physical processes, one could compute all the structure in the universe (leaving aside practical problems like chaos and the uncertainty principle). Turning the problem around, we may hope to work backward from observations to determine $\phi(\vec{x})$. Once this is determined, nothing more can be learned about the initial conditions for structure formation.

Clearly, it is impossible in practice to infer the complete initial conditions or to compute exactly the present state of the universe even if the initial conditions were known. However, even if we are capable only of very limited forward computations or backward extrapolations from observations, these give us some means of comparing predictions with observations and thereby testing theories for the origin of fluctuations. So, what do theories of the early universe predict for the initial conditions $\phi(\vec{x})$?

The first important point is this: No scientific theory attempts to predict $\phi(\vec{x})$ exactly. No theory attempts to predict that the Virgo cluster should be where it is, with its mass and distribution of galaxies, with our own Local Group in its suburbs, with M31 falling now toward the Milky Way, etc. All fluctuation theories are *stochastic*. In other words, theories predict only the *statistical properties* of $\phi(\vec{x})$. Thus, a complete theory might predict the mean abundance of clusters and of galaxies and the mean correlations of galaxies but not the specific locations of galaxies and clusters in our universe.

Given that theories predict only the statistical properties of the initial fluctuations $\phi(\vec{x})$, we must frame statistical hypotheses. Our starting point is a statistical restatement of the Cosmological Principle: $\phi(\vec{x})$ is a *homogeneous and isotropic random field*. By random field we mean that $\phi(\vec{x})$ for our universe is a random realization from a statistical ensemble.

Just as a full description of a random variable is provided by its probability distribution, a full description of the random field $\phi(\vec{x})$ (a set of random variables, one for each point in space) is given by the probability distribution functional $p[\phi]$ with measure (infinitesimal probability)

$$d\mu[\phi(\vec{x})] = p[\phi]\,[D\phi(\vec{x})] \equiv \lim_{N\to\infty} p(\phi_1, \phi_2, \dots, \phi_N) \prod_{k=1}^{N} d\phi_k , \qquad (49)$$

where $\phi_k \equiv \phi(\vec{x}_k)$ and $p(\phi_1, \phi_2, \dots, \phi_N)$ is the finite-dimensional probability distribution for the ϕ_k at N points in space. The limit may be made well-defined (at least, for the cases of interest to us) by specifying the field on a regular three-dimensional lattice of spacing Δx in a periodic cube of length L and then taking the limits $\Delta x \to 0$ and $L \to \infty$; the corresponding measure is called Wiener measure. For finite Δx and L the field may equally well be specified by its Fourier series expansion. (For some calculations it may be necessary to apply a cutoff at high spatial frequencies, i.e., to work with the smoothed field, which is most easily done in the Fourier domain.) In the continuum limit, the probability distribution depends on the field $\phi(\vec{x})$ at all points in space so that $p[\phi]$ is a functional (a function of the random field) giving the infinite-dimensional probability density for the random field $\phi(\vec{x})$. For a homogeneous and isotropic random field $\phi(\vec{x})$, translation or rotation of the coordinates leaves the distribution invariant. Thus, homogeneity is equivalent to the condition $p[\phi(\vec{x} + \vec{a})] = p[\phi(\vec{x})]$ for any constant $\vec{a}$. For a mathematical treatment of random fields see Adler (1981) or Vanmarcke (1983). An introduction to functional measure and functional integrals (in the context of quantum mechanics) is given by Feynman & Hibbs (1965); for more advanced treatments in quantum field theory see Amit (1984), Glimm & Jaffe (1987), Rivers (1987), Negele & Orland (1988) and Ramond (1989). While the practicing cosmologist need not be an expert in quantum field theory, some familiarity with the basics of functional differentiation and integration is helpful for understanding the statistics of random fields.

   Our task boils down to a classic problem of statistical inference. Given observations of the present matter distribution, how can we infer the probability distribution functional $p[\phi]$ for the initial conditions? So stated, the problem appears simple, but it is complicated by two major difficulties. First, how does one relate the present distribution of galaxies (including positions, velocities and other observable properties) to the initial fluctuation field $\phi(\vec{x})$? Linear theory is part of the answer but far from all of it; also needed (and lacking) are a full theory of nonlinear gravitational clustering and a good understanding of the physics of galaxy formation. In the following we will drastically oversimplify this problem by assuming implicitly that observations of galaxies can give some information about the present density fluctuation field $\delta(\vec{x})$ and that this may be related in some way to the initial conditions. A second problem is the choice of statistics. Given measurements of a sample of $\phi(\vec{x})$, what are good statistics (functions of the observations) to use for testing theories? The statistics should be sufficiently powerful to distinguish among rival theories and sufficiently robust so that they can be measured accurately from observational data.

Before discussing specific statistical measures we must mention one more general statistical issue, the ergodic problem. Observations probe only the spatial distribution in one realization of $\phi(\vec{x})$. Theory, on the other hand, specifies the probability distribution over an ensemble. How are the two related? Implicit in practically all theoretical discussions is the *Ergodic Hypothesis* for homogeneous random fields: *Ensemble averages equal spatial averages taken over one realization of the random field.* Note that, in contrast with the custom of statistical mechanics, our Ergodic Hypothesis refers to the spatial distribution of the random field at a fixed time rather than to the time evolution of the system.

The Ergodic Hypothesis for random fields has been proved for Gaussian random fields (for which $p[\phi]$ is a Gaussian functional and the finite-dimensional distributions are all multidimensional Gaussians) with a continuous power spectrum (Adler 1981). We will discuss the power spectrum below. Essentially, the Ergodic Hypothesis requires spatial correlations to decay sufficiently rapidly with increasing separation so that there exist many statistically independent volumes in one realization. These conditions are met in a large class of theories of initial conditions, including quantum fluctuations produced during inflation (Brandenberger 1985). If the Ergodic Hypothesis is valid, then all the information present in the complete distribution functional $p[\phi]$ is available from a single sample of $\phi(\vec{x})$ over all space. While it is impossible in principle to measure a complete sample in a universe with particle horizons, we still have the hope that by measuring $\phi(\vec{x})$ over a sufficiently large volume arbitrarily precise tests of theories can be made.

Peebles (1980) combines the assumptions of the statistical Cosmological Principle (homogeneous and isotropic random fluctuation fields), the weak Cosmogonical Principle (small fluctuations initially and today on the Hubble scale) and the Ergodic Hypothesis and calls the resulting set of assumptions the *fair sample hypothesis*. We shall adopt this set of hypotheses in the following.

This section will focus on the statistical description of general random processes rather than on the specific properties of the initial conditions. The statistical tools that will be introduced are sufficiently powerful so that they may be applied equally well to the initial fluctuation field $\phi(\vec{x})$ or to the present distribution of mass or galaxies. We will not attempt to infer the initial conditions; describing the present galaxy distribution alone is a challenging problem.

5.1 Basic statistics of random processes

We now discuss some of the statistics used by cosmologists to characterize the spatial distribution of matter and galaxies. By the term statistic we mean a function of observational data taken in a realization of a random process (the mass or galaxy distribution in our universe).

At the outset it is necessary to distinguish between a discrete distribution such as galaxies (each one treated as a point) and a continuous distribution like the density fluctuation field $\delta(\vec{x})$. A random distribution of discrete points is called a random point process. Continuous random fields and point processes are particular cases of random processes. The spatial statistics that we shall discuss come in continuous and discrete versions (for random fields and point processes, respectively) and we will show the correspondence between the two.

It is an observational fact that luminosity is concentrated into discrete objects — galaxies. Theoretically, the distribution of mass is expected to be macroscopically continuous (albeit with concentrations where luminous galaxies form), while the distribution of luminosity is effectively discrete. Simple phenomenological models, called biasing schemes, have been proposed to relate the distribution of relative mass density fluctuations in linear (or even nonlinear) theory, $\delta(\vec{x})$, to the distribution of galaxies (Kaiser 1984; Dekel & Rees 1987). However, the true relation between galaxies and the mass distribution is unknown, and we will not discuss biasing schemes further. Lacking this information, we cannot, with much confidence, use observations of galaxies to infer the statistics of the continuous mass density field. In spite of our ignorance about galaxy formation, however, it is still useful to discuss the statistics of both discrete and continuous distributions and to make a formal connection between the two.

There are several ways to relate discrete and continuous distributions. A straightforward approach, which will be adopted here, is to define the number density field for a point process as a sum over Dirac delta functions:

$$n(\vec{x}) = \bar{n}\left[1 + \delta(\vec{x})\right] = \sum_i \delta^{\mathrm{D}}(\vec{x} - \vec{x}_i) , \quad \left\langle \sum_i \delta^{\mathrm{D}}(\vec{x} - \vec{x}_i) \right\rangle = \bar{n} , \quad (50)$$

where $\bar{n}$ is the mean density, δ^{D} is the Dirac delta function, and the angle brackets denote an average over the statistical ensemble of point processes. The Dirac delta functions may be replaced by smooth kernels if a spatially continuous field is desired; Scherrer & Bertschinger (1991) considered the statistical properties of continuous fields defined in this way.

Equation (50) allows us to generate a continuous distribution from a discrete one, but not the other way around. The simplest method to generate a discrete distribution from a continuous one is to employ the Poisson process with density (in statistical jargon, rate) given by the realization $n(\vec{x}) = \bar{n}\left[1 + \delta(\vec{x})\right]$ of the continuous random process. In other words, given a realization $n(\vec{x})$, one generates a point process by randomly placing a particle in each volume element d^3x with probability $n(\vec{x})\,d^3x$, with independent probabilities for all volume elements. Correlations are built into the point process through correlations in the density field $n(\vec{x})$. Layzer (1956) introduced the Poisson model for galaxy clustering, which has been developed more extensively by Peebles (1980) and Fry (1985). Note that the outcome is a double stochastic process, with one level of randomness coming from the random field $n(\vec{x})$ and a second from the Poisson sampling.

Table 1.

CONTINUOUS	DISCRETE
2-point density correlation function $\xi(r_{12}) = \langle \delta(\vec{x}_1)\,\delta(\vec{x}_2)\rangle$ with $r_{12} \equiv |\vec{x}_1 - \vec{x}_2|$, $\langle \delta(\vec{x})\rangle = 0$	**Pair distribution function** $dP(\vec{x}_1, \vec{x}_2) = \bar{n}^2 d^3x_1 d^3x_2 \,[1 + \xi(r_{12})]$ = probability for 2 particles in d^3x_1, d^3x_2
2-point velocity correlation function $\langle \vec{v}(\vec{x}_1)\,\vec{v}(\vec{x}_2)\rangle = $ $\psi_{\parallel}(r_{12})\,\vec{n}\vec{n} + \psi_{\perp}(r_{12})\left(\overleftrightarrow{1} - \vec{n}\vec{n}\right)$ with $\vec{n} \equiv (\vec{x}_1 - \vec{x}_2)/r_{12}$	**Pair velocity statistics** $v_{\parallel} = \langle v_{12}\rangle = \langle \vec{n}\cdot\vec{v}_{12}\rangle,$ $\sigma_{\parallel}^2 = \langle v_{12}^2\rangle - \langle v_{12}\rangle^2, \quad \sigma_{\perp}^2 = \langle |\vec{n}\times\vec{v}_{12}|^2\rangle$ with $\vec{v}_{12} = \vec{v}_1 - \vec{v}_2$
N-point density correlation functions $\xi^{(N)}(\vec{x}_1,\ldots,\vec{x}_N) = \langle \delta(\vec{x}_1)\cdots\delta(\vec{x}_N)\rangle_c$	**N-point correlation functions** cluster expansion for $\xi^{(N)}(\vec{x}_1,\ldots,\vec{x}_N)$
1-point PDF of smoothed fields $dP(\tilde{\delta}) = p(\tilde{\delta})\,d\tilde{\delta},$ $\tilde{\delta}(\vec{x}) = \int d^3x'\, W(\vec{x} - \vec{x}')\,\delta(\vec{x}')$	**Counts in cells** $P_N(V)$ = probability that a randomly placed cell of volume V contains N objects
Moments of $\tilde{\delta}$ $\langle \tilde{\delta}^k\rangle = \int \tilde{\delta}^k\, dP(\tilde{\delta})$	**Moments of counts in cells** $\langle N^k\rangle = \sum N^k P_N(V)$
N-point PDFs of smoothed fields $dP(\tilde{\delta}_1,\ldots,\tilde{\delta}_N) = p(\tilde{\delta}_1,\ldots,\tilde{\delta}_N)\,d^N\tilde{\delta}$	**N-point distribution function** $dP(\vec{x}_1,\ldots,\vec{x}_N)$

Table 1 lists several of the statistics used to characterize the spatial distribution of mass and of galaxies. For each statistic a continuous version and its discrete counterpart are listed. Table 1 is far from complete but it does include most of the popular statistics, including several (the last 4 sets) that are, in a certain sense, complete. (As we will see, subject to a few caveats they completely characterize the statistical properties of the spatial distribution of matter or particles.) Where expectation values are indicated, they may be interpreted (under the Ergodic Hypothesis) either as ensemble averages or spatial averages. Similarly, probability distributions may either be interpreted over the ensemble or over independent spatial volumes in one realization. The subscript c on the N-point density correlation function indicates the cumulant (also known as the connected part or the irreducible correlation function), which is explained below. When the statistics are measured over a finite volume in one realization of the random process they are called sample statistics and they may be regarded as estimators for the true (ensemble-averaged) quantities. The following sections assume some familiarity with basic statistics. Readers who are not fluent in statistics will find Lindgren (1976) to be a good introductory text. For mathematical introductions to random fields see Doob (1953) and Yaglom (1962). Several of our continuous statistics are discussed more fully by Monin and Yaglom (1971).

5.2 2-point correlation function

Let us discuss the contents of Table 1 in more detail, beginning first with the 2-point density correlation function and the pair correlation function, both denoted $\xi(r_{12})$. One should be careful to distinguish the discrete and continuous cases, for $\xi(r_{12})$ has a different meaning in the two cases.

In the continuous case, $\xi(r_{12}) = \langle \delta(\vec{x}_1)\,\delta(\vec{x}_2)\rangle$ is, in the language of statistics, the autocovariance function of the field $\delta(\vec{x})$ [in statistics the term correlation function is actually reserved for $\xi(r_{12})/\xi(0)$] but we will call it a correlation function in keeping with standard practice in statistical physics. The correlation function is a measure of the degree of spatial correlation of the density fluctuations $\delta(\vec{x})$. Clearly, for a homogeneous and isotropic random process ξ can depend only on the distance between the two points.

In the discrete case, correlations are defined (Table 1) in terms of probabilities for finding discrete objects (treated as point particles) at specified points. For a homogeneous random point process the probability that one particle lies in the infinitesimal volume d^3x is the 1-point distribution function $dP(\vec{x}) = \bar{n}\,d^3x$, where $\bar{n}$ is the mean number density (independent of $\vec{x}$). For a Poisson process, the probability for finding one particle in d^3x_1 and another in d^3x_2 (with $\vec{x}_1 \neq \vec{x}_2$) is $\bar{n}^2 d^3x_1 d^3x_2$. Thus, $\xi(r)$ is proportional to the excess probability (relative to Poisson) that two different objects are separated by a distance r.

To relate the discrete and continuous cases, we use (50) to give a continuous field from the point process. The 2-point correlation function of the number density field is

$$\langle n(\vec{x}_1)\,n(\vec{x}_2)\rangle = \bar{n}^2\left[1 + \langle\delta(\vec{x}_1)\,\delta(\vec{x}_2)\rangle^{\mathrm{d}}\right]$$

$$= \left\langle \sum_i \sum_j \delta^{\mathrm{D}}(\vec{x}_1 - \vec{x}_i)\,\delta^{\mathrm{D}}(\vec{x}_2 - \vec{x}_j)\right\rangle = \bar{n}^2\left[1 + \xi(r_{12})\right] + \bar{n}\delta^{\mathrm{D}}(\vec{x}_1 - \vec{x}_2)\;.$$

$$(51)$$

A superscript d is used to indicate that the density field is derived from a discrete point process. The correlation function $\xi(r_{12})$ is the discrete pair correlation function. The first equality of (51) follows immediately from (50) with $\langle\delta\rangle = 0$. The last equality is obtained by multiplying (51) by $d^3x_1 d^3x_2$ so that it gives $\langle N_1 N_2\rangle$, where N_i is the number of objects in the volume element d^3x_i. Because d^3x_i is an infinitesimal, N_i is either 0 or 1. The expectation value $\langle N_1 N_2\rangle$ is computed using the pair distribution function (Table 1), unless $\vec{x}_1 = \vec{x}_2$ (i.e., the same particle is counted twice), in which case $\langle N_1^2\rangle = \langle N_1\rangle$ and the 1-point distribution function is used instead, giving the Dirac delta function term. Thus, the density correlation function for the point process is

$$\langle\delta(\vec{x}_1)\,\delta(\vec{x}_2)\rangle^{\mathrm{d}} = \xi(r_{12}) + \bar{n}^{-1}\delta^{\mathrm{D}}(\vec{x}_1 - \vec{x}_2)\;. \qquad (52)$$

In fact, the delta function should be added to the expression for the pair distribution function in Table 1 (but not to the continuous density correlation function) unless the restriction is imposed that $\vec{x}_1 \neq \vec{x}_2$ or that the particles be distinct. The Dirac delta function contribution to the correlation function arises from discreteness; it is not present in the continuous case. Aside from this delta function, the continuous and discrete 2-point correlation functions agree.

It is also easy to relate the continuous and discrete cases starting from the former. Suppose that we have a realization of the random field $\delta(\vec{x})$, with corresponding number density $n(\vec{x}) = \bar{n}(1 + \delta)$ for some mean density $\bar{n}$. Adopting the Poisson model, the probability for two particles to be in d^3x_1 and d^3x_2 (for $\vec{x}_1 \neq \vec{x}_2$) is

$$dP(\vec{x}_1, \vec{x}_2) = n(\vec{x}_1)\,n(\vec{x}_2)\,d^3x_1 d^3x_2 = \bar{n}^2\,d^3x_1 d^3x_2\,[1 + \delta(\vec{x}_1)]\,[1 + \delta(\vec{x}_2)]\ . \tag{53}$$

So far this has been done for a given realization of $\delta(\vec{x})$; now we average over the ensemble of random fields (but not over the Poisson process). Using $\langle \delta(\vec{x}) \rangle = 0$, we recover the pair distribution function of Table 1, with the continuous density correlation function equaling the pair correlation function, $\langle \delta(\vec{x}_1)\,\delta(\vec{x}_2) \rangle = \xi(r_{12})$, for $r_{12} \neq 0$. When the two points are coincident, $dP(\vec{x}_1, \vec{x}_1) = dP(\vec{x}_1) = n(\vec{x}_1)\,d^3x_1$ and so we must add the extra Poisson variance occurring when the same particle is counted twice. The result agrees with (52).

5.3 2-point velocity statistics

Correlation functions are not restricted to the density field; they may be evaluated for any random field, even a vector field, e.g., the peculiar velocity field $\vec{v}(\vec{x})$. Statistics of the velocity field are particularly important because they contain information about the dynamical processes (e.g., gravity) responsible for inducing the motions.

First we consider the continuous case. By isotropy, $\langle \vec{v}(\vec{x}) \rangle = 0$. The 2-point velocity correlation function $\langle \vec{v}(\vec{x}_1)\,\vec{v}(\vec{x}_2) \rangle$ is a tensor. By isotropy, this tensor must be expressible (Table 1) in terms of a longitudinal correlation function $\psi_\parallel(r_{12}) = \langle (\vec{n} \cdot \vec{v}_1)(\vec{n} \cdot \vec{v}_2) \rangle$ and a transverse correlation function $\psi_\perp(r_{12}) = \langle (\vec{t} \cdot \vec{v}_1)(\vec{t} \cdot \vec{v}_2) \rangle$ where $\vec{n} = (\vec{x}_1 - \vec{x}_2)/r_{12}$ is the radial unit vector and $\vec{t}$ is any transverse unit vector ($\vec{n} \cdot \vec{t} = 0$). For an irrotational velocity field, one can show that $\psi_\parallel(r) = d\,[r\psi_\perp(r)]/dr$. Górski (1988) introduced the velocity correlation function in cosmology and evaluated it theoretically for several models of the initial conditions. Górski et al. (1989) applied this statistic to measurements of the peculiar velocity and gravity fields, finding it to be a useful diagnostic for large-scale structure.

The velocity correlations are somewhat different in the discrete case. The reason is that in the continuous case points $\vec{x}_1$ and $\vec{x}_2$ may be at any point in space while in the discrete case the points are required to coincide with particles. In the former case the points are equally likely to be in regions of high and low density while in the latter case the points are necessarily in regions of high (in fact, infinite) density. Pair correlations involving positions and velocities therefore modify the expectation values in the discrete case. In both cases isotropy demands $\langle \vec{v}(\vec{x}) \rangle = 0$. However, while $\langle \vec{n} \cdot (\vec{v}_1 - \vec{v}_2) \rangle$ vanishes in the continuous case (where $\vec{x}_1$ and $\vec{x}_2$ are considered fixed and no conditions are placed on the densities at these points while the ensemble average is taken), it is generally nonzero in the discrete case.

To see more clearly how this difference arises, consider a sample of a point process with given positions and velocities and with phase space density (using $\vec{v}$ in place of the conjugate momentum)

$$f(\vec{x}, \vec{v}) = \sum_i \delta^D(\vec{x} - \vec{x}_i) \, \delta^D(\vec{v} - \vec{v}_i) \, . \tag{54}$$

We may define a continuous velocity field $\vec{v}(\vec{x})$ in the usual fluid limit by averaging over the 1-particle phase space distribution in a small volume around $\vec{x}$. The spatial average is achieved by integrating over a normalized "window function" $W(\vec{x})$ satisfying $\int d^3x \, W(\vec{x}) = 1$:

$$\vec{v}(\vec{x}) = \frac{\int d^3x' \, W(\vec{x} - \vec{x}') \int d^3v' \, f(\vec{x}', \vec{v}') \vec{v}'}{\int d^3x' \, W(\vec{x} - \vec{x}') \int d^3v' \, f(\vec{x}', \vec{v}')} = \frac{\vec{F}(\vec{x})}{\tilde{n}(\vec{x})} \, , \tag{55}$$

where $\vec{F}(\vec{x}) \equiv \sum_i W(\vec{x} - \vec{x}_i) \vec{v}_i$ is the smoothed particle flux density and $\tilde{n}(\vec{x}) \equiv \sum_i W(\vec{x} - \vec{x}_i)$ is the smoothed number density. The mean (ensemble-averaged) velocity field may now be defined either by the volume-weighted average $\langle \vec{v}(\vec{x}) \rangle = \langle \vec{F}/\tilde{n} \rangle$ or by the number-weighted average $\langle \vec{F} \rangle / \langle \tilde{n} \rangle$. The latter definition is required in the discrete case with no smoothing because the velocity is undefined except at the positions of particles. However, the former definition is acceptable if smoothing is used or the mass distribution is continuous; it is natural in the continuous case to average the velocity (rather than the mass flux) because, under the Ergodic Hypothesis, the spatial average equals the ensemble average. In any case, both averages vanish by isotropy.

A difference arises, however, for the relative velocity of pairs:

$$\vec{v}_1 - \vec{v}_2 = \frac{\vec{F}(\vec{x}_1)}{\tilde{n}(\vec{x}_1)} - \frac{\vec{F}(\vec{x}_2)}{\tilde{n}(\vec{x}_2)} = \frac{\vec{F}(\vec{x}_1) \, \tilde{n}(\vec{x}_2) - \vec{F}(\vec{x}_2) \, \tilde{n}(\vec{x}_1)}{\tilde{n}(\vec{x}_1) \, \tilde{n}(\vec{x}_2)} \, . \tag{56}$$

For a continuous fluid the expectation value $\langle \vec{v}_1 - \vec{v}_2 \rangle = \langle \vec{v}_1 \rangle - \langle \vec{v}_2 \rangle$ is computed using the first equation in (56). With either volume-weighting or number-weighting, the expectation value vanishes because $\langle \vec{v}_1 \rangle = \langle \vec{v}_2 \rangle = 0$. However, in the discrete case (without prior smoothing to make continuous fields) pair measurements require that there be particles at both $\vec{x}_1$ and $\vec{x}_2$ so that the second of (56) must be used, in the pair-weighted form $\langle \vec{v}_1 - \vec{v}_2 \rangle = \langle \vec{F}_1 \, \tilde{n}_2 - \vec{F}_2 \, \tilde{n}_1 \rangle / \langle \tilde{n}_1 \, \tilde{n}_2 \rangle$. (The positions are treated as fixed and only those samples that have particles in *both* $d^3 x_1$ and $d^3 x_2$ contribute to the ensemble average in the discrete case.) Now isotropy does not require $\langle \vec{F}(\vec{x}_1) \, \bar{n}(\vec{x}_2) \rangle = 0$; there exists a preferred direction, $\vec{x}_1 - \vec{x}_2$. Consequently, the mean radial relative velocity of pairs need not vanish. Indeed, the gravitational force each particle exerts on the other induces a nonzero mean.

Similarly, the expectation value $\langle \vec{v}(\vec{x}_1) \, \vec{v}(\vec{x}_2) \rangle$ differs depending on whether it is computed in the volume-weighted form $\langle \vec{F}_1 \vec{F}_2 / \bar{n}_1 \bar{n}_2 \rangle$ or in the pair-weighted form $\langle \vec{F}_1 \vec{F}_2 \rangle / \langle \bar{n}_1 \bar{n}_2 \rangle$, the latter being necessary for a discrete process with no smoothing. The tensor must decompose in either case into longitudinal and transverse parts as before, but the results will depend on how the averages are taken. Moreover, if a continuous velocity field is defined by smoothing a discrete distribution, moments of the velocity field will depend on the window function used for the smoothing.

These complications and differences between the various weighting schemes have a practical significance for galaxy pair velocities. In observational samples galaxies are generally given equal weight (excluding those galaxies that are too faint or small to be present in the sample). Theoretically, it may be preferable to give each galaxy a weight proportional to its mass. The different weights may lead to a statistical "velocity bias" of galaxy pair velocities when equal weighting is used for all galaxy pairs, relative to the pair velocities obtained by weighting each pair by the product of masses (Bertschinger & Gelb 1991). Such a bias was noticed in N-body simulations by Carlberg, Couchman & Thomas (1990), who proposed that the difference arose from dynamical friction of galaxies in clusters rather than being a statistical bias. The origin of this bias has not yet been fully understood. While a great deal of attention has focused on "biasing" of the galaxy spatial distribution relative to the mass, surprisingly little attention has been paid to bias of velocity statistics. This is an important oversight, especially considering that both position and velocity information are needed for dynamical estimates of Ω_m via, e.g., the cosmic virial theorem (Davis & Peebles 1983).

In the discrete case, the velocity and correlation statistics are related dynamically by the BBGKY hierarchy equations of kinetic theory. We will not discuss these equations here, although, in principle, they offer important tests of the gravitational clustering paradigm. See Peebles (1980) for details.

5.4 N-point correlation functions

The N-point correlation functions are a generalization of the 2-point correlations discussed above. (We will consider higher-order correlations of only the density.) First we consider the discrete case. The correlation functions are defined by the irreducible parts of the N-point distribution functions (Table 1). The expansion of these distributions in irreducible correlation functions is called a cluster expansion (Pathria 1972; Huang 1987). The motivation for this expansion is the fact that if our N points are divided up into several subsets that are so widely separated from each other as to be independent, then the N-point distribution function factors into products of lower-dimensional distributions, one for each independent subset. Therefore, the N-point distribution function may be approximated by a sum of products of lower-dimensional distribution functions, one for each partition of the N points. The remainder gives the irreducible N-point correlation function. Put another way, the irreducible N-point correlation function is the part of the N-point distribution that cannot be obtained from lower-order irreducible correlation functions. In statistical mechanics, the cluster expansion forms the basis for approximation schemes because clustering is generally very weak and the higher-order correlation functions are small. In a self-gravitating system, however, for small separations clustering is strong and the higher-order correlation functions are large. The irreducible correlation functions are, nevertheless, of great interest.

Note that because the N-point probability distribution is symmetric under exchange of any pair of arguments (the probabilities do not distinguish particles), the N-point correlations are also fully symmetric. Also, translational invariance implies that $\xi^{(N)}$ depends only on $N-1$ difference vectors $\vec{x}_i - \vec{x}_{i+1}$ ($i = 1$ to $N-1$). We assume that, unless otherwise noted, all N points are distinct. In this section we will use $\xi^{(2)}$ for the 2-point correlation function and we will use subscripts to indicate the arguments. We will drop the $\vec{x}$ from the arguments of the probability distributions below. The one-point correlation function is defined in the discrete case by $\xi^{(1)} \equiv 1$.

The cluster expansion is best illustrated by example. The expansion of the 2-point distribution is (Table 1) $dP(\vec{x}_1, \vec{x}_2) = \bar{n}^2 d^3x_1 d^3x_2 [1 + \xi_{12}^{(2)}]$. As advertized, the 2-point correlation function is the residual probability not accounted for by a product of 1-point probabilities. Similarly, the 3-point distribution function has cluster expansion

$$dP(1,2,3) = \left(\prod_{k=1}^{3} \bar{n}\, d^3x_k \right) \left[1 + \xi_{12}^{(2)} + \xi_{23}^{(2)} + \xi_{31}^{(2)} + \xi_{123}^{(3)} \right] . \qquad (57)$$

When the three points are moved to infinity, only the first term in brackets remains, while, when one point is moved to infinity, one of the 2-point correlation functions remains. With four points the combinatorics becomes more complicated but the expansion is similar:

$$dP(1,2,3,4) = \left(\prod_{k=1}^{4} \bar{n}\, d^3x_k \right) \left[1 + \xi_{12}^{(2)} + \xi_{13}^{(2)} + \xi_{14}^{(2)} + \xi_{23}^{(2)} + \xi_{24}^{(2)} + \xi_{34}^{(2)} \right.$$
$$\left. + \xi_{12}^{(2)} \xi_{34}^{(2)} + \xi_{13}^{(2)} \xi_{24}^{(2)} + \xi_{14}^{(2)} \xi_{23}^{(2)} + \xi_{123}^{(3)} + \xi_{124}^{(3)} + \xi_{134}^{(3)} + \xi_{234}^{(3)} + \xi_{1234}^{(4)} \right] .$$
$$(58)$$

The cluster expansion gives the N-point distribution as a sum of terms, one for each partition of the set of N points $\vec{x}_1, \ldots, \vec{x}_N$. A partition of a set

S is the set of disjoint subsets $\{T_i\}$, whose union gives the whole: $\cup T_i = S$. For example, there are two partitions of the set $\{1,2\}$: $\{\{1\},\{2\}\}$ and $\{1,2\}$. Similarly, there are four partitions of the set $\{1,2,3\}$: $\{\{1\},\{2\},\{3\}\}$, $\{\{1,2\},\{3\}\}$, $\{\{1,3\},\{2\}\}$, $\{\{2,3\},\{1\}\}$ and $\{1,2,3\}$. The correspondence between these partitions and the cluster expansions should be clear: each subset in a partition corresponds to an irreducible correlation function. A subset containing only one point gives $\xi^{(1)} \equiv 1$. The product of all irreducible correlations is taken for each partition. Because of the symmetry of the correlation functions, only one product is taken for each distinct partition. For example, the partitions $\{\{1,3\},\{2\}\}$ and $\{\{2\},\{3,1\}\}$ are not considered distinct: the terms $\xi_{13}^{(2)}\xi_{2}^{(1)}$ and $\xi_{2}^{(1)}\xi_{31}^{(2)}$ are identical and we avoid double-counting.

The N-point correlation functions, together with the mean density, completely specify the spatial statistics of a homogeneous and isotropic random point process. This follows because the N-point distributions, which themselves provide a complete statistical description, are built up entirely from these ingredients [e.g., (57), (58)]. However, one should not conclude from this that the hierarchy of galaxy N-point correlation functions is sufficient to completely characterize the initial conditions that gave rise to cosmological structure. Not only is half of phase space missing (peculiar velocities), we do not even know the relation between galaxies and the mass.

In the continuous case, the irreducible correlation functions are defined as the cumulants (also known as irreducible moments, connected moments and semi-invariants) of the density fluctuation field $\delta(\vec{x})$. Cumulants arise frequently in statistics (Stuart & Ord 1987). They are defined recursively by expanding the moments of $\delta(\vec{x})$ in a series of products in a way analogous to the discrete cluster expansion (Ma 1985). The cumulant expansion of the 2- and 3-point correlations of $\delta(\vec{x})$ are (indicating the point $\vec{x}_i$ by subscript i)

$$\langle \delta_1 \delta_2 \rangle = \langle \delta_1 \rangle_c \langle \delta_2 \rangle_c + \langle \delta_1 \delta_2 \rangle_c \; , \tag{59}$$

$$\langle \delta_1 \delta_2 \delta_3 \rangle = \langle \delta_1 \rangle_c \langle \delta_2 \rangle_c \langle \delta_3 \rangle_c$$
$$+ \langle \delta_1 \delta_2 \rangle_c \langle \delta_3 \rangle_c + \langle \delta_2 \delta_3 \rangle_c \langle \delta_1 \rangle_c + \langle \delta_3 \delta_1 \rangle_c \langle \delta_2 \rangle_c + \langle \delta_1 \delta_2 \delta_3 \rangle_c \; . \tag{60}$$

The irreducible correlation functions of $\delta(\vec{x})$ are defined to be the cumulants: $\xi^{(N)} \equiv \langle \delta_1 \cdots \delta_N \rangle_c$. Note that $\langle \delta \rangle_c \equiv \langle \delta \rangle = 0$ so that $\langle \delta_1 \delta_2 \rangle_c = \langle \delta_1 \delta_2 \rangle$ and $\langle \delta_1 \delta_2 \delta_3 \rangle_c = \langle \delta_1 \delta_2 \delta_3 \rangle$. However, we retain the $\langle \delta \rangle_c$ terms in (59) and (60) to make clear the correspondence with the cluster expansion. Like the cluster expansion, the cumulant expansion is a sum over partitions. The equivalence of the two expansions becomes clear [cf. (57) and (60)] when we recall that $\xi^{(1)} \equiv 1$ in the discrete case while $\xi^{(1)} \equiv \langle \delta \rangle_c = 0$ in the continuous case.

We can relate the discrete and continuous N-point correlation functions using either (50) or the Poisson model, as we did for the 2-point correlation functions. Consider first the cumulants of the density field for a point process. The procedure here is to compute moments as in (51), integrating the number density field over infinitesimal volumes $d^3x_1 \cdots d^3x_N$ and using the N-point distribution functions to compute expectation values. The cumulants are then obtained by subtracting off the contributions from lower-order connected correlation functions as in (59) and (60). Care is needed to properly handle the cases when some or all of the points are identical. We leave it as an exercise for the reader to show that, including the discreteness contributions arising from identical points, the 3- and 4-point irreducible density correlation functions for the point process are

$$\langle \delta_1 \delta_2 \delta_3 \rangle_c^d = \xi_{123}^{(3)} + \bar{n}^{-1} \left[\delta_{12}^D \xi_{23}^{(2)} + \delta_{23}^D \xi_{31}^{(2)} + \delta_{31}^D \xi_{12}^{(2)} \right] + \bar{n}^{-2} \delta_{12}^D \delta_{23}^D , \quad (61)$$

$$\langle \delta_1 \delta_2 \delta_3 \delta_4 \rangle_c^d = \xi_{1234}^{(4)} + \bar{n}^{-1} \left[\delta_{12}^D \xi_{134}^{(3)} + \delta_{13}^D \xi_{124}^{(3)} + \delta_{14}^D \xi_{123}^{(3)} + \delta_{23}^D \xi_{124}^{(3)} + \right.$$
$$\left. \delta_{24}^D \xi_{123}^{(3)} + \delta_{34}^D \xi_{123}^{(3)} \right] + \bar{n}^{-2} \left[\delta_{12}^D \delta_{34}^D \xi_{13}^{(2)} + \delta_{13}^D \delta_{24}^D \xi_{12}^{(2)} + \delta_{14}^D \delta_{23}^D \xi_{12}^{(2)} + \right. \quad (62)$$
$$\left. \delta_{12}^D \delta_{23}^D \xi_{14}^{(2)} + \delta_{12}^D \delta_{24}^D \xi_{13}^{(2)} + \delta_{13}^D \delta_{34}^D \xi_{12}^{(2)} + \delta_{23}^D \delta_{34}^D \xi_{12}^{(2)} \right] + \bar{n}^{-3} \delta_{12}^D \delta_{23}^D \delta_{34}^D .$$

Dirac delta functions are given for identical pairs of points, e.g., $\delta_{12}^D \equiv \delta^D(\vec{x}_1 - \vec{x}_2)$, with a correlation function left over for the distinct points. Note the similarity with the cluster and cumulant expansions. Each term in (61) and (62) corresponds to a partition, with $M - 1$ delta functions for each subset of M (identical) points and one connected correlation function for the whole partition (corresponding to disjoint subsets). For example, the term $\delta_{12}^D \xi_{134}^{(3)}$ in the 4-point cumulant corresponds to the partition $\{\{1,2\},\{3\},\{4\}\}$. Scherrer & Bertschinger (1991) give the general result for the Nth-order cumulant [their (2.13) and (2.14), with the mass-dependence removed and with their window function f replaced by a delta function].

The reader may wish to show that (61) and (62) are identical with the result obtained by applying the Poisson model to the continuous density field $\delta(\vec{x})$, as in (53). Thus, the point process may be represented as a continuous density field through a sum over delta functions [(50)], or, the continuous field may be Poisson sampled to give a point process. The same relation between discrete and continuous correlation functions obtains in either case. We conclude that the irreducible N-point cumulant of the density fluctuation field is identical with the discrete N-point correlation function of the corresponding point process aside from discreteness terms needed when some or all of the N points are identical. Note that these discreteness terms are more than just the Poisson cumulants corresponding to N identical points [the last terms in (61) and (62)]. In the discrete case, all $N-1$ lower-order irreducible correlation functions contribute to the N-point density correlation function.

5.5 Cumulant expansion theorem

The N-point correlation functions are closely related to the the probability distribution functional $d\mu[\delta(\vec{x})]$ for a continuous density field and to the probability distribution for counts in cells in the discrete case. The connection is made by a mathematical result known as the cumulant expansion theorem. This theorem finds widespread use in statistical mechanics (e.g., Ma 1985) and in quantum field theory (e.g., Negele & Orland 1988), where it is known as the linked-cluster theorem. The cumulant expansion theorem is important because it simplifies many statistical calculations. In this section we therefore present the theorem in several practical forms.

It is useful to develop the ideas of the cumulant expansion theorem starting with a simple statistical system: a single random variable X, with continous probability distribution $dP(X)$. The moments of this distribution are defined by $\langle X^n \rangle = \int X^n \, dP(X)$. The moments do not always exist; the Cauchy distribution $dP/dX = [\pi(1 + X^2)]^{-1}$ is a well-known example of a distribution some of whose moments do not exist. When the moments do exist, they may be obtained from a Taylor series expansion of the moment generating function, defined by

$$M(s) \equiv \langle e^{-sX} \rangle = \int e^{-sX} \, dP(X) = \sum_{n=0}^{\infty} \frac{(-s)^n}{n!} \langle X^n \rangle . \qquad (63)$$

Expanding the exponential establishes the proof. Note that $M(-is)$, called the characteristic function, is the Fourier transform of the differential probability distribution function. We use the real transform for convenience in order to work with real-valued functions. The moment generating function exists for $\mathrm{Re}\, s \geq 0$ even though it may have no Taylor series expansion about $s = 0$. Under very general conditions, the characteristic function $M(-is)$ determines the distribution function and vice versa (Stuart & Ord 1987).

The cumulant expansion theorem states that the logarithm of the moment generating function is the cumulant generating function. For the univariate distribution of (63), this relation, which we will refer to as the univariate cumulant expansion theorem, is

$$\ln M(s) = \langle e^{-sX} - 1 \rangle_c \equiv \sum_{n=1}^{\infty} \frac{(-s)^n}{n!} \langle X^n \rangle_c . \qquad (64)$$

The second equality defines the connected part of $\langle e^{-sX} - 1 \rangle$ by the series expansion over cumulants; note that the zeroth-order cumulant is excluded. The proof of (64) is straightforward but tedious (Ma 1985). Basically, by exponentiating (64) and regrouping terms, one finds a correspondence between terms contributing to nth-order moments of X (e.g., $\langle X^2 \rangle_c \langle X \rangle_c^2$ for $n = 4$) and partitions of n points (e.g., $\{\{1,2\},\{3\},\{4\}\}$). Summing over partitions (carefully counting the number of ways that each partition can arise) reproduces the cumulant expansion for $\langle X^n \rangle$.

Equation (64) can be generalized to a p-dimensional multivariate distribution $dP(X_1, X_2, \ldots, X_p)$ with moment generating function $M(s_1, s_2, \ldots, s_p)$. The result is the multivariate cumulant expansion theorem:

$$\ln M(s_1, s_2, \ldots, s_p) \equiv \ln \left\langle \exp\left[-\sum_{k=1}^{p} s_k X_k\right] \right\rangle$$

$$= \left\langle \exp\left[-\sum_{k=1}^{p} s_k X_k\right] - 1 \right\rangle_c \equiv \sum_{N=1}^{\infty} \frac{(-1)^N}{N!} \left[\prod_{k=1}^{N} \sum_{l_k=1}^{p} s_{l_k}\right] \langle X_{l_1} \cdots X_{l_N}\rangle_c .$$

$$(65)$$

The proof of (65) is left as an exercise for the reader. The cumulants follow at once from derivatives of $\ln M$:

$$\langle X_1^{n_1} X_2^{n_2} \cdots X_p^{n_p}\rangle_c = (-1)^{n_{tot}} \left[\frac{\partial^{n_{tot}} \ln M(s_1, s_2, \ldots, s_p)}{\partial s_1^{n_1} \partial s_2^{n_2} \cdots \partial s_p^{n_p}}\right]_{s_1 = s_2 = \cdots = s_p = 0},$$

$$(66)$$

where $n_{tot} \equiv n_1 + n_2 + \cdots n_p$. Likewise, the moments $\langle X_1^{n_1} X_2^{n_2} \cdots X_p^{n_p}\rangle$ follow from derivatives of M.

We now generalize the multivariate cumulant expansion theorem to the infinite-dimensional probability distribution for the random field $\delta(\vec{x})$. With infinitely many variables, the sums in (65) become integrals and the expectation values are functional integrals. The moment generating functional is

$$Z[s(\vec{x})] \equiv \left\langle \exp\left[-\int d^3 x \, s(\vec{x}) \, \delta(\vec{x})\right] \right\rangle =$$

$$\int d\mu[\delta(\vec{x})] \exp\left[-\int d^3 x \, s(\vec{x}) \, \delta(\vec{x})\right] .$$

$$(67)$$

Students of statistical physics or quantum field theory will recognize $Z[s]$ as the partition functional or vacuum-to-vacuum transition amplitude in the presence of a source $s(\vec{x})$ (Amit 1984, Rivers 1987). Equation (65) now becomes the functional cumulant expansion theorem or linked-cluster theorem:

$$\ln Z[s(\vec{x})] = \left\langle \exp\left[-\int d^3 x \, s(\vec{x}) \, \delta(\vec{x})\right] - 1 \right\rangle_c$$

$$\equiv \sum_{N=1}^{\infty} \frac{(-1)^N}{N!} \left[\prod_{k=1}^{N} \int d^3 x_k \, s(\vec{x}_k)\right] \langle \delta(\vec{x}_1) \cdots \delta(\vec{x}_N)\rangle_c .$$

$$(68)$$

Using the fact that $\langle \delta(\vec{x}_1) \cdots \delta(\vec{x}_N)\rangle_c = \xi^{(N)}(\vec{x}_1, \ldots, \vec{x}_N)$, we see that the linked-cluster theorem gives $\ln Z[s]$ as a series expansion over the N-point irreducible correlation functions.

The analog of (66) for an infinite-dimensional distribution involves functional derivatives of $\ln Z[s]$. Consider a functional $F[s(\vec{x})]$. The functional derivative is defined by extending the concept of partial derivatives:

$$\delta F = \lim_{\delta s \to 0} \{F[s(\vec{x}) + \delta s(\vec{x})] - F[s(\vec{x})]\} \equiv \int d^3x \, \frac{\delta F}{\delta s} \, \delta s(\vec{x}) \,, \qquad (69)$$

where $\delta s(\vec{x})$ is an arbitrary variation of the field $s(\vec{x})$. Functional differentiation of (68) gives the irreducible correlation functions:

$$\xi^{(N)}(\vec{x}_1, \ldots, \vec{x}_N) = (-1)^N \left[\frac{\delta^N \ln Z[s(\vec{x})]}{\delta s(\vec{x}_1) \cdots \delta s(\vec{x}_N)} \right]_{s=0}. \qquad (70)$$

The functional cumulant expansion theorem shows that one may reconstruct the cumulant generating functional $\ln Z[s]$ from the N-point correlation functions when the latter exist (as they do in general for nonzero pair separations). The probability distribution functional $p[\delta]$ then follows from the inverse functional Fourier transform of the characteristic functional $Z[-is]$. Therefore, in principle the hierarchy of N-point correlation functions completely specifies the statistics of the homogeneous and isotropic random field $\delta(\vec{x})$, just as they fully specify the statistical properties of a homogeneous and isotropic random point process.

The cumulant expansion theorem also has a form appropriate for a discrete random point process. We define $Z_{\rm d}[s(\vec{x})]$ in a manner similar to (67), using (50) for $\delta(\vec{x})$:

$$Z_{\rm d}[s(\vec{x})] \equiv \left\langle \exp\left[-\int d^3x \, s(\vec{x}) \, \delta(\vec{x})\right] \right\rangle =$$

$$\exp\left[\int d^3x \, s(\vec{x})\right] \left\langle \exp\left[-\sum_i \frac{s(\vec{x}_i)}{\bar{n}}\right]\right\rangle. \qquad (71)$$

The sum is taken over the entire set of points in the point process. Now there are two different ways to proceed in deriving the discrete cumulant expansion theorem; both involve expanding the exponential in (71). One may either use the discrete N-point distribution functions [(53), (57), (58), etc.] to evaluate moments of products of $s(\vec{x}_i)$, taking care when one or more points in the products are identical, or one treats $\delta(\vec{x})$ as a continuous field, yielding (68). We adopt the latter procedure. Now, for a discrete distribution, the cumulants of δ are not the discrete N-point correlation functions, because the latter neglect the discreteness contributions discussed in Sect. 5.4. Rather, the cumulants of $\delta(\vec{x})$ are given by (52), (61), (62), etc.; note that these equations involve partition sums analogous to (59) and (60). It is straightforward to show that the N-point cumulants in (68) sum to give an exponential. The final result, referred to as the discrete cumulant expansion theorem, is

$$\ln Z_{\rm d}[s(\vec{x})] = \int d^3x \, s(\vec{x}) +$$

$$\sum_{N=1}^{\infty} \frac{(-1)^N}{N!} \left\{ \prod_{k=1}^{N} \int d^3x_k \, \bar{n} \left[1 - e^{-s(\vec{x}_k)/\bar{n}}\right]\right\} \xi^{(N)}(\vec{x}_1, \ldots, \vec{x}_N) \,, \qquad (72)$$

with $\xi^{(1)} \equiv 1$. Aside from the extra discreteness term $\int d^3x\, s(\vec{x})$, (72) is equivalent to (68) for the continuous case with the replacement of s by $\bar{n}[1 - \exp(-s/\bar{n})]$.

It is easy to generalize (72) to the case in which each discrete object contributes some kernel function $f(\vec{x} - \vec{x}_i)$ to the density instead of a delta function (Scherrer & Bertschinger 1991). The density fluctuation field [(50)] is simply convolved with $f(\vec{x})$. So, then, are the cumulants. Thus, in the cumulant expansion theorem one simply replaces $\xi^{(N)}(\vec{x}_1, \ldots, \vec{x}_N)$ with $\int d^3x_1' \cdots d^3x_N' \, f(\vec{x}_1 - \vec{x}_1') \cdots f(\vec{x}_N - \vec{x}_N')\, \xi^{(N)}(\vec{x}_1', \ldots, \vec{x}_N')$.

Equation (72) may also be derived using the Poisson model, starting from a continuous random field $\delta(\vec{x})$ and using $n(\vec{x}) = \bar{n}[1 + \delta(\vec{x})]$ as the rate function for Poisson sampling (Fry 1985). For this derivation we divide space into small cells of volume d^3x_k, with the kth cell containing n_k objects. The moment generating function [(71)] includes the factor

$$\left\langle \exp\left[-\sum_i \frac{s(\vec{x}_i)}{\bar{n}}\right]\right\rangle = \left\langle \prod_k e^{-s_k n_k/\bar{n}}\right\rangle = \prod_k \left\langle e^{-s_k n_k/\bar{n}}\right\rangle \qquad (73)$$

where i labels the objects and k labels the cells, with $s_k = s(\vec{x}_k)$. Note that we have assumed a Poisson process (independent probabilities for different cells) to interchange the product and expectation values. Now we average over the Poisson process for the kth cell, using the fact that, to first order in d^3x_k, the probability that the cell contains one object is $n(\vec{x}_k)\, d^3x_k$ and the probability that the cell contains no objects is $1 - n(\vec{x}_k)\, d^3x_k$. Taking the limit of infinitely many small cells, the result of this averaging is

$$\left\langle \exp\left[-\sum_i \frac{s(\vec{x}_i)}{\bar{n}}\right]\right\rangle = \exp\left\{-\int d^3x \left[1 - e^{-s(\vec{x})/\bar{n}}\right] n(\vec{x})\right\} . \qquad (74)$$

Next, we average over the continuous random field $n(\vec{x}) = \bar{n}[1 + \delta(\vec{x})]$. Note that (74) is in the form of (67) for $Z[S(\vec{x})]$ with $S = \bar{n}[1 - \exp(-s/\bar{n})]$ and with δ replaced by $1 + \delta$. Except for $\langle 1 + \delta \rangle = 1$, the cumulants of $1 + \delta$ equal those of δ. So, from (71) we recover (72), with $\xi^{(N)}$ equaling the N-point correlation function of the continuous density fluctuation field $\delta(\vec{x})$ except for $\xi^{(1)} \equiv 1$.

5.6 Smoothed density distribution, counts in cells, and moments

Using the cumulant expansion theorem, we can obtain formal expressions for the moments and the probability distribution of the mass contained in a volume V, as well as related quantities. Estimates of these quantities are easier to obtain from observations than are the correlation functions yet, as we will see, they contain equivalent information.

We begin with the continuous case. Given a density fluctuation field $\delta(\vec{x})$, we define a smoothed field (Table 1) $\tilde{\delta}(\vec{x})$ by convolution with a window function $W(\vec{x})$ normalized by $\int d^3x\, W(\vec{x}) = 1$. We would like to compute the probability distribution, or at least the moment generating function, for $\tilde{\delta}(\vec{x})$ at one point in space. For a homogeneous random process, the 1-point probability distribution function $p(\tilde{\delta})$ is independent of $\vec{x}$. Therefore it suffices to consider $\tilde{\delta}(\vec{x})$ for $\vec{x} = 0$. From the discussion after (63), $p(\tilde{\delta})$ is fully specified by its moment generating function $M(s)$ which, from (67), equals the moment generating functional $Z[S(\vec{x})]$ for the random field $\delta(\vec{x})$, evaluated with $S(\vec{x}) = sW(-\vec{x})$ (here s is a scalar argument, not a field):

$$M(s) \equiv \left\langle e^{-s\tilde{\delta}} \right\rangle = \int e^{-s\tilde{\delta}} p(\tilde{\delta})\, d\tilde{\delta} =$$

$$\left\langle \exp\left[-s \int d^3x\, W(-\vec{x})\, \delta(\vec{x}) \right] \right\rangle = Z[sW(-\vec{x})] .$$

(75)

Employing the univariate and functional cumulant expansion theorems [(64) and (68)], derivatives of $\ln M(s)$ give the cumulants of $\tilde{\delta}$ as the volume-averaged correlation functions weighted by the window function W:

$$(-1)^N \left(\frac{d^N \ln M}{ds^N} \right)_{s=0} = \langle \tilde{\delta}^N \rangle_c = \left[\prod_{k=1}^{N} \int d^3x_k\, W(-\vec{x}_k) \right] \xi^{(N)}(\vec{x}_1, \ldots, \vec{x}_N).$$

(76)

Equation (76) can also be obtained directly, without using functionals, from the definitions $\tilde{\delta}(0) = \int d^3x\, W(-\vec{x})\delta(\vec{x})$ and $\langle \delta(\vec{x}_1) \cdots \delta(\vec{x}_N) \rangle_c = \xi^{(N)}(\vec{x}_1, \ldots, \vec{x}_N)$. In any case, the moments of the smoothed density allow us to reconstruct the moment generating function $M(s)$ from its Taylor series expansion. Note that even if the moments do not exist, the moment generating function does exist for Re $s \geq 0$. The inverse Fourier transform of $M(-is)$ (the characteristic function) gives the probability distribution function $p(\tilde{\delta})$. Coles & Jones (1991) discuss a useful model for the 1-point probability distribution function of the density fields produced by nonlinear clustering, the lognormal distribution.

Note that, for a given window function, $p(\tilde{\delta})$ depends on all of the correlation functions but does not contain enough information to reconstruct them. However, if the window function is varied, p and its moment generating function M may be considered to be functionals of $W(\vec{x})$: $M(s) = Z[sW(-\vec{x})]$. Thus, all statistical information about the random field $\delta(\vec{x})$ is contained in the moments $\langle \tilde{\delta}^N \rangle$ or the distribution function $p(\tilde{\delta})$ when all possible (normalized) window functions are considered. If one prefers to avoid functionals, complete information is also given for fixed W_V by the complete set of N-point probability distribution functions of the smoothed density field, $dP(\tilde{\delta}_1, \ldots, \tilde{\delta}_N)$ (Table 1).

A commonly-employed window function is the top-hat window function for volume V, $W_V(\vec{x}) = V^{-1}$ if $\vec{x} \in V$ and $W_V = 0$ otherwise. With this choice, $\tilde{\delta} = M/\bar{M} - 1$, where M is the mass in volume V and $\bar{M} = \bar{\rho}V$ is the mean mass. Note that the shape of the volume, as well as the total volume, affects the integrals over the correlation functions [(76)] and therefore the probability distribution $p(\tilde{\delta})$.

Consider next the discrete case with the top-hat window function $W_V(\vec{x})$. Now the smoothed density is not a continuous random variable; rather, $\tilde{\delta} = N/\bar{N} - 1$ is discrete, where N is the number of points in volume V and $\bar{N} = \bar{n}V$ is the mean number. Equation (75) must be modified accordingly:

$$M(s) \equiv \left\langle e^{-s\tilde{\delta}} \right\rangle = e^s \left\langle e^{-sN/\bar{N}} \right\rangle = e^s \sum_{N=0}^{\infty} P_N(V) e^{-sN/\bar{N}} = Z_{\mathrm{d}}\left[sW_V(-\vec{x})\right],$$

$$(77)$$

where $P_N(V)$, known as the distribution of counts in cells, is the probability that the volume V contains exactly N objects.

The discrete probability distribution P_N is fully characterized by its probability generating function, $\mathcal{P}(t) \equiv \sum_{N=0}^{\infty} P_N t^N$. Note that $\mathcal{P}(t) = \langle t^N \rangle = t^{\bar{N}} M(-\bar{N} \ln t) = M'(-\ln t)$ where $M(s)$ is the moment generating function for $\tilde{\delta} = N/\bar{N} - 1$ and $M'(s) = \langle \exp(-sN) \rangle$ is the moment generating function for N [cf. (63)]. Using (72) for $Z_{\mathrm{d}}\left[sW_V(-\vec{x})\right]$ and defining $t = \exp(-s/\bar{N})$, from (77) we obtain an important and useful expression for $\mathcal{P}(t)$ in terms of volume averages of the N-point correlation functions:

$$\mathcal{P}(t) \equiv \sum_{N=0}^{\infty} P_N(V) t^N = \exp\left\{ \sum_{N=1}^{\infty} \frac{(t-1)^N}{N!} \left[\prod_{k=1}^{N} \int_V \bar{n}\, d^3x_k \right] \xi^{(N)}(\vec{x}_1, \ldots, \vec{x}_N) \right\}$$

$$(78)$$

where $\xi^{(N)}$ is the discrete N-point correlation function, with $\xi^{(1)} = 1$. A derivation similar to ours has been given by Balian & Schaeffer (1989a) (see also Carruthers 1991). Note that we did not assume the Poisson model; we assumed a point process throughout. Application of the discrete cumulant expansion theorem has made the derivation simple. The moments of counts in cells (Table 1) follow simply from differentiating $\mathcal{P}(t) = \langle t^N \rangle$:

$$\langle N^k \rangle = \sum_{N=0}^{\infty} P_N(V) N^k = \left[\frac{d^k \mathcal{P}}{d(\ln t)^k} \right]_{t=1}.$$

$$(79)$$

Similarly, the cumulants of counts in cells follow from replacing $\mathcal{P}$ with $\ln \mathcal{P}$ in (79).

Equation (78) can also be derived by applying the Poisson model to a continuous density field $n(\vec{x})$ and then averaging over the ensemble of random fields $n(\vec{x}) = \bar{n}[1 + \delta(\vec{x})]$. Using $s(\vec{x}) = -\bar{N}W_V(\vec{x})\ln t$ in (74), we find that the probability generating function is given by an average over the continuous random field $n(\vec{x})$ (Layzer 1956, 1975; Peebles 1980; Fry 1985):

$$\mathcal{P}(t) = \left\langle \exp\left[(t-1) \int_V d^3x\, n(\vec{x}) \right] \right\rangle.$$

$$(80)$$

Application of the functional cumulant expansion theorem [(68)] for the random field $n(\vec{x})$ now gives (78) [see the discussion following (74)].

The probability distribution of counts in cells, $P_N(V)$, has long been used to describe galaxy clustering (e.g., Hubble 1934; Zwicky 1957; Shane & Wirtanen 1967; for recent work see Efstathiou *et al.* 1990; Maurogordato & Lachièze-Rey 1991; and Vogeley, Geller & Huchra 1991). The counts distribution may be estimated from observations by taking independent volumes to be independent samples of the distribution; theoretically, the distribution is over the ensemble of random point processes. The cells may be either elements of solid angle on the sky or, when redshifts are available, three-dimensional volumes in redshift space (with redshift taking the place of distance). Both the volume and shape of the cells may be varied to obtain information about the clustering of objects.

It is possible to define the density distribution for discrete objects using window functions other than top-hats. If the window function is smooth, $p(\tilde{\delta})$ is no longer discrete. Nevertheless, $p(\tilde{\delta})$ can be measured, and, considered as a functional of $W(\vec{x})$, it contains full information on all of the N-point correlation functions and, therefore, provides a complete statistical description of a homogeneous and isotropic point process. (In practice, sampling fluctuations arising from the finite size of surveys prevents a complete determination of the functional.) Saunders *et al.* (1991) measured the low-order cumulants of $\tilde{\delta}$ using Gaussian window functions of several radii. Coles & Frenk (1991) discuss the third moment — the skewness — in detail. As these authors discuss, some care should be exercised when trying to obtain information about the N-point correlation functions from $p(\tilde{\delta})$ because discreteness introduces nonvanishing cumulants of all orders. The correct generating function for $p(\tilde{\delta})$ for any window function is given by (77), with $W_V(-\vec{x})$ replaced by the desired function. However, the resulting expression for $p(\tilde{\delta})$ is cumbersome even for discrete sampling of Gaussian random field (with $\xi^{(N)} = 0$ for $N > 2$).

The special case $P_0(V)$, the probability that a cell of volume V contains no objects, is called the void probability function. The void probability function, considered as a function of the volume V, contains essentially complete statistical information about the random point process. From (78), we have $\mathcal{P}(0) = P_0(V)$ and thus

$$\ln P_0(V) = \sum_{N=1}^{\infty} \frac{(-\bar{N})^N}{N!}\, \bar{\xi}_N(V)\,,$$

$$\bar{\xi}_N(V) \equiv \left[\prod_{k=1}^{N} \frac{1}{V} \int_V d^3 x_k \right] \xi^{(N)}(\vec{x}_1, \ldots, \vec{x}_N)\,,$$

(81)

where $\bar{N} = \bar{n}V$ and $\bar{\xi}_N(V)$ is the volume-averaged N-point correlation function ($\bar{\xi}_1 = 1$). Equation (81) follows as a simple consequence of the cumulant expansion theorem. This important formula was first derived in a different manner by White (1979); see also Fry (1986) and Otto *et al.* 1986.

¿From (78), we see that when the mean density is treated as a variable, the probability generating function is a function of $(t-1)\bar{n}$. Using this fact, one can show that the void probability function is a generating function for the other count probabilities when the number density is varied with the volume fixed (White 1979; Balian & Schaeffer 1989a):

$$P_N(V) = \frac{(-\bar{n})^N}{N!} \frac{\partial^N}{\partial \bar{n}^N} P_0(V) . \tag{82}$$

In practice, the mean density is fixed by nature and cannot be varied. However, a sample may be diluted by including each object with probability $f < 1$. The result of this dilution (as the reader may verify for an exercise) is simply to replace $\bar{n}$ with $\bar{n}f$ in all of our formulas. Thus, $\bar{n}$ may be replaced with f in (82), allowing P_N to be determined from partial derivatives of $P_0(V, f)$ with respect to the dilution factor f. However, this is not the best way in practice to estimate P_N for $N > 0$ because finite-sized samples have sampling fluctuations, making the derivatives noisy. It is better to determine all desired $P_N(V)$ directly from cell counts. For a given V, from the counts in cells distribution $P_N(V)$ one can, in principle, determine the volume-averaged correlations, $\bar{\xi}_N(V)$, through the probability generating function $\mathcal{P}(t)$.

When the volume is varied $\ln P_0(V)$ becomes a generating functional for the correlation functions themselves [cf. (68), (70)]:

$$\xi^{(N)}(\vec{x}_1, \ldots, \vec{x}_N) = (-\bar{n})^N \left[\frac{\delta^N \ln P_0[W_V(\vec{x})]}{\delta W_V(\vec{x}_1) \cdots \delta W_V(\vec{x}_N)} \right]_{W_V = 0} . \tag{83}$$

Here $P_0(V)$ is considered a functional of the top-hat window function, $P_0[W_V(\vec{x})]$. It is not rigorously correct to write (83) with functional derivatives with respect to $W_V(\vec{x})$ because the window function for $P_0(V)$ must always have the top-hat form. However, for small volumes (i.e., window functions close, in the space of functions, to $W_V = 0$), the top-hat window functions are sufficiently complete so that the $\xi^{(N)}$ may be determined in principle from measurements of $P_0(V)$ for all possible small volumes. Sampling fluctuations would unfortunately defeat any such attempt in practice.

5.7 Scale-invariance of the distribution of counts in cells

As we have seen, the counts in cells distribution $P_N(V)$ (or its moments), considered as a function of N and a functional of V (i.e., of the top-hat window function W_V), or the generating function $\mathcal{P}(t)$ (a function of t and a functional of W_V), contains complete information about the statistics of a homogeneous and isotropic random point process. This information can be just as unwieldy as the full set of N-point correlation functions or N-point distribution functions. However, nonlinear gravitational clustering may lead to a simplification. It has long been noted that the 3-point correlation function is well approximated by a symmetrized product of two-point correlation functions (Peebles & Groth 1975): $\xi_{123}^{(3)} = Q[\xi_{12}\xi_{23} + \xi_{23}\xi_{31} + \xi_{31}\xi_{12}]$, where $Q \approx 1$ is independent of scale (note that we have dropped the superscript from the 2-point correlation function). Similarly, the 4-point correlation function may be written as a sum of products of three 2-point correlation functions (Fry & Peebles 1978). This hierarchical factorization of the correlation functions is consistent with, and possibly even implied by, nonlinear gravitational clustering on small scales (Davis & Peebles 1977; Peebles 1980, Sect. 73; Hamilton 1988).

In an important paper, White (1979) conjectured that the N-point correlation functions may all be composed of sums of products of $(N - 1)$ 2-point correlation functions, and he pointed out that this leads to a simple scaling law for $\ln P_0/\bar{N}$ which we will derive below. Investigating the (two-dimensional) Zwicky catalog, Sharp (1981) found evidence for this scaling. Theorists soon suggested several hierarchical clustering models that exhibit such behavior (Carruthers & Minh 1983; Schaeffer 1984; Saslaw & Hamilton 1984; Fry 1985); they showed that initial nonlinear evolution of a Gaussian random field leads to this scaling if $\Omega = 1$ (Fry 1984; Bernardeau 1992); and they confirmed scaling behavior in fully nonlinear N-body simulations (Davis *et al.* 1985; Fry & Melott 1985; Bouchet, Schaeffer & Davis 1991). Meanwhile, evidence for the predicted scaling accumulated in observational work (Hamilton, Saslaw & Thuan 1985; Bouchet & Lachièze-Rey 1986; Maurogordato & Lachièze-Rey 1987, 1991; Fry *et al.* 1989; Alimi, Blanchard & Schaeffer 1990; but Vogeley, Geller & Huchra 1991 showed that the scaling breaks down on large scales). If this scaling of $P_N(V)$ is correct, it represents a leap in understanding and in treating the complexity of fully nonlinear gravitational clustering.

Balian & Schaeffer (1989a) synthesized key results concerning the scaling behavior of counts in cells, postulating and deriving many analytic results that have since been confirmed in N-body simulations (Bouchet, Schaeffer & Davis 1991). Their work, although rather mathematical, is required reading for anyone seeking to describe or understand nonlinear gravitational clustering. We follow their treatment in deriving the key scaling relations using the N-point correlation functions.

The 2-point correlation function is well-approximated by $\xi(r) = (r_0/r)^\gamma$ with $r_0 \approx 5\,h^{-1}\,\mathrm{Mpc}$ and $\gamma = 1.8$ (Peebles 1980), at least for $r \lesssim 2r_0$. If the N-point correlation function is proportional to a symmetrized product of $(N - 1)$ 2-point functions, then these functions obey the following scaling property (Balian & Schaeffer 1989a):

$$\xi^{(N)}(\lambda\vec{x}_1, \ldots, \lambda\vec{x}_2) = \lambda^{-\gamma(N-1)}\xi^{(N)}(\vec{x}_1, \ldots, \vec{x}_2) \ . \tag{84}$$

This condition is sufficient to establish scaling of the counts in cells, as we show by evaluating the void probability function [(81)]. Consider a volume of fixed shape but with a scale parameter l; e.g., a cube of length l. We may consider other shapes, but in any case we require a scale parameter l to vary the volume without changing the shape of the cell. The volume-averaged 2-point correlation function is $\bar{\xi} \equiv V^{-2} \int d^3 x_1 \int d^3 x_2\, \xi(r_{12}) \equiv (l_0/l)^\gamma$ where $l_0 \propto r_0$ is a constant (independent of l but dependent on the shape of the volume). Now, it follows from (84) that the l-dependence of the volume-averaged N-point correlation is $\bar{\xi}_N \propto l^{-\gamma(N-1)}$, implying

$$\bar{\xi}_N(l) = [\bar{\xi}(l)]^{N-1} S_N \,, \tag{85}$$

where S_N is independent of l but may depend on N and on the shape of the cell. From (81) and (85), we conclude that the logarithm of the void probability function, considered as a function of l, may be written

$$\ln P_0(l) = -\bar{N}\, \sigma(N_c)\,, \quad N_c \equiv \bar{N}\bar{\xi}\,, \quad \sigma(x) \equiv \sum_{N=1}^{\infty} \frac{(-1)^{N-1}}{N!} S_N\, x^{N-1} \,, \tag{86}$$

where $\bar{N} \propto l^3$ and $\bar{\xi} \propto l^{-\gamma}$ depend explicitly on l but the functional form of $\sigma(x)$ depends only on the shape of the volume and not on l. We have introduced the conventional notation N_c for the mean number of neighbors, in excess of Poisson, of an object in the cell.

We conclude that, if the N-point correlation functions are scale-invariant [(84)], the void probability function is determined by the scaling function $\sigma(N_c) = -\ln P_0/\bar{N}$. This is a significant simplification because, even for cells of a fixed shape, P_0 depends on both the cell size l and the dilution factor f (equivalently, the number density $\bar{n}$), while σ depends on the single parameter $N_c \propto f\, l^{3-\gamma}$. Moreover, it is plausible that the functional form of $\sigma(N_c)$ depends very weakly on the shape of the volume (Schaeffer 1984; Bouchet, Schaeffer & Davis 1991). If so, $\sigma(N_c)$ and $\bar{\xi}(l)$ completely determine the void probability function, with all of the dependence on the window function being relegated to the simple functional $\bar{\xi}[W_V]$. There is still more beauty here: from (78) it follows that the generating function for all of the cell counts is also determined by σ:

$$\mathcal{P}(t) = \exp\left\{ -\bar{N}\sigma[(1-t)N_c] \right\} \,. \tag{87}$$

Thus, *all* of the counts in cells distributions $P_N(V)$ depend in a simple way on V. From (87), it follows that the scaling function $\sigma(N_c)$ may be determined in several different ways: (1) from the l-dependence of $P_0(l)$ or any other P_N; (2) from the f-dependence (where f is the dilution factor for the number density of objects) of P_0 or any other P_N for fixed l (such that $N_c \gg 1$ and $\bar{\xi} \gg 1$); or (3) from the N-dependence of $P_N(l)$ for fixed l (such that $N_c \gg 1$ and $\bar{\xi} \gg 1$). In short, a complete statistical description of a random point process obeying the scaling hypothesis may be given by just two functions, $\xi(r)$ and $\sigma(N_c)$.

Scaling of the distribution of counts in cells is not just a mathematical nicety; it has several important implications. First, it implies that there is persistent substructure as one goes to smaller and smaller scales (Peebles 1980, Sect. 60). Viewed on a length scale l, the mean number of particles in a clump is $N_c(l) = \bar{n}V\bar{\xi}$. Thus, the mean density in a clump of size l is $\bar{n}\bar{\xi}(l) = \bar{n}(l_0/l)^\gamma \propto l^{-\gamma}$. The number density of clumps is $\sim 1/(V\bar{\xi})$ and the fraction of volume occupied by these clumps is $\sim \bar{\xi}^{-1} = (l/l_0)^\gamma$ — most of the mass is in the clumps, even on small scales. This is inconsistent with the picture that gravitational clustering entirely erases all substructure (Press & Schechter 1974; White & Rees 1978) and produces a set of nonoverlapping objects, each with a density profile $n(r) \propto r^{-\gamma}$. The hierarchical scaling of the N-point correlation functions implies that we must have a statistically scale-invariant distribution of subclumps within clumps. One may question the *degree* of subclustering, but not its existence. Observationally, the scale-invariant description seems to hold down to scales of $\sim 10\ h^{-1}$ Mpc; perhaps dissipation destroys the hierarchy on small scales or perhaps there is continuing substructure in invisible dark matter. The reader should be aware, however, that the existence of subclustering in galaxy groups and in galaxies is still controversial.

The existence of substructure implies that the mass distribution of clumps $n(M)\,dM$ is dependent on the length scale, or density contrast, for which these clumps are defined. Many workers have tried to derive the mass distribution of clumps, beginning with Press & Schechter (1974). The approach of Press & Schechter supposes that substructure is erased so that a density peak on one scale will not contain density peaks of smaller scale. The scale-invariant clustering paradigm concludes otherwise; the distribution of clump masses $n(M, l)\,dM$ is also a function of the resolution scale l. Balian & Schaeffer (1989a) make the plausible assumption that $n(M, l)$ scales similarly to the distribution of counts in cells. However, this is still insufficient to give the mass distribution of objects like galaxies or galaxy clusters. A relation between the mass and size is needed: $M = M(l)$. The mass-size relation cannot be determined within the context of the scaling paradigm but depends rather on poorly understood dissipative processes. Recently, Bernardeau & Schaeffer (1991) have shown that, with reasonable mass-radius relations for galaxies and clusters, the scaling paradigm can account well for the galaxy luminosity function and for cluster mass multiplicity functions. While their results do not give a complete explanation of the observed distributions, their success demonstrates that, using scale invariance, analytical progress can be made toward understanding the difficult problem of galaxy formation.

The existence of scale-invariant substructure does not imply that the matter distribution is a fractal, for we have focussed on only the dense regions. As Balian & Schaefer (1989a) argue, for large N_c scale invariance implies $\sigma(N_c) \propto N_c^{-\omega}$. Scale invariance does not determine the exponent ω (aside from imposing the constraints $0 \leq \omega \leq 1$), but observations and numerical simulations suggest $\omega \approx 0.5 \pm 0.1$ (Balian & Schaeffer 1989a; Alimi, Blanchard & Schaeffer 1990; Bouchet, Schaeffer & Davis 1991). Now, provided that P_0 is not very small, the fraction of occupied cells of size l is $1 - P_0(l) \approx \bar{N}\sigma(N_c) \propto l^{3-\omega(3-\gamma)}$. By definition, the Hausdorff (capacity) dimension, is $D_0 \equiv 3 - d\ln(1-P_0)/d\ln l$. The scale-invariant distribution is, over a range of length scales, a bifractal, with two distinct fractal dimensions (Balian & Schaeffer 1989b): the correlation dimension $D_2 = 3 - \gamma$ and the Hausdorff dimension $D_0 = \omega(3 - \gamma)$. On larger scales, it appears that the galaxy distribution may be a multifractal, with a continuous range of fractal dimensions (Jones *et al.* 1988; Martinez *et al.* 1990).

Finally, the scaling paradigm opens up new ways to attack the problem of fully nonlinear gravitational clustering. Note that all of our discussion of counts in cells has been based on a static description of clustering at a fixed time. However, gravitational clustering is a dynamic process. Up to now, analytical attacks on the dynamical clustering problem have either been based on toy models such as the evolution of isolated objects or on the exceedingly difficult BBGKY hierarchy equations (Davis & Peebles 1977). By replacing volume averages of all of the N-point correlation functions with the scaling function $\sigma(N_c)$ plus the 2-point correlation function $\xi(r)$, scale-invariance may simplify the BBGKY hierarchy sufficiently so that analytical solutions can be obtained with fewer approximations than were made by Davis & Peebles (1977). However, one must add back the missing half of phase space: momentum (or velocity). It remains to be seen whether an approach based on scale-invariance can give a satisfactory dynamical description of clustering. Although scale-invariance is probably only approximate, it appears to offer a powerful paradigm for describing and understanding galaxy clustering.

5.8 Power spectrum

5.8.1 Definitions

Our spatial statistics have so far been based on the density field $\delta(\vec{x})$. However, it is often convenient to work with the Fourier transform,

$$\hat{\delta}(\vec{k}) \equiv \int \frac{d^3x}{(2\pi)^3} \, e^{-i\vec{k}\cdot\vec{x}} \, \delta(\vec{x}) \; ; \quad \delta(\vec{x}) = \int d^3k \, e^{i\vec{k}\cdot\vec{x}} \, \hat{\delta}(\vec{k}) \; . \qquad (88)$$

Warning: many authors adopt different conventions for where to place the factors of 2π and for the signs in the exponents. The reader who prefers a different convention is responsible for inserting all of needed factors of 2π into the following formulas. All consistent definitions of the forward and inverse transforms are based on the well-known representation of the Dirac delta function:

$$\delta^D(\vec{k}) = (2\pi)^{-3} \int d^3x \, e^{\pm i\vec{k}\cdot\vec{x}} \,. \tag{89}$$

Note that $\hat{\delta}(\vec{k})$ and $\delta^D(\vec{k})$ have units of volume.

The power spectrum, known more fully as the power spectral density of the random field $\delta(\vec{x})$, is defined from the expectation value of the two-point function in Fourier space, as follows:

$$\left\langle \hat{\delta}(\vec{k}_1)\,\hat{\delta}(\vec{k}_2) \right\rangle \equiv P(k_1)\,\delta^D(\vec{k}_1 + \vec{k}_2) \,. \tag{90}$$

For the present we restrict ourselves to continuous density fields. As the reader should verify using the definition of the two-point density correlation function (Table 1) plus (88) and (89), the power spectrum and the 2-point correlation function are Fourier transform pairs (Wiener-Khintchine theorem):

$$\xi(r) = \int d^3k \, e^{i\vec{k}\cdot\vec{x}} \, P(k) \,, \quad P(k) = \int \frac{d^3x}{(2\pi)^3} \, e^{-i\vec{k}\cdot\vec{x}} \, \xi(r) \,, \quad r \equiv |\vec{x}| \,. \tag{91}$$

The reader will discover that the Dirac delta function in (90) is required because of translational invariance, $\langle \delta(\vec{x}_1)\,\delta(\vec{x}_2) \rangle = \xi(|\vec{x}_1 - \vec{x}_2|)$. Similarly, isotropy implies that $P(k)$ depends only on the magnitude of the wavevector $\vec{k}$.

The power spectral density is well-defined for almost all homogeneous random fields; nevertheless, there is a great deal of confusion in the large-scale structure literature about its definition. Note that $P(k)$, being a *spectral density*, has units of volume (inverse of volume in wavenumber space). Many workers, following Peebles (1980), prefer to avoid integrals and to work instead with a periodic volume and discrete transforms, requiring, for consistency, extra factors of the volume of the fundamental cube (the "volume of the universe"). Others make outright blunders:

$$P(k) = \left\langle |\delta_{\vec{k}}|^2 \right\rangle \quad \text{(false)} \,, \tag{92a}$$

where $\delta_{\vec{k}}$ is supposed to be a Fourier component of $\delta(\vec{x})$, although whether the continuous or discrete Fourier transform is intended is often unclear. Equation (92a) is false in either case, for $P(k)$ must have units of volume. (To those who think otherwise: Do you insert the volume of the universe into your integrals over the power spectrum? If not, your units are inconsistent. If you do insert the appropriate volume factors, your units are consistent but then your P depends on the size of the volume; see below.) The correct relation follows from (90) together with the fact that $\hat{\delta}(-\vec{k}) = \hat{\delta}^*(\vec{k})$ because $\delta(\vec{x})$ is real:

$$\left\langle \hat{\delta}(\vec{k}_1)\,\hat{\delta}^*(\vec{k}_2) \right\rangle = P(k_1)\,\delta^D(\vec{k}_1 - \vec{k}_2) \quad \text{(true)} \,. \tag{92b}$$

5.8.2 Discrete fourier transforms

Why is there so much confusion over the definition of the power spectrum? Probably because of the casual use of discrete Fourier transforms. The term discrete refers here to the representation of a continuous function by its values on a regular lattice and not to a discrete random point process. Consider a periodic cube of length L and suppose that the density is evaluated on a lattice of spacing $\Delta x = L/n$ where n is an integer (often taken to be a power of 2 to take advantage of the speed of the fast Fourier transform algorithm). Because the density is periodic in the spatial domain, it is also discrete in the Fourier domain, with lattice spacing $\Delta k = 2\pi/L$. The discrete Fourier transform pair [cf. (88)] is

$$\delta_{\vec{k}} \equiv n^{-3} \sum_{\vec{j}} e^{-i\vec{k}\cdot\vec{x}_{\vec{j}}} \delta(\vec{x}_{\vec{j}}) \,, \quad \delta(\vec{x}_{\vec{j}}) = \sum_{\vec{k}} e^{i\vec{k}\cdot\vec{x}_{\vec{j}}} \delta_{\vec{k}} \,, \tag{93}$$

where $\vec{x}_{\vec{j}} = (j_x\vec{e}_x + j_y\vec{e}_y + j_z\vec{e}_z)\Delta x$ and $0 \le j_x, j_y, j_z < n$; $\vec{k} = (l_x\vec{e}_x + l_y\vec{e}_y + l_z\vec{e}_z)\Delta k$ is similarly discretized, with $0 \le l_x, l_y, l_z < n$. The sums are taken over all n^3 values of the indices. The factor of n^{-3} in (93) may be moved to the inverse transform (from the Fourier to the spatial domain), or it may be split between the forward and inverse transforms, but it must be present somewhere. The reader may verify that $\delta_{\vec{k}}$ and $\delta(\vec{x}_{\vec{j}})$ form an exact transform pair. Comparing (88) and (93), one sees that $\delta_{\vec{k}}$ is proportional to a discrete approximation of the continuous transform:

$$\hat{\delta}(\vec{k}) = \lim_{\Delta k \to 0} \left(\frac{\Delta x}{2\pi}\right)^3 n^3 \delta_{\vec{k}} = \lim_{\Delta k \to 0} (\Delta k)^{-3} \delta_{\vec{k}} \,. \tag{94}$$

Note that $\delta_{\vec{k}}$ is dimensionless, in contrast with the continuous Fourier transform $\hat{\delta}(\vec{k})$. The reader may be bothered that the limit in (94) is divergent, but there is no problem here. The Dirac delta function behaves in a similar manner. Using a discrete approximation to (89), one obtains

$$\delta^{\mathrm{D}}(\vec{k}) = \lim_{\Delta k \to 0} \left(\frac{L}{2\pi}\right)^3 \delta^{\mathrm{K}}_{\vec{k},0} = \lim_{\Delta k \to 0} (\Delta k)^{-3} \delta^{\mathrm{K}}_{\vec{k},0} \,, \tag{95}$$

where $\delta^{\mathrm{K}}_{\vec{k},0}$ is the Kronecker delta: $\delta^{\mathrm{K}}_{\vec{k},0} = 0$ unless $\vec{k} = 0$, with $\delta^{\mathrm{K}}_{0,0} = 1$. Now, combining (94) and (95) for finite Δk with (92b), we obtain

$$\langle |\delta_{\vec{k}}|^2 \rangle \approx P(k)(\Delta k)^3 \,, \tag{96}$$

where $P(k)$ is the power spectral density. Note that $(\Delta k)^3$ is the volume element associated with one mode (harmonic) of the discrete Fourier transform. Therefore, the dimensionless quantity $\langle |\delta_{\vec{k}}|^2 \rangle$ is the *power per mode*. Lest the reader conclude that (92a) has been vindicated, he or she should consider what happens to the power per mode for a fixed $\vec{k}$ as the size of the cube is changed. The power per mode depends on $L = 2\pi/\Delta k$ (vanishing in the limit $L \rightarrow \infty$) while the power spectrum (i.e., spectral density) is independent of L. The correct way to define the power for a homogeneous random field in an infinite universe is with the power spectral density $P(k)$.

5.8.3 Importance of the power spectrum

There are several reasons for the importance of the power spectrum. First, it (or its Fourier transform, the correlation function) completely specifies the statistics of a homogeneous and isotropic Gaussian random field. By definition, the N-point probability distribution functions for a Gaussian random field are all multivariate normal (Gaussian) distributions, with co-variance matrix $\langle \delta(\vec{x}_i)\,\delta(\vec{x}_j) \rangle = \xi(|\vec{x}_i - \vec{x}_j|)$. The probability distribution functional [(49)] is therefore a Gaussian functional. Representing the density fluctuation field by its Fourier transform, the distribution functional is (Bertschinger 1987)

$$ p[\hat{\delta}(\vec{k})] \propto \exp\left[-\frac{1}{2} \int d^3k\, \frac{|\hat{\delta}(\vec{k})|^2}{P(k)} \right] , \tag{97} $$

which is consistent with (92b). In Fourier transform space, the covariance matrix is diagonal [(97) has a single integral over wavevectors, not two], leading to considerable simplification of calculations performed in Fourier transform space. On account of the reality constraint $\hat{\delta}(-\vec{k}) = \hat{\delta}^*(\vec{k})$, the integral in (97) should be taken over half of Fourier space (e.g., $k_z \geq 0$), with the integrand doubled for compensation. Equation (97) says that the Fourier components of $\delta(\vec{x})$ are independently normally distributed, or, $(\mathrm{Re}\,\hat{\delta}, \mathrm{Im}\,\hat{\delta})$ has an independent two-dimensional circular Gaussian distribution for each wavevector $\vec{k}$. Thus, generating realizations of a Gaussian random field on a periodic cubic lattice of length L is easy: the real and imaginary parts of $\delta_{\vec{k}}$ for $k_z \geq 0$ are independent normal random variables with zero mean and with variance $(1/2)P(k)(2\pi/L)^3$ [(96)]. For $k_z < 0$ the components follow from the reality constraint. It is also true that, writing $\delta_{\vec{k}} = |\delta_{\vec{k}}|\exp(i\phi_{\vec{k}})$, the phases are uniformly distributed ($0 \leq \phi_{\vec{k}} < 2\pi$) and the moduli have a Rayleigh distribution (Lindgren 1976), $dP(|\delta_{\vec{k}}|) = \sigma^{-2}\exp(-|\delta_{\vec{k}}|^2/2\sigma^2)|\delta_{\vec{k}}|\,d|\delta_{\vec{k}}|$ for $|\delta_{\vec{k}}| > 0$ with $\sigma^2 = P(k)(2\pi/L)^3$. [For the Rayleigh distribution, $\exp(-|\delta_{\vec{k}}|^2/2\sigma^2)$ is uniformly distributed between 0 and 1. Thus, two independent normal random variables may be generated from two independent uniformly distributed random variables, one for the phase and one for the modulus; this is known as the Box-Muller method.] One often hears a Gaussian random field described, incompletely, as a "random-phase distribution" and one sometimes hears it said, incorrectly, that the magnitude $|\delta_{\vec{k}}|$ has a Gaussian distribution.

Gaussian random fields are very popular with cosmologists, for two reasons. The first one is that the inflation theory predicts that the initial density fluctuation field is a Gaussian random field. (Energy density fluctuations arise from the vacuum fluctuations of a weakly coupled, hence nearly free, scalar field. All the connected Green's functions vanish aside from the 2-point function — the Feynman propagator — leading to a Gaussian functional for the vacuum-to-vacuum transition amplitude and for the scalar field fluctuations; see Brandenberger 1985.) The second reason for the popularity of Gaussian fields is ease of calculation. Because the Fourier transform of a Gaussian is also Gaussian, the moment-generating functional [(67)] for a Gaussian random field is a Gaussian, implying that the cumulant generating functional [(68)] is quadratic. Consequently, all of the N-point correlation functions of a Gaussian random field $\delta(\vec{x})$ vanish except for the 2-point correlation function, greatly simplifying statistical calculations. Many local properties of a Gaussian random field can be calculated analytically in terms of the power spectrum $P(k)$; see Bardeen et $al.$ (1986) for a detailed discussion.

Another reason for the importance of the power spectrum is that it measures the mean-square amplitude of density fluctuations as a function of wavelength $\lambda = 2\pi/k$. The variance of the smoothed density is

$$\tilde{\sigma}_\delta^2 \equiv \left\langle \tilde{\delta}^2(0) \right\rangle = \int d^3x_1 \, W(-\vec{x}_1) \int d^3x_2 \, W(-\vec{x}_2) \, \left\langle \delta(\vec{x}_1) \delta(\vec{x}_2) \right\rangle$$
$$= \int d^3k \, \widetilde{W}^2(\vec{k}) \, P(k) \,, \quad \widetilde{W}(\vec{k}) \equiv \int d^3x \, e^{i\vec{k}\cdot\vec{x}} W(\vec{x}) \,, \tag{98}$$

where $W(\vec{x})$ is a normalized window function ($\int d^3x \, W = 1$) and $\widetilde{W}$ is $(2\pi)^3$ times its Fourier transform. This particular definition is convenient because $\widetilde{W}(0) = 1$ and, if the density field is not smoothed, $\widetilde{W}(\vec{k}) = 1$ for all $\vec{k}$. From (98), we see that $P(k)d^3k$ is the contribution to the variance of δ from waves with wavevectors in the volume element d^3k about $\vec{k}$. (With a window function, convolution in the spatial domain becomes multiplication in the Fourier domain, illustrating the well-known convolution theorem.) Often this result is stated as follows: $d\sigma_\delta^2/d\ln k = 4\pi k^3 P(k)$ is the contribution to the variance of δ per logarithmic interval of k. This is true whether or not $\delta(\vec{x})$ is a Gaussian random field. As workers are increasingly beginning to realize (Baumgart & Fry 1991; Peacock 1991; see also Hamilton et $al.$ 1991), $P(k)$ is a powerful statistic for characterizing both large-scale and small-scale structure.

5.8.4 Normalization of the power spectrum

Recall that, for the growing mode of linear perturbation theory, the peculiar velocity and gravitational potential perturbations are related to the density fluctuation field: $\delta(\vec{x}) \approx -\vec{\nabla} \cdot \vec{v}/[aHf(\Omega_m)]$ [(46)] and $\nabla^2\phi = (3/2)\Omega_m H^2 a^2 \delta$ [(26)]. It follows that the variances of δ, $\vec{v}$ and ϕ are all proportional to the power spectrum:

$$\frac{d\sigma_\delta^2}{d\ln k} = 4\pi k^3 P(k) , \qquad \frac{d\sigma_v^2}{d\ln k} = 4\pi (aHf)^2 kP(k) ,$$

$$\frac{d\sigma_\phi^2}{d\ln k} = 4\pi \left(\frac{3}{2} \Omega_m H^2 a^2\right)^2 \frac{P(k)}{k} . \qquad (99)$$

The fluctuations of spacetime curvature are proportional to $\phi(\vec{x})$ [(6)]. The most natural primordial fluctuations are scale-invariant constant curvature fluctuations, with $d\sigma_\phi^2/d\ln k$ =constant, or $P(k) \propto k$ (Harrison 1970; Peebles & Yu 1970; Zel'dovich 1972). Inflation predicts an initial spectrum $P(k, a_i)$ of essentially this form (with a logarithmic correction). Evolution of the fluctuations after the inflation epoch modifies the spectrum. While the fluctuations are still linear the modification is simply a linear filtering, with transfer function $T(k, a) = P(k, a)/P(k, a_i)$. (Waves grow at different rates depending on the relation between the wavelength, the Jeans length and the Hubble length; cf. Sect. 3.1). For a discussion of the transfer function in various models, see Efstathiou (1990).

Although inflation generically predicts the shape of the initial power spectrum, the overall amplitude is effectively a free parameter that cannot be calculated with any confidence because it is highly model-dependent. Therefore, it is meaningless to speak of a standard theory based on inflation (such as the "standard cold dark matter theory") without specifying the amplitude of $P(k)$. It is possible to normalize a theoretical power spectrum to agree with the observed power spectrum of galaxy clustering at some wavelength, but it is by no means certain that all matter clusters the same way as do galaxies. Therefore, the normalization of theoretical power spectra must be specified some other way. Conventionally, the normalization of a theoretical spectrum is performed using linear theory, i.e., the normalization is given for the power spectrum extrapolated from high redshift to the present day assuming linear theory. For nonlinear calculations, the power spectrum is then scaled back to a high redshift and nonlinear evolution is carried forward to the present day.

It is conventional to state the normalization of the linear power spectrum in one of four ways. The first and most common way (Peebles 1982) is in terms of the rms relative fluctuation of mass in a sphere of radius $R = 8\ h^{-1}$ Mpc, i.e., the standard deviation of the density smoothed with a spherical top-hat window function of radius R:

$$\sigma_R^2 = \int d^3k\, W_{\mathrm{TH}}^2(kR)P(k) , \qquad W_{\mathrm{TH}}(x) = \frac{3}{x^3}(\sin x - x\cos x) . \qquad (100)$$

It is easy to verify that $W_{\mathrm{TH}}(kR)$ is $(2\pi)^3$ times the Fourier transform of the top-hat window function for a sphere of radius R. Normalization now corresponds to specifying σ_R for some R. For $R = 8\ h^{-1}$ Mpc, the rms relative mass fluctuation is denoted σ_8. Galaxy number counts give $\sigma_8 \approx 1$ (Peebles 1982); if galaxies trace mass on scales of several Mpc and if nonlinear effects do not modify σ_8 appreciably, then the normalization $\sigma_8 = 1$ is reasonable. Even if one prefers a different normalization, it is still convenient to specify that normalization using σ_8. For example, simple models of biased galaxy formation suggest that the galaxy correlation function may be b^2 times that of the mass, with $b > 1$ (Kaiser 1984). If so, the correct normalization for the power spectrum of mass density fluctuations would then be $\sigma_8 = b^{-1}$. Warning: many authors invoke the "bias parameter" b when they mean nothing more than a linear normalization $\sigma_8 = b^{-1}$. If one is normalizing the power spectrum, it may be better to avoid using the term bias parameter, because no subject in the large-scale structure literature is more confused than "bias."

A second normalization method is also based on galaxy clustering, through the integral of the correlation function:

$$J_3(R) \equiv \int_0^R \xi(r)\,r^2\,dr = \frac{R^3}{3} \int d^3k\,W_{\mathrm{TH}}(kR)P(k)\ . \tag{101}$$

Estimates from the Center for Astrophysics redshift survey give $J_3(10\ h^{-1}\,\mathrm{Mpc}) = 277\,(h^{-1}\,\mathrm{Mpc})^3$ and $J_3(25\ h^{-1}\,\mathrm{Mpc}) = 780\,(h^{-1}\,\mathrm{Mpc})^3$ (Davis & Peebles 1983). Note that the σ_8 and J_3 normalizations are sensitive to the power on scales near $10\ h^{-1}$ Mpc but are insensitive to much longer wavelengths.

A more physical normalization is based on fluctuations of the microwave background radiation. On large angular scales, aside from the dipole fluctuation due to our motion relative to the comoving frame (or, possibly, to an entropy fluctuation with wavelength exceeding the Hubble length), the fluctuations are expected theoretically to be produced by the Sachs-Wolfe effect. The fluctuations arise from the gravitational redshift microwave background photons suffer as they climb out of potential wells present at recombination (the precursors of large-scale structure). Because the gravitational potential varied from place to place, the amount of redshift depends on the direction in the sky. The result (Sachs & Wolfe 1967) is a microwave background anisotropy

$$\frac{\Delta T}{T}(\vec{n}) = \frac{1}{3}\frac{\Delta\phi(\vec{x})}{c^2}\ , \tag{102}$$

where $\vec{n}$ is a unit vector giving the direction and $\Delta\phi(\vec{x})$ is the fluctuation in gravitational potential at comoving position $\vec{x}$ on the last-scattering surface (recombination layer). Using the Robertson-Walker metric [(6)], one can show that an angle θ corresponds to a comoving distance at recombination of $(4c/\Omega_0 H_0)\sin(\theta/2)$ (or $105\,\Omega_0^{-1}\,h^{-1}$ Mpc for $1°$). The angular correlation function of the fluctuations is, excluding the dipole and unobservable monopole terms,

$$\left\langle \frac{\Delta T}{T}(\vec{n}_1)\frac{\Delta T}{T}(\vec{n}_2)\right\rangle \equiv C(\theta) \equiv \frac{1}{4\pi}\sum_{l=2}^{\infty}(2l+1)\,a_l^2\,P_l(\cos\theta)\ ,\qquad (103)$$

where $\cos\theta = \vec{n}_1\cdot\vec{n}_2$ and P_l is the Legendre polynomial of order l. (Warning: there are several different definitions of the expansion coefficients in the literature. Equation 103 is consistent with Peebles 1982.) Using (102) and the Poisson equation to relate ϕ to δ, we get the Sachs-Wolfe contribution to $C(\theta)$:

$$C(\theta) = \left(\frac{\Omega_0 H_0^2}{2c^2}\right)^2\int d^3k\,k^{-4}P(k)\frac{\sin kR}{kR}\ ,\qquad R = \frac{4c}{\Omega_0 H_0}\sin(\theta/2)\ .\quad (104)$$

Although we have retained the dependence on Ω_0, (104) is invalid in spatially curved models ($\Omega_0 \neq 1$) on large angular scales ($\theta \gtrsim \Omega_0\,|1-\Omega_0|^{-1/2}$) because the plane wave decomposition of the density fluctuations breaks down on scales comparable to or larger than the curvature distance. Using the orthonormality of the Legendre polynomials, $\int_{-1}^{1}dx\,P_m(x)P_n(x) = 2\delta_{mn}^{\rm K}/(2m+1)$ and the integral $\int_0^1 dx\,P_l(1-2x^2)\sin(Ax) = (A/2)\,j_l^2(A/2)$, we obtain the Sachs-Wolfe expansion coefficients

$$a_l^2 = 4\pi\left(\frac{\Omega_0 H_0^2}{2c^2}\right)^2\int d^3k\,k^{-4}P(k)\,j_l^2\left(\frac{2ck}{\Omega_0 H_0}\right)\ ,\qquad l\gg\pi\Omega_0^{-1}|1-\Omega_0|^{1/2}\ ,$$
$$(105)$$

where $j_l(x)$ is the spherical Bessel function of order l. Equation (105) agrees with (11) of Peebles (1982) with $P(k) = (4\pi)^{-1}|\delta_k|^2$. [Peebles defines the power spectrum, $|\delta_k|^2$ in his notation, with an extra factor of 4π from his (10).] Note that the contribution to the rms anisotropy from a_l is $(\Delta T/T)_l = [(2l+1)/4\pi]^{1/2}\,a_l$.

There are two commonly used conventions for power spectrum normalization on the Hubble scale. The first (Peebles 1982) is the quadrupole coefficient a_2; the dominant contributions to the integral in (105) come from $k \sim \Omega_0 H_0/2c$. No quadrupole anisotropy has yet been detected; the best upper limit is $(\Delta T/T)_2 < 3\times 10^{-5}$ or $a_2 < 5\times 10^{-5}$ from the COBE satellite (Smoot et al. 1991). An alternative large-scale normalization (Abbott & Wise 1984) is based on the rms gravitational potential fluctuation [(99)], nondimensionalized as follows:

$$\epsilon_{\rm H}^2 \equiv (9\pi\Omega_0^2 c^4)^{-1}\left(\frac{d\sigma_\phi^2}{d\ln k}\right)_{k_{\rm H}} = k_{\rm H}^3 P(k_{\rm H})\ ,\qquad k_{\rm H}\equiv H_0/c\ .\qquad (106)$$

For the scale-invariant constant-curvature fluctuations predicted by infla-
tion, $P(k) \propto k$ on large scales so that $d\sigma_\phi^2/d\ln k$ is independent of k. Ignoring the change in slope of the power spectrum arising from the small-scale
transfer function (valid for $l \ll 50$ for cold dark matter) and using the
quadrature $\int_0^\infty dx\, j_l^2(x)/x = [2l(l+1)]^{-1}$, we get

$$a_l \approx \frac{2\pi\Omega_0}{\sqrt{2l(l+1)}}\, \epsilon_H\, , \qquad \pi\Omega_0^{-1}|1-\Omega_0|^{1/2} \ll l \ll 50\, . \tag{107}$$

For $\Omega_0 = 1$, $a_2 = 1.814\,\epsilon_H = 1.585\,(\Delta T/T)_2$.

The σ_8 and J_3 normalizations measure power on scales of $\sim 10\,h^{-1}\,\mathrm{Mpc}$
while the a_2 and ϵ_H normalizations measure power on scales $\sim 10^4\,h^{-1}\,\mathrm{Mpc}$.
Relating the two requires a power spectrum spanning the full range. We
will illustrate the normalization and relate the various conventions with the
$\Omega_0 = 1$ cold dark matter model (Peebles 1982; Blumenthal et al. 1984; Davis
et al. 1985). Using the transfer function of Bardeen et al. (1986) with the
scale-invariant constant curvature spectrum for a universe dominated by
cold dark matter, the present linear power spectrum is

$$P(k) = \epsilon_H^2 \left(\frac{kc}{H_0}\right) T^2(k)\, ,$$

$$T(q) = \frac{(2.34q)^{-1}\ln(1+2.34q)}{[1 + 3.89q + (16.1q)^2 + (5.46q)^3 + (6.72q)^4]^{1/4}}\, , \tag{108}$$

where $q = k/(\Omega_0 h^2\,\mathrm{Mpc}^{-1})$ and $H_0 = 100\,h\,\mathrm{km\,s^{-1}\,Mpc^{-1}}$. For the cold
dark matter model with $\Omega_0 = 1$ and $h = 0.5$, using (100), (101), (105) and
(106) one obtains

$$J_3(10) = 398\,\sigma_8^2\, , \qquad J_3(25) = 520\,\sigma_8^2\, ,$$

$$\epsilon_H = 4.08 \times 10^{-6}\sigma_8\, , \qquad a_2 = 7.31 \times 10^{-6}\,\sigma_8\, , \tag{109}$$

where R is measured in units of $h^{-1}\,\mathrm{Mpc}$ and J_3 is measured in units of
$(h^{-1}\,\mathrm{Mpc})^3$. For comparison, Davis & Peebles (1983) measure $J_3(10) = 277$
and $J_3(25) = 780$. Efstathiou (1990) adopts the normalization $J_3(10) = 270$,
corresponding to $\sigma_8 = 0.82$. Also, the power per $\ln k$ of the large-scale gravitational potential fluctuations is $d\sigma_\phi^2/d\ln k = 9\pi\epsilon_H^2 c^4 = (2.17 \times 10^{-5}\sigma_8 c^2)^2$.
Note that for $\sigma_8 \lesssim 1$ the quadrupole moment is well below the COBE limit
($a_2 < 5 \times 10^{-5}$, 95% confidence level). Even taking account of better limits
on smaller scales, Bond et al. (1991) and Vittorio et al. (1991) conclude
that, for the reasonable baryon density parameter $\Omega_{b,0} = 0.03$, present microwave background anisotropy limits impose the constraint $\sigma_8 \lesssim 1.5$ (95%
c.l.). Thus, occasional news reports notwithstanding, the cold dark matter
model is not eliminated based on the isotropy of the microwave background
radiation. This model does have serious problems in other respects, but that
is not the subject of these notes.

5.8.5 Power spectrum for a point process

Our previous discussion of the power spectrum has assumed that $\delta(\vec{x})$ is a continuous random field. Now we consider the power spectrum for a point process with mean number density $\bar{n}$. Using (50) for the density field, the Fourier transform of the density fluctuation field is

$$\hat{\delta}(\vec{k}) = \frac{1}{\bar{n}\,(2\pi)^3} \sum_j e^{-i\vec{k}\cdot\vec{x}_j} - \delta^{\mathrm{D}}(\vec{k}) \,, \tag{110}$$

where the sum is taken over the point process and the Dirac delta function is needed so that $\langle\hat{\delta}\rangle = 0$. Now, using (89)–(91) together with the pair distribution function (Table 1), we get

$$\left\langle \hat{\delta}(\vec{k}_1)\,\hat{\delta}(\vec{k}_2) \right\rangle^{\mathrm{d}} = \left[P(k_1) + \frac{1}{(2\pi)^3\bar{n}} \right] \delta^{\mathrm{D}}(\vec{k}_1 + \vec{k}_2) \,,$$

$$P(k) \equiv \int \frac{d^3x}{(2\pi)^3}\, e^{-i\vec{k}\cdot\vec{x}}\, \xi(r) \,. \tag{111}$$

Note that for a point process, $P(k)$ is defined as the Fourier transform of the 2-point correlation function and is not the same as the 2-point density function in Fourier transform space. We see that the latter includes the discreteness contribution $(2\pi)^{-3}\bar{n}^{-1}$. This result also follows from the Fourier transform of (52). The reader may also verify that the same result is obtained using the Poisson model. Note that the discreteness contribution corresponds to a power per mode of $\bar{N}^{-1}$ for a cube of length L ($\bar{N} = \bar{n}L^3$), yielding the expected variance in the relative mass fluctuation in the volume. In the pure Poisson case, where all N-point correlation functions vanish, $P = 0$ and only the discreteness contribution remains, having the value expected from Poisson statistics. Note that the discreteness contribution is independent of $\vec{k}$. A random process with constant power spectrum is called white noise, and a power per mode of $\bar{N}^{-1}$ is often called the white noise level. However, a Poisson point process has non-vanishing higher-order correlations of density, dependent on $\bar{n}$ [cf. (61)], that distinguish it from a Gaussian white noise random field.

5.8.6 N-point spectra: bispectrum, trispectrum, etc.

Just as the 2-point correlation function may be generalized to the N-point correlation function, the power spectrum (the 2-point function in Fourier transform space, in the continuous case) may be generalized to N-point spectra. In the continuous case, the N-point spectrum is defined by the irreducible part of the N-point correlation function in Fourier transform space, or, equivalently, as the N-dimensional Fourier transform of the N-point correlation function (Fry & Melott 1985; Baumgart & Fry 1991):

$$P^{(N)}(\vec{k}_1,\ldots,\vec{k}_N)\,\delta^{\mathrm{D}}(\vec{k}_1+\cdots+\vec{k}_n)=\left\langle\hat{\delta}(\vec{k}_1)\cdots\hat{\delta}(\vec{k}_2)\right\rangle_{\mathrm{c}}$$

$$=\left[\prod_{j=1}^{N}\int\frac{d^3x_j}{(2\pi)^3}\,e^{-i\vec{k}_j\cdot\vec{x}_j}\right]\xi^{(N)}(\vec{x}_1,\ldots,\vec{x}_N)\,. \tag{112}$$

The Dirac delta function is present because of translational invariance: the N-point correlation function depends only on $(N-1)$ differences $\vec{x}_j-\vec{x}_{j+1}$ so that one of the volume integrals gives a delta function [(89)]. For $N=2$, $P_{12}^{(2)}=P(k_1)$ (with the constraint $\vec{k}_2=-\vec{k}_1$) and (112) agrees with (90). For $N=3$, $P_{123}^{(3)}=B(\vec{k}_1,\vec{k}_2,\vec{k}_3)$ is the bispectrum while $N=4$ gives the trispectrum (Baumgart & Fry 1991).

The statistical advantages of the N-point correlation functions are also present in the N-point spectra, with the added advantage that spectra more naturally distinguish large and small scales. From (112) it follows that the hierarchical factorization of the N-point correlation functions is also present for the N-point spectra. For example, $\xi_{123}^{(3)}=Q[\xi_{12}\xi_{23}+\xi_{23}\xi_{31}+\xi_{31}\xi_{12}]$ implies $P_{123}^{(3)}=Q[P_1P_2+P_2P_3+P_3P_1]$ (where $P_j=P(\vec{x}_j)$ and we require $\vec{k}_1+\vec{k}_2+\vec{k}_3=0$), with the same value of Q (if Q is constant). Note that $P^{(N)}$ has the units of $(N-1)$ volume factors, which is consistent with its factorization into sums of products of $(N-1)$ 2-point spectra.

Most of the results that were derived in Sect. (5.5)–(5.7) can be adopted straightforwardly to the Fourier domain. For example, the functional cumulant expansion theorem [(68)] carries over directly with $\int d^3x\,s(\vec{x})$ being replaced by $\int d^3k\,\hat{s}(\vec{k})$ and with $\delta(\vec{x})$ by $\hat{\delta}(\vec{k})$. The extension to a point process is also clear: the N-point spectra are the Fourier transforms of the N-point correlation functions; the N-point cumulant of $\hat{\delta}(\vec{k})$ includes discreteness contributions from lower-order spectra just like the N-point correlation function of $\delta(\vec{x})$ in the spatial domain [cf. (61), (62)]. For counts in cells analysis using a top-hat window function it is preferable to work in the spatial domain. For smooth density distributions (or a smooth window function convolving a point process) it may be preferable to work in the Fourier domain.

6 Conclusions

We have considered two small parts of the theory of large-scale structure: linear perturbation theory in the Newtonian approximation and the statistical characterization of random processes. While there are few outstanding unsolved problems in these areas, there are many technical details that should be mastered by a would-be practitioner. For example, our discussion of the cumulant expansion theorem makes clear how N-point correlation functions are related to probability distributions for the density and for counts in cells (especially the void probability function), and we have paid special attention to the relation between discrete and continuous density distributions. There are surprisingly few comprehensive pedagogical presentations of this sort of material in the literature; most treatments leave out the details. To quote a well-known MIT professor of mechanical engineering (Woodie Flowers), God is in the details. Students, take note. Attention to details may reduce the frequency of false reports about the death of Big Bang cosmology and other foolishness.

ACKNOWLEDGEMENTS. This work was supported by NSF grant AST90-01762. This paper is also being published in the proceedings of the International School of Astrophysics "D. Chalonge", *Current Topics in Astrofundamental Physics* (1992, ed. N. Sanchez & A. Zichichi).

References

Abbott, L. F. & Wise, M. B. 1984: ApJ **282** L47

Adler, R. J. 1981: The Geometry of Random Fields (New York: Wiley)

Alimi, J.-M., Blanchard, A. & Schaeffer, R. 1990: ApJ **349** L5

Amit, D. J. 1984: Field Theory, the Renormalization Group, and Critical Phenomena (Singapore: World Scientific)

Balian, R. & Schaeffer, R. 1989a: A&A **220** 1

Balian, R. & Schaeffer, R. 1989b: A&A **226** 373

Bardeen, J. M., Bond, J. R., Kaiser, N. & Szalay, A. S. 1986: ApJ **304** 15

Baumgart, D. J. & Fry, J. N. 1991: ApJ **375** 25

Bernardeau, R. 1992: preprint

Bernardeau, R. & Schaeffer, R. 1991: A&A **250** 23

Bertschinger, E. 1987: ApJ **323** L103

Bertschinger, E. 1990: in Particle Astrophysics: The early universe and cosmic structures, ed. Alimi, J. M., Blanchard, A., Bouquet, A., Martin de Volnay, F. & Tran Thanh Van, J. (Gif sur Yvette: Editions Frontières), p. 411

Bertschinger, E. & Dekel, A. 1989: ApJ **336** L5

Bertschinger, E., Dekel, A., Faber, S. M., Dressler, A. & Burstein, D. 1990: ApJ **364** 370

Bertschinger, E. & Gelb, J. M. 1991: Comp. in Phys. **5** 164

Blumenthal, G. B. 1988: in TASI-87 Proceedings of the Theoretical Advanced Studies Institute in Particle Physics,; ed. Slansky, R. & West, G. B. (Singapore: World Scientific)

Blumenthal, G. R., Faber, S. M., Primack, J. R. & Rees, M. J. 1984: Nature **311** 517

Bouchet, F. R. & Lachièze-Rey, M. 1986: ApJ **302** L37

Bouchet, F. R., Schaeffer, R. & Davis, M. 1991: ApJ **383** 19

Bond, J. R., Efstathiou, G., Lubin, P. M. & Meinhold, P. R. 1991: Phys. Rev. Lett. **66** 2179

Brandenberger, R. H. 1985: Rev. Mod. Phys. **57** 1

Carlberg, R. G., Couchman, H. M. P. & Thomas, P. A. 1990: ApJ **352** L29

Carruthers, P. 1991: ApJ **380** 24

Carruthers, P. & Minh, C. C. 1983: Phys. Lett. **131B** 116

Coles, P. & Frenk, C. S. 1991: MNRAS **253** 727

Coles, P. & Jones, B. 1991: MNRAS **248** 1

Davis, M., Efstathiou, G., Frenk, C. S. & White, S. D. M. 1985: ApJ **292** 371

Davis, M. & Peebles, P. J. E. 1983: ApJ **267** 465

Dekel, A. 1992: in Astrofundamental Physics, ed. Sanchez, N. & Zichichi, A. (Singapore: World Scientific), in press

Dekel, A., Bertschinger, E. & Faber, S. M. 1990: ApJ **364** 349

Dekel, A. & Rees, M. J. 1987: Nature **326** 455

Doob, J. L. 1953: Stochastic Processes (New York: Wiley)

Efstathiou, G. 1990: in Physics of the Early Universe, ed. Peacock, J. A., Heavens, A. F. & Davies, A. T. (Bristol: IOP), p. 361

Efstathiou, G., Kaiser, N., Saunders, W., Lawrence, A., Rowan-Robinson, M., Ellis, R. S. & Frenk, C. S. 1990: MNRAS **247** 10p

Efstathiou, G. & Silk, J. 1983: Fund. Cosm. Phys. **9** 1

Feynman, R. P. & Hibbs, A. R. 1965: Quantum Mechanics and Path Integrals (New York: McGraw-Hill)

Fry, J. N. 1984: ApJ **279** 499

Fry, J. N. 1985: ApJ **289** 10

Fry, J. N. 1986: ApJ **306** 358

Fry, J. N., Giovanelli, R., Haynes, M. P., Melott, A. L. & Scherrer, R. J. 1989: ApJ **340** 11

Fry, J. N. & Melott, A. L.1985: ApJ **292** 395

Fry, J. N. & Peebles, P. J. E. 1978: ApJ **221** 19

Giavalisco, M., Mancinelli, B., Mancinelli, P. & Yahil, A. 1991: preprint

Glimm, J., and Jaffe, A. 1987: Quantum Physics: A Functional Integral Point of View (New York: Springer-Verlag)

Górski, K. 1988: ApJ **332** L7

Górski, K., Davis, M., Strauss, M. A., White, S. D. M. & Yahil, A. 1989: ApJ **344** 1

Gurbatov, S. N., Saichev, A. I. & Shandarin, S. F. 1989: MNRAS **236** 385

Guth, A. H. 1981: Phys. Rev. **D23** 347

Hamilton, A. J. S. 1988: ApJ **332** 67

Hamilton, A. J. S., Kumar, P., Lu, E. & Mathews, A. 1991: ApJ **374** L1

Hamilton, A. J. S., Saslaw, W. C. & Thuan, T. X. 1985: ApJ **297** 37

Harrison, E. R. 1970: Phys. Rev. **D1** 2726

Heath, D. 1977: MNRAS **179** 351

Huang, K. 1972: Statistical Mechanics (New York: Wiley)

Hubble, E. 1934: ApJ **79** 8

Jones, B. J. T., Martinez, V. J., Saar, E. & Einasto, J. 1988: ApJ **332** L1

Kaiser, N. 1984: ApJ **284** L9

Kaiser, N. 1990: Contemp. Phys. **31** 149

Kaiser, N., Efstathiou, G., Ellis, R., Frenk, C., Lawrence, A., Rowan-Robinson, M. & Saunders, W. 1991: MNRAS **252** 1

Kodama, H. & Sasaki, M. 1984: Prog. Theo. Phys. Suppl. **78** 1

Kodama, H. & Sasaki, M. 1986: Int. J. Mod. Phys. **A1** 265

Layzer, D. 1956: AJ **61** 383

Layzer, D. 1975: in Galaxies and the Universe, ed. Sandage, A., Sandage, M. & Kristian, J. (Chicago: Chicago University Press)

Lindgren, B. W. 1976: Statistical Theory (New York: Macmillan)

Ma, S.-K. 1985: Statistical Mechanics (Philadelphia: World Scientific)

Martinez, V. J., Jones, B. J. T., Domínguez-Tenreiro, R. & Van de Weygaert, R. 1990: ApJ **357** 50

Maurogordato, S. & Lachièze-Rey 1987: ApJ **320** 13

Maurogordato, S. & Lachièze-Rey 1991: ApJ **369** 30

Monin, A. S. & Yaglom, A. M. 1971: Statistical Fluid Mechanics, vols. 1 & 2 (Cambridge: MIT Press)

Mukhanov, V. F., Feldman, H. A. & Brandenberger, R. H. 1991: preprint

Negele, J. W. & Orland, H. 1988: Quantum Many-Particle Systems (Redwood City: Addison-Wesley)

Nusser, A., Dekel, A., Bertschinger, E. & Blumenthal, G. R. 1991: ApJ **379** 6

Otto, S., Politzer, H. D., Preskill, J. & Wise, M. 1986: ApJ **304** 62

Pathria, R. K. 1972: Statistical Mechanics (Oxford: Pergamon)

Peacock, J. A. 1991: MNRAS **253** 1p

Peacock, J. A. 1992: these proceedings

Peebles, P. J. E. 1976: ApJ **205** 318

Peebles, P. J. E. 1980: The Large-Scale Structure of the Universe (Princeton: Princeton University Press)

Peebles, P. J. E. 1982: ApJ **263** L1

Peebles, P. J. E. 1984: ApJ **284** 439

Peebles, P. J. E. & Groth, E. J. 1975: ApJ **196** 1

Peebles, P. J. E. & Yu, J. T. 1970: ApJ **162** 815

Press, W. H. & Schechter, P. 1974: ApJ **187** 425

Ramond, P. 1989: Field Theory: A Modern Primer (Redwood City: Addison-Wesley)

Rivers, R. J. 1987: Path Integral Methods in Quantum Field Theory (Cambridge: Cambridge University Press)

Sachs, R. K. & Wolfe, A. M. 1967: ApJ **147** 73

Saslaw, W. C. & Hamilton, A. J. S. 1984: ApJ **276** 13

Saunders, W. et al. 1991: Nature **349** 32

Schaeffer, R. 1984;A&A: **134** L15

Scherrer, R. J. & Bertschinger, E. 1991: ApJ **381** 349

Shandarin, S. F. & Zel'dovich, Ya. B. 1989: Rev. Mod. Phys. **61** 185

Shane, C. D. & Wirtanen, C. A. 1967: Publ. Lick Obs. **22** part 1

Sharp, N. A. 1981: MNRAS **195** 857

Stuart, A. & Ord, J. K. 1987: Kendall's Advanced Theory of Statistics, Vol. 1 (London: Charles Griffin)

Silk, J. 1974: ApJ **193** 525

Strauss, M. A., Davis, M., Yahil, A. & Huchra, J. P. 1990: ApJ **361** 49

Vanmarcke, E. 1983: Random Fields: Analysis and Synthesis (Cambridge: MIT Press)

Vittorio, N., Meinhold, P., Muciaccia, P. M., Lubin, P. & Silk, J. 1991: ApJ **372** L1

Vogeley, M. S., Geller, M. J. & Huchra, J. P. 1991: ApJ **382** 44

Weinberg, S. 1972: Gravitation and Cosmology; (New York: Wiley)

White, S. D. M. 1979: MNRAS **186** 145

White, S. D. M. & Rees 1978: MNRAS **183** 341

White, S. D. M. 1990: in Physics of the Early Universe, ed. Peacock, J. A., Heavens, A. F. & Davies, A. T. (Bristol: IOP), p. 1

Whitham, G. B. 1974: Linear and Nonlinear Waves (New York: Wiley)

Yaglom, A. M. 1962:An Introduction to the Theory of Stationary Random Functions (Englewood Cliffs: Prentice-Hall)

Yahil, A. 1990: in Particle Astrophysics: The early universe and cosmic structures, ed. Alimi, J. M., Blanchard, A., Bouquet, A., Martin de Volnay, F. & Tran Thanh Van, J. (Gif sur Yvette: Editions Frontières), p. 483

Yahil, A. 1992: in Astrofundamental Physics, ed. Sanchez, N. & Zichichi, A. (Singapore: World Scientific), in press

Zel'dovich, Ya. B. 1970: A&A **5** 84

Zel'dovich, Ya. B. 1972: MNRAS **160** 1p

Zel'dovich, Ya. B. & Novikov, I. D. 1983: Relativistic Astrophysics, vol. 2 (Chicago: University of Chicago Press)

Zwicky, F. 1957: Morphological Astronomy (Berlin: Springer-Verlag)

Clusters of Galaxies as Probes for the Large Scale Structure

Hans Böhringer and Gerda Wiedenmann

Max-Planck-Institut für Extraterrestrische Physik, D-8046
Garching, Germany

Abstract: The large efforts that are presently undertaken in optical redshift surveys and X-ray surveys of clusters of galaxies will make them increasingly important tracers for the study of the large scale structure in the Universe. Compared to galaxies, clusters are very "biased" objects and therefore provide a different statistical view on the density distribution. In the present paper we explore how information on the mass or X-ray luminosity function of clusters, the evolution of clusters with redshift, and the spatial distribution of clusters can be used to study the large scale structure and the cosmological initial conditions of the Universe.

1 INTRODUCTION

Our present knowledge on the large scale structure of the matter distribution in the Universe has almost exclusively been derived from the investigation of the spatial distribution of galaxies (e.g. Geller and Huchra, 1990; Saunders et al., 1991). Hereby it is still a matter of debate how closely the galaxy distribution actually follows the density distribution of all matter (Yahil et al., 1991; Davis et al., 1991). Alternative tracers of the large scale structure like quasars, quasar absorption line systems, and clusters of galaxies have hardly provided any additional insight up to date. Also the unsuccessful search for fluctuations in the cosmological microwave background does only yield some contraints on the cosmological models but does not provide direct information on the density distribution in the early Universe (Vittorio et al., 1991; Smoot et al., 1991).

In the hierarchical sequence of structures of increasing order of integration, clusters of galaxies following single galaxies, are probably the last clearly defined step. They are the largest units which have decoupled from the Hubble flow, have collapsed and are approaching a quasi equilibrium state characterized by virial equilibrium. Clusters are close enough to this

virial state that many of them can well be described to first order by an
equlibrium configuration like the King model. All structers of larger size,
like superclusters, sheets, filaments, and voids, are far from being describable by a proper equlibrium state. In that sense clusters of galaxies are the
largest objects with a characteristic constitution of their own and they may
be the largest units which favourably can be investigated as tracers of the
large scale structure. The reason why one has not made much use of these
objects so far is the sparcity and imprecission of the currently available data.

This situation is now rapidly changing. The use of automatic scanning
machines for astronomical photographic plates, like AMP and COSMOS
(Maddox *et al.*, 1990; Heydon-Dumbleton *et al.*, 1989) to produce large
galaxy and cluster catalogues, the increased velocity with which redshifts
can be obtained and the possibility to detect many clusters independently
in X-ray surveys (as described in the second paper) will offer the possibility
to use clusters of galaxies to study the mass distribution in the Universe in
a similar fashion as with galaxies but at an even larger scale.

In this paper we therefore like to investigate how clusters of galaxies can
be used to study the large scale strucuctre of the Universe and to obtain
information on the cosmological initial conditions. Four types of data sets
are of interest in this respect:

   i) Statistics of the internal structure in clusters

   ii) The mass spectrum of clusters of galaxies

   iii) The evolution of the mass spectrum of clusters of galaxies

   iv) The correlation in the spatial distribution of clusters of galaxies

Increasingly larger scales are probed by these data from top to bottom,
starting at a scale of about 1 Mpc up to several 100 Mpc depending on the
depth of the cluster survey.

The large scale structure as observable today is closely tied to the form of
the initial conditions that prevailed in the early Universe. Since the process
of the formation of structure in a cosmological model is rather complicated
the initial conditions can not easily be deconvolved from the observable
structure. The observations are rather used as a test for the different cosmological models. A large number of cosmological scenarios have been proposed in the literature. But we will limit the discussion here to the most
conservative and popular models: the global dynamics of the Universe will
be described by Friedmann-Lemaître cosmologies, the seed inhomogenities
in the very early Universe will be assumed to be scale free Gaussian density
fluctuations (which for example applies to density fluctuations in a thermal
gas or quantum fluctuations in the inflationary epoch, but does not apply
to topological defects of space-time like cosmic strings, walls or textures);
most of the mass of the Universe today is assumed to be in the form of yet
unidentified, very weakly interacting "dark matter", and we will also think
of clusters of galaxies primarily as deep gravitational potential wells made
up mostly of dark matter which constitutes the "missing mass" in clusters

deduced from observations. The galaxies and the hot plasma in the clusters are merely seen as tracers of the gravitational potential.

The theory of the formation and spatial distribution of clusters of galaxies is part of the models on the formation of struture in the Universe and a number of excellent reviews were written about this topic as for example Peebles, 1980; Efstathiou and Silk, 1983; Szalay, 1987; Longair, 1989; White, 1989; Efstathiou, 1989. Part of the theory will also be the subject of some other lectures at this conference in particular the contribution by John Peacock. Therefore we will place more emphasis on a discussion of the interpretation of the observable parameters of clusters of galaxies in the context of the theoretical models. Section 2 provides the basic cosmological formalism used in the paper. Section 3 is concerned with the formation of clusters and the characterization of protoclusters in the underlying density fluctuation field. The mass function of clusters is derived in section 4 and compared to the observed temperature function of a statistical sample of clusters. Section 5 gives a brief description of the structure of the intracluster medium and the observable X-ray properties. In section 6 observations on the evolution of the X-ray luminosity function of galaxy clusters are discussed, and section 7 provides an outlook on the use of the more refined theory of peak statistics. Section 8 shows what can be learned from a correlation analysis of the spatial distribution of clusters of galaxies and section 9 provides a conclusion.

For numerical examples we will prefer a value for the Hubble constant of $H_o = 50h_{50}$ km s^{-1} Mpc^{-1} with $h_{50} = 1$ which is most commonly used in observational X-ray astronomy, and we will use h_{50} as a scaling parameter where it is useful.

2 COSMOLOGICAL FRAMEWORK

In this section we briefly review the basic equations describing the dynamics of a homogeneous Universe and the evolution of density perturbations. This will also serve to set up the terminology used in the following analysis. A derivation of these relations can be found in standard textbooks (e.g. Mc Vittie, 1963; Weinberg, 1972; Peebles, 1971,1980) and the review articles mentioned above.

The expansion of the Universe is governed by the following equations (assuming that the cosmological constant is zero) :

$$\dot{a}^2 = H_o^2 \left(1 - \Omega_o + \frac{\Omega_o}{a^\epsilon} \right) \qquad (1)$$

where a is the scaling parameter of the Universe defined such that $a(t = 0) = 1$ (for the present epoch). The Hubble constant, H, and the density parameter, Ω, are defined as

$$H = \frac{\dot{a}}{a} \quad \text{and} \quad H^2 \Omega = \frac{8\pi}{3} G\rho \tag{2}$$

H_o and Ω_o are the values at the present time. In eqn. (1) $\epsilon = 1$ for a matter dominated Universe and $\epsilon = 2$ for the radiation dominated Universe. Thus the expansion is characterized by a scaling parameter $a(t)$ which for a flat Universe is $\propto t^{2/3}$ for matter and $\propto t^{1/2}$ for radiation while for an open Universe with $\Omega \ll 1$ free expansion ($a \propto t$) prevails (For $\Omega_o = 1$ the transition from the radiation dominated dynamics to the matter dominated dynamics of the Universe occurs at $z \sim 10^4$ when $a \sim 9.5 \cdot 10^{-5}$).

The evolution of the density fluctuations with small amplitude, $\Delta = \frac{\delta\rho}{\rho}$, is characterized by a growing and a decaying mode. If pressure and damping effects are neglected for a moment, the time dependence of the growing mode is given by

$$\Delta \propto t^{2/3} \quad \text{for matter and} \quad \Omega = 1$$

$$\Delta \propto t \qquad \text{for radiation and} \quad \Omega = 1 \tag{3}$$

$$\Delta = const. \qquad\qquad \text{for} \quad \Omega \ll 1$$

In the radiation dominated era two further effects modify the power spectrum of the density fluctuations. The radiation pressure prevents the growth of fluctuations on a scale of the size of the Horizon of the Universe. And if most of the matter density is made up by dark matter, density fluctuations are washed out due to diffusion of the dark matter particles over a scale length comparable to the Horizon radius as long as the particles are relativistic. The epoch when the various proposed dark matter particles become non-relativistic leads to a distinction between "cold", "warm", and "hot" dark matter (CDM, WDM, HDM). The redshifts for the transitions are $z \sim 4 \cdot 10^{12}$, $z \sim 4 \cdot 10^6$, and $z \sim 4 \cdot 10^4$ and the corresponding masses contained within the Horizon are $M < M_\odot$, $M \sim 10^{10} M_\odot$, and $M \sim 10^{15} M_\odot$, respectively.

For a quantitative analysis of the evolution of the density fluctuation spectrum describable by a Gaussian field one usually chooses a Fourier representation. The fluctuation field is then statistically completely described by a power spectrum, $P(k)$, and can be represented by

$$\Delta(\mathbf{x}) = 2 \frac{\sqrt{V}}{(2\pi)^3} \sum_{l,m,n} \sqrt{|P(k)|}_{l,m,n} \; \cos\left(\frac{2\pi}{L}[lk_x + mk_y + nk_z] + \phi_{l,m,n}\right),$$

$$\tag{4}$$

where the sum is restricted to the upper half of k-space and $\sqrt{V}/(2\pi)^3$ is a normalization factor. (One can find different normalization factors for the Fourier representation in the literature; we have chosen a normalization where $P(k)$ has the dimension of a density in k-space, that is $(length)^3$, while $\Delta(x)$, $\sigma(R)$, and $\xi(r)$ as defined below remain dimensionless).

Now we are actually interested in the variation of the mass or mean density in regions of the size of protoclusters. This can be investigated mathematically by means of a filtering of the original Gaussian field with a filter of the size of protoclusters. Popular filters for this purpose are Gaussian filters

$$\Delta(x)_{R_f} = \int \frac{1}{\left(2\pi R_f^2\right)^{3/2}} \exp\left(-\frac{|x-x'|^2}{2R_f^2}\right) \Delta(x')dx'^3 \qquad (5)$$

with the corresponding modification of the power spectrum

$$P(k)_{R_f} = P(k)\, W^2(k, R_f) \text{ with } W(k, R_f) = \exp(-\frac{k^2 R_f^2}{2}) \qquad (6)$$

and top-hat filters

$$\Delta(x)_{R_t} = \frac{3}{4\pi R_t^3} \int \Theta\left(1 - \frac{|x-x'|}{R_t}\right) \Delta(x')dx'^3 \qquad (7)$$

with the effective power spectrum

$$P(k)_{R_t} = P(k)\, W^2(k, R_t) \text{ with } W(k, R_t) = \frac{3(\sin k R_t - k R_t \cdot \cos k R_t)}{(k R_t)^3} \qquad (8)$$

The variance of the density in the filtered field is then given by

$$< \Delta(R)^2 >= \sigma_o(R)^2 = \frac{1}{(2\pi)^3} \int W(k, R)^2 P(k) 4\pi k^2 dk \qquad (9)$$

If $P(k)$ is approximated by a power law with exponent n one gets

$$< \Delta(R)^2 > \propto R^{-(n+3)} \qquad (10)$$

independent of the sort of filter used.

Let us now turn to the origin and evolution of the density fluctuation spectrum. The most popular assumption about the power spectrum of the density fluctuations that prevails in the very early Universe just after inflation is

$$P(k)_{init} \propto k \qquad (11)$$

which is often termed Zel'dovich spectrum (after Zel'dovich 1972). From the early epoch to the post-recombination era the fluctuation field grows linearly, independent of the wavelength but it is modified due to the radiation pressure and due to diffusion as mentioned above. This modification is in practice usually expressed in the form of a transfer function, $T(k)$, that modifies the intial power spectrum in the following way (see e.g. Bond and Efstathiou, 1984; Peebles, 1982; Bond et al., 1982; Bond and Szalay 1983)

$$P(k) = T(k)^2 \cdot P(k)_{init} \tag{12}$$

where $P(k)$ is the power spectrum after recombination. Fig. 1 shows for example the post recombination power spectra for CDM, WDM, and HDM for a Zel'dovich initial spectrum and transfer functions taken from Bardeen, Bond, Kaiser, and Szalay (1986, BBKS in the following). The asymptotic behavior of the CDM power spectrum is characterized by $P(k) \propto k$ at small k and $P(k) \propto k^{-3}$ at large k.

3 THE COLLAPSE OF OVERDENSE REGIONS AND THE FORMATION OF CLUSTERS

An overdense region with a mean density larger than the critical density of the Universe, $\rho > \rho_{crit}$, is bound to collapse eventually. The overdense region can be considered like a small Universe that obeys eqn. (1). The dynamics is equivalent to the laws of Newtonian mechanics (as long as the size of the overdense region is much smaller than the Horizon). The collaps time for an overdensity is only dependent on the amplitude

$$t_{coll} \propto \frac{R}{<v>} \propto \frac{R}{<\dot{v}>} \propto \frac{R^3}{\Delta M(r)} \propto \Delta \tag{13}$$

Thus eqn. (13) tells us that objects form simultaneously on all scales for a power spectrum with $P(k) \propto k^{-3}$. For flatter power spectra small objects form first. This is the case for all the regions of the power spectra shown in Fig. 1 except for the steep cut-off regions in the WDM and HDM spectra. We will in the following only consider purely hierarchical clustering models where similar to the CDM model objects form first on the smallest scales and then aggregate into successively larger units (in the interesting size range from 1 to several 100 Mpc).

If we approximate the overdense region (protocluster) by a sphere with homogeneous density, the dynamics of the protocluster can also be described by eqn. (1) (we will replace the parameter a here by the local scaling parameter b for the overdense region). From eqn. (1) one finds that the maximum expansion of the protocluster when it decouples from the Hubble flow and turns around is $b_t = \Omega_i/(\Omega_i - 1)$ where Ω_i is the ratio of the mean density of the protocluster to the critical density at an initial time t_i. The time at

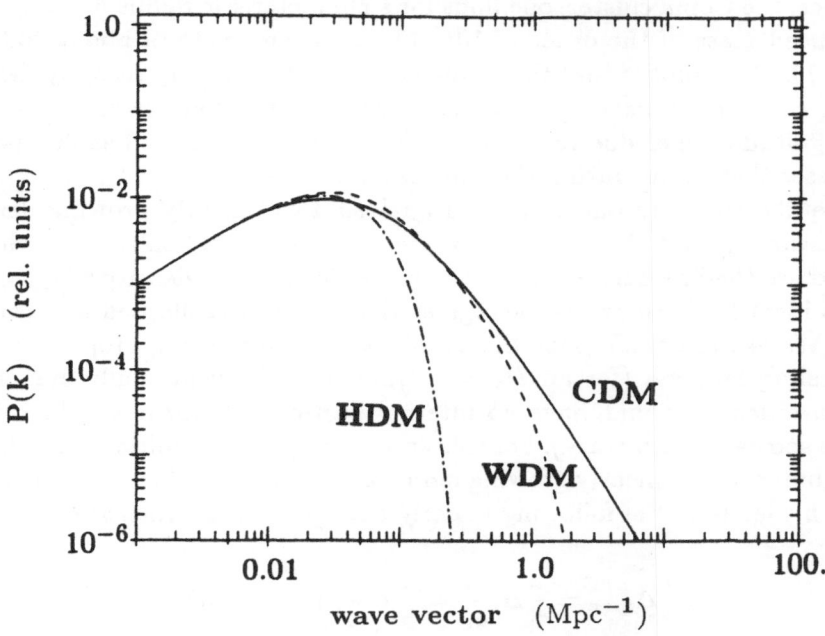

Fig. 1. Post recombination power spectra of the density fluctuation field for a Zel'dovich initial spectrum and hot (HDM), warm (WDM), and cold dark matter (CDM) - (for $h_{50} = 1$).

the turn around can be found by integrating the differential equation (1) from $b = 0$ to $b = b_t$ which yields

$$t_{turn} = \frac{\pi}{2} \left(\frac{b_t^3}{\Omega_i H_i^2} \right)^{1/2} = \frac{\pi}{2} \frac{\Omega_i}{H_i} \left(\frac{1}{\Omega_i - 1} \right)^{3/2} \tag{14}$$

The recollapse of the protocluster takes again as long as the expansion and leads to an object with a finite radius in virial equilibrium. The collapse is characterized by a "thermalization" process termed "violent relaxation" described by Lynden-Bell (1967). One special feature of violent relaxation is that it leaves the ensemble of mass particles of the protocluster in a state where all particles have the same velocity distribution independent of their mass. This is close to what is actually observed for the galaxies in clusters.

For the case of $\Omega_o = 1$ the density at turn around is higher by a factor of $\left(\frac{3\pi}{4} \right)^2 \sim 5.55$ than that of the background medium and assuming that the mean radius decreases by a factor of 2 during violent relaxation one finds an overdensity of the collapsed region of ~ 177 at $t = t_{rel} = 2 \cdot t_{turn}$.

If one takes a look at one of the best studied fairly relaxed clusters of galaxies, the Coma cluster, one finds for a characteristic radius of 3 Mpc and a deduced mass of the order of 10^{15} $M_\odot$ an overdensity of about 200 (for $h_{50} = 1$). This implies that the Coma cluster collapsed very recently. Galaxies for comparison have characteristic overdensities that are several orders of magnitude larger due to an earlier formation time as well as dissipative processes that occur during their formation.

We like to apply our statistical analysis to a linearly growing density field assuming that the small amplitude approximation holds. Thus what is needed are the linearely extrapolated values for the overdensity at t_{turn} and t_{rel} to identify those overdense regions that will have collapsed at a certain time. We use the result that the overdensities are growing proportional to the scaling factor a (for an $\Omega_o = 1$ Universe). Conventionally one starts with overdensities that have no internal particular velocities (which is of course somewhat arbitrary). For this case only 3/5 of the initial protocluster contributes to the linearly growing mode (as shown by Peebles, 1980, chapter 15). This leads to the following linearly extrapolated overdensities

$$\Delta_{turn} = \frac{3}{5} \Delta_i \frac{a_t}{a_i} = \frac{3}{5} \left(\frac{3\pi}{4} \right)^{2/3} \sim 1.06$$

$$\text{and} \quad \Delta_{rel} = \frac{3}{5} \left(\frac{3\pi}{2} \right)^{2/3} \sim 1.69 \tag{15}$$

Therefore we follow the custom and identify those peaks for which $\Delta \geq 1.69$ as collapsed and virialized clusters.

The effects occuring when the assumption of spherical symmetry in the collapse is relaxed have been studied extensively by N-body simulations (e.g. White, 1976,1977, Cavaliere et al., 1986, Evrard, 1986,1987). Fig. 2 shows two time series of N-body simulations by Schindler (1991) with 500 particles that were initially Poisson distributed in an overdense sphere in model (b) and distributed in a more homogenous way in model (a) achieved by choosing a Poisson distribution with a finite minimal separation between particles (with a space filling factor of the avoiding regions of $\sim 30\%$). In model (a) the collapse occurs in a fairly homogeneous way while in model (b) strong clumping is already present at turn around. Models with Poisson initial conditions of the kind of model (b) often break up into two or three major subclumps before the final merge, and formation of the tidely bound subclusters can prolong the collapse up to a factor of five over the value calculated above (Cavaliere et al., 1986; Schindler, 1991). The time delay can easily be explained by energetic reasons; the large amount of binding energy that goes into the subclustering results in a lowering of the bond of the cluster as a whole and increases the effective turn around radius. One should therefore expect a spread in the collaps times for peaks with the same size and the same mean overdensity in the filter region.

N-body simulations also show the interesting effect that different collapse geometries can lead to different cluster configurations in the virialized state. Schindler (1991) compared the results of simulations with distinctly different initial conditions: nearly homogeneous collapse, central collapse followed by steady infall, and the collision of two major subclusters. The results for the virialized state for the first two models are schown in Fig. 3. The collision model closely resembles the homogeneous collapse model. The density distribution of the final configuration in both cases can nicely be described by the following King density profile with a cut-off radius $r_t \sim 30 - 50 r_c$ (where r_c is the core radius of the cluster).

$$S(R) = S_o \left(\frac{1}{\sqrt{1 + \left(\frac{R}{r_c} \right)^2}} - \frac{1}{\sqrt{1 + \left(\frac{r_t}{r_c} \right)^2}} \right)^2 \tag{16}$$

Here S(R) is the surface mass density in the cluster (the corresponding formula for the volume density is more complicated and can be found in King (1962)).

The steady infall model on the contrary has a more compact core with a smaller core radius and a very extended halo. Precise data to very large radii would be necessary to trace this cluster shapes in real observations. Such data are hardly available. Therefore extra care has to be taken in a theoretical interpretation of the distribution of observed core radii in clusters of galaxies. Would all collapse models yield the same final configuration, the deduced ages of clusters with equal mass would increase with decreasing core radius, for example. The differences in the final configurations for varying initial conditions make this interpretation dangerous.

Observationally cluster masses can not easily be determined out to radii larger than 3 Mpc since there are hardly enough galaxy positions and velocities or X-ray emission at larger radii. Therefore it is comforting to know that 80 to 90 % of the total cluster mass is contained within 3Mpc for a generic cluster with $r_c \sim 250$ kpc and $r_t/r_c = 30-50$. For clusters that were formed in a way similar to the steady infall model, the core radii are considerably smaller and the ratios r_t/r_c are much larger and therefore 3 Mpc is also a good radius limit for the mass determination (see Fig. 3). Clusters of galaxies presumably are not relaxed at radii larger than 3 Mpc and thus one should not consider the above numbers as correction parameters but rather as systematic uncertainties in the determination of the total cluster mass.

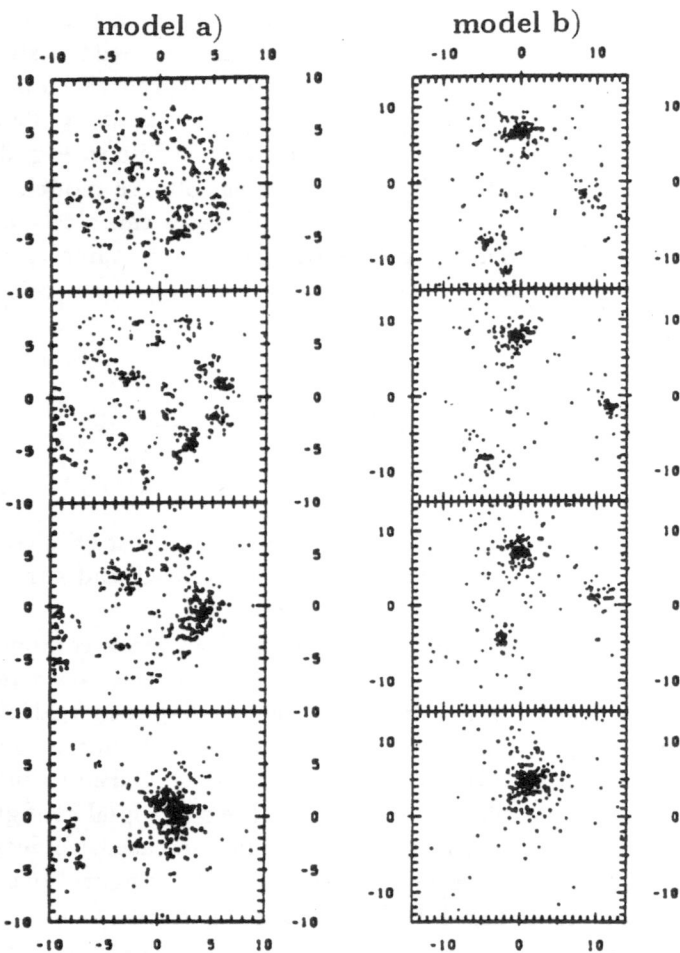

Fig. 2. N-body simulations of the formation of clusters of galaxies by Schindler (1991). Shown are the projected distributions of the 500 particles representing the cluster for subsequent configurations. Model a) was started with a Poisson distribution in a sphere with a lower limit for the particle separation and model b) had pure Poissonian initial conditions in a sphere. The time intervals are 0.55 free fall times for a) and 2.7 free fall times for b). The formation of a virialized cluster takes much longer in the more inhomogeneous model b).

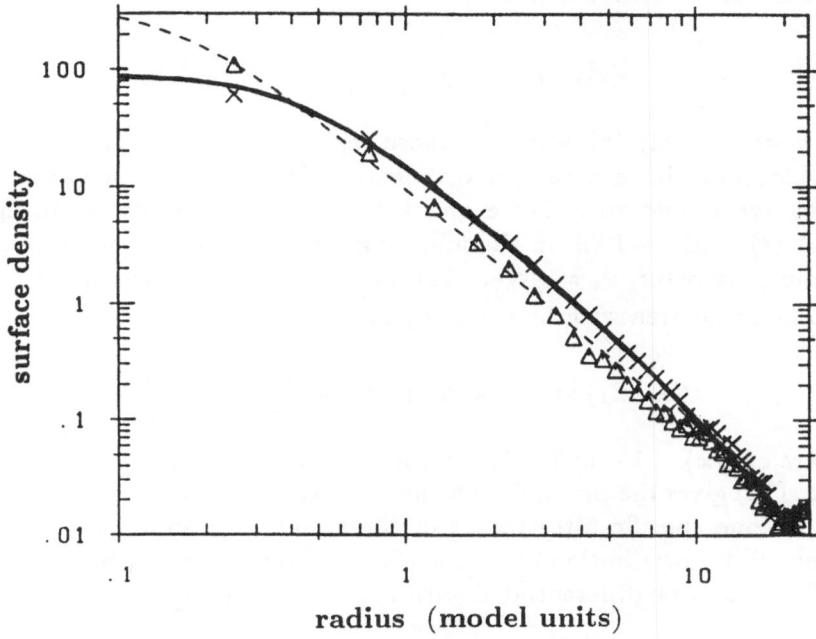

Fig. 3. Comparison of the final configurations of two N-body simulations of the cluster formation for different idealized initial conditions: nearly spherical collapse (crosses) and steady infall (triangles). Shown are the projected surface densities and fits of the empirical King profiles to the data points. The steady infall model has a very small core radius and an extended halo.

4 THE MASS FUNCTION OF CLUSTERS DERIVED FROM THE GLOBAL DENSITY DISTRIBUTION ("Press-Schechter Models")

For the hierachical clustering models described above a simple approach to derive the mass function of the forming objects has been developed first by Press and Schechter (1974) and has been applied to the current problem and to the problem of galaxy formation by many authors (e.g. Perrenod, 1980; Kaiser, 1986; Evrard, 1989; Peacock and Heavens, 1990; Henry and Arnaud, 1991; Bower, 1991; Henry et al., 1991). The mass function can in turn be used to derive the distribution functions of other physical cluster parameters that can be observed.

The basic point in this approach is, that in a random Gaussian field filtered on the scale R_f the quantity $\Delta(R_f)$ is also Gaussian distributed (due to the central limit theorem):

$$p(\Delta, R_f) = \frac{1}{\sqrt{2\pi}\sigma(R_f)}\ e^{-\frac{\Delta^2}{2\sigma(R_f)^2}} \tag{17}$$

Now we are only interested in those regions of the filtered fluctuation field which are above a certain overdensity limit, $\Delta_\star$, at an initial time t_i. They are bound to collapse and form virialized objects at an epoch when $\Delta_\star(t) = \Delta_c \sim 1.69$. In the following we will also use the normalised threshold parameter, $\nu_\star = \frac{\Delta_\star}{\sigma(R_f)}$. The probability for a given point in the field to be in an area with height $> \Delta_\star$ is given by

$$P(\Delta, R_f) = \int_{\Delta_\star}^{\infty} p(\Delta, R_f)d\Delta = \frac{1}{2}\mathrm{erfc}\left(\frac{\nu_\star}{\sqrt{2}}\right) \tag{18}$$

where $\mathrm{erfc}(x) = 1 - \mathrm{erf}(x)$ is the complementary error function.

Eqn. (18) gives the probability to find an arbitrary point in an overdense region for one specific filter radius or filter mass. To obtain the differential probability distribution of overdensities for regions with given mass M, $P(\Delta, R_f)$ has to be differentiated with respect to $M(R_f)$

$$Q(\nu_\star, M) = -\frac{dP(\nu_\star, R_f)}{dM} = -\frac{1}{\sqrt{2\pi}}\ \frac{\nu_\star}{\sigma(R_f)}\ e^{-\frac{\nu_\star^2}{2}}\ \frac{d\sigma(R_f)}{dM} \tag{19}$$

Using $M = 4/3\pi\rho R_f^3$ (which is correct for a top-hat filter; for other filters one may integrate the area under the filter function to calculate $M(R_f)$) one obtains

$$Q(M, \nu_\star) = -\frac{1}{3\sqrt{2\pi}}\ \nu_\star\ \frac{1}{M}\ e^{-\frac{\nu_\star^2}{2}}\ \frac{dlog\sigma(R_f)}{dlog R_f} \tag{20}$$

While this formula gives the volume fraction occupied by the objects in a given mass interval, the number density is obtained by multiplying with $\frac{\rho}{M}$ (where ρ is the mean density of the Universe)

$$n(M, \nu_\star) = -\frac{\sqrt{2}}{3\sqrt{\pi}}\ \nu_\star\ \frac{\rho}{M^2}\ e^{-\frac{\nu_\star^2}{2}}\ \frac{dlog\sigma(R_f)}{dlog R_f} \tag{21}$$

An extra factor of two has been added to eqn. (21). Omitting that factor of two, the integration of eqn. (21) from $\nu_\star = 0$ to infinity would lead to only half of the mass in the Universe that could be incorporated into objects which is an uncomforting result. Therefore Press and Schechter argued in their original paper that the remaining mass is somehow consumed by later accretion onto the compact objects formed. To account for that they added a factor of two to eqn. (21) which was shown by Bond et al.(1990) to be correct for a sharp-k-space filter. In a more rigorous approach this problem has to be solved in connection with the "cloud - in - cloud problem" (see below).

Unfortunately one cannot compare these theoretical results directly to observations, since a reliable mass function of a statistical sample of clusters of galaxies has not been determined yet. An available data set that comes closest to a mass function is the X-ray temperature function of a statistical sample of clusters (Edge and Stewart, 1991a,b; Henry and Arnaud, 1991). In the paper by Henry and Arnaud the temperature function is already compared to the predictions of the Press Schechter theory and we will follow their derivation to illustrate the application of this method.

For the derivation of the relation between the mass and temperature function one can refer to the virial theorem

$$\sigma_r^2 = \frac{1}{3} < v_{gal}^2 > \propto \frac{GM}{r_{turn}} \propto \frac{GM}{R_f}(1+z_f) \tag{22}$$

where $M(R_f) = 4\pi/3\rho_b R_f^3$ is the cluster mass, ρ_b is the mean background density, z_f the formation epoch of the cluster, and σ_r is the line-of-sight velocity dispersion of the cluster galaxies. σ_r is usually related to the intracluster plasma X-ray temperature by an empirical parameter β through

$$\beta = \frac{\sigma_r^2 \mu m_p}{kT} \tag{23}$$

where μm_p is the mean particle mass. To determine the proportionality constant in these relations Henry and Arnaud refer to the results of N-body simulations including the hydrodynamics of the intracluster medium by Evrard (1990) who finds

$$M_{15} = \left(\frac{kT}{4.03\text{keV}}(1+z_f)^{-1}\right)^{3/2} h_{50}^{-1} \tag{24}$$

(where M_{15} is the cluster mass in units of 10^{15} M$_\odot$) which with a proper evaluation of the constants in the above relations corresponds to $\beta \sim 1.2$.

If the post-recombination power spectrum of the density fluctuation field is approximated by a power law in the wavelength region relevant for clusters

$$P(k) = k_o^{-(n+3)} k^n \tag{25}$$

(where k_o is a constant in units of the wavenumber) the mass function and temperature function of clusters of galaxies can be calculated analytically. The mass function can be obtained by means of eqn. (9) and (21). Henry and Arnaud chose a top-hat filter - eqn. (8) - to smooth the fluctuation field on cluster scales. And using the relation $n(T) = n(M(T))\frac{dM}{dT}$ together with eqn. (24) yields the temperature function which can be found in Henry and Arnaud (1991).

Henry and Arnaud used a χ^2 test to determine the power spectrum parameters n and k_o that provide the best fit to the observational data. They also applied a convolution with a Gaussian of 10% width to the theoretical temperature function to account for a spread in the mass-temperature relation (24) similar to what was found in N-body simulations by Evrard (1990). The best fitting parameters are

$$n = -1.7 \begin{bmatrix} +0.65 \\ -0.35 \end{bmatrix} \qquad k_o = 0.0225 \begin{bmatrix} +0.0075 \\ -0.0125 \end{bmatrix} h_{50} \text{ Mpc}^{-1} \qquad (26)$$

Fig. 4 shows the theoretically determined temperature function for the above fitting parameters in comparison to the observational data. The observed cluster sample contains 25 objects with known temperatures and the data were binned such that each bin contains five clusters. The cluster temperatures for the sample span a range from about 3 to more than 8 keV. It is interesting to consider the statistical properties of the corresponding overdensities. The 3 keV clusters have a mass of about $6 \cdot 10^{14}$ $M_\odot$ and correspond to 2.5σ overdensities, while the 8 keV clusters have masses around $3 \cdot 10^{15}$ $M_\odot$ and can be identified with rare objects of an amplitude higher than 3σ. These high σ amplitudes were the justification for the linear approximation for the dynamics and statistics of overdensities in the fluctuation field. When the overdensities become nonlinear the main field is still close to linear evolution.

Also shown in Fig. 4 is the theoretical result for the temperature function for the same parameters n and k_o if a Gaussian filter - eqn. (6) - is chosen as filter function. It shows how sensitive the results are to the choice of the filter function.

5 THE INTRACLUSTER MEDIUM AND ITS ORIGIN

In the last section we used the temperature of the intracluster plasma as derived from X-ray observations as a means to estimate the mass of the cluster. To understand how reliably one can use observable parameters of the intracluster plasma some knowledge about the constitution and origin of the plasma in clusters is required. Detailed investigations of some nearby clusters in X-rays show that the plasma content in rich clusters is around 10% and may even reach up to 30% of the estimated virial mass of clusters. The uncertainty of this ratio is caused mainly by the possible errors in the virial mass estimate.

The Perseus cluster for example which is well studied may be a case of a very gas rich cluster. Recently obtained X-ray data from Spacelab-2 (Eyles et al., 1991) show that the total mass of Perseus may be only about half of what has been deduced previously from optical data (Kent and Sargent,

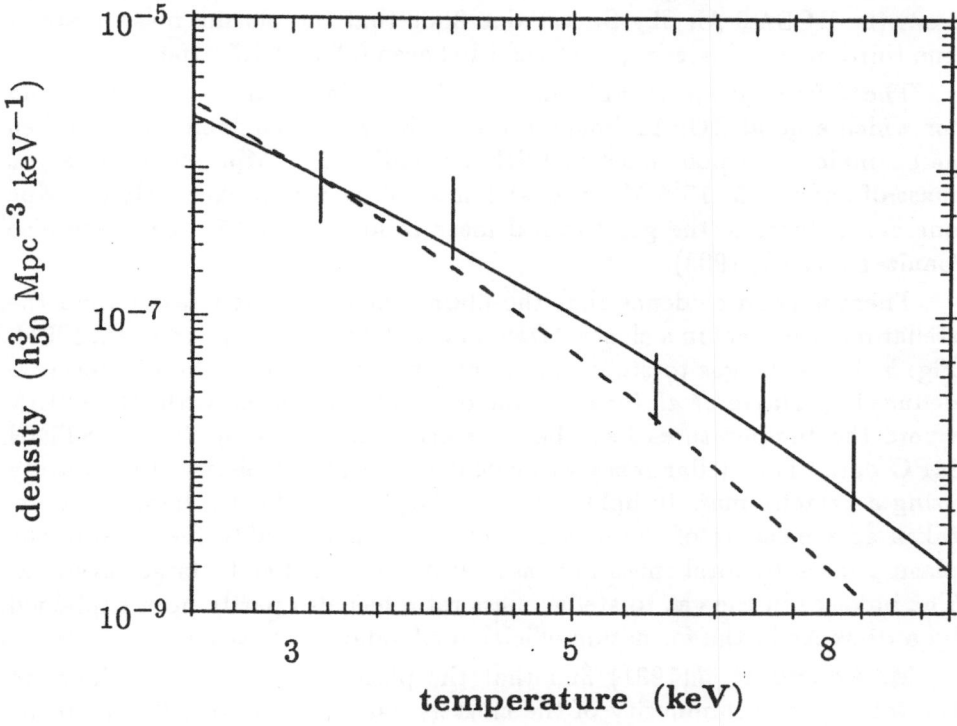

Fig. 4. Temperature function of clusters of galaxies with experimental data points adapted from Henry and Arnaud (1991) and the fit of the Press-Schechter model result (solid line). Also shown is the Press-Schechter result obtained for the same power spectrum but Gaussian filtering of the fluctuation field (dashed line).

1983). The new X-ray instruments like Spacelab-2 (coded mask telescope) and Spartan (thin scanning collimated proportional counter; e.g. Snyder *et al.*, 1990) provide simultaneous information on the temperature and density distribution of the intracluster plasma even though they don't provide the spatial resolution of an X-ray telescope. By means of a hydrostatic approximation one can use the plasma as tracer to deduce the gravitational potential of the cluster. This is a more direct approach than using galaxy redshifts. The Perseus cluster is a very striking and complex system in this respect. Recent ROSAT observations show that the X-ray temperature is not azimuthally symmetric in the cluster but indicate that the cluster may have suffered a recent merger event (Schwarz *et al.*, 1991). This may be the reason why the mass estimate using the galaxy velocity dispersion on the assumption of complete relaxation gives a too high value of the cluster mass. The gas mass can be very reliably determined with an X-ray telescope such as ROSAT. Combining the mass estimate of Eyles *et al.* and imaging data from the ROSAT All Sky Survey one finds that the plasma mass is about one third of the cluster mass at radii between 0.5 and 1.5 Mpc.

The galaxy group AWM7 that contains a bright central cD galaxy and for which a good ROSAT image from the Survey is available can be taken as example for a poor cluster. Within a radius of 1 Mpc one finds a gas mass of about $2.5 \cdot 10^{13}$ $M_\odot$ and within a radius of approximately 0.4 Mpc one can determine the gas to total mass ratio to about 7 - 20% (see also Canizares *et al.*, 1983).

There is good evidence that the plasma mass increases faster than the stellar mass content in a cluster (David *et al.*, 1990a; M. Arnaud *et al.*, 1991). Fig. 5 shows the gas to stellar mass ratio as a function of the temperature deduced by David *et al.* for a sample of clusters observed with EINSTEIN where the temperatures have been determined by use of the EINSTEIN MPC data. The stellar mass was calculated from the observed luminosity using a galactic mass to light ratio of 8 $M_\odot/L_\odot$. If the gas temperature is taken as a measure of the clusters mass then presumably the ratio of the plasma mass to total mass increases also from smaller to larger systems. The increase in the gas to stellar mass ratio has also partly been explained by a decrease in the formation efficiency of galaxies, however.

M. Arnaud *et al.*(1991) find that the plasma mass is proportional to the total galaxy luminosity of the clusters with a power of 1.9 and to the combined luminosity of E and SO galaxies with a power of 1.5. If one uses a generous value for the mass to light ratio in galaxies of 10 $M_\odot/L_\odot$ one finds that the gas to stellar mass ratio increases from about 1 in poor clusters and groups to up to 5 or 6 in rich clusters (M. Arnaud *et al.*, 1991). These more detailed results confirm the above conclusions that more massive clusters seem to be more gas rich. At the same time there is an indication that the relative iron abundance decreases with the gas mass probably implying that rich clusters contain a larger fraction of primordial gas.

If spectroscopically determined plasma temperatures are compared to the galaxy velocity dispersions by means of the β parameter defined in eqn. (23) one usually finds that β scatters around 1. Thus the gas is about as "hot" as the galaxies. Therefore the gas can have acquired its presently observed temperature from the release of the potential energy during the formation of the cluster and the thermalization of the galactic winds. The cooling time of the gas is larger than the Hubble time except for the central regions ($\sim$ 100 kpc) of a large part of the clusters which consequently have a cooling flow. Thus most of the cluster gas is still at the formation temperature.

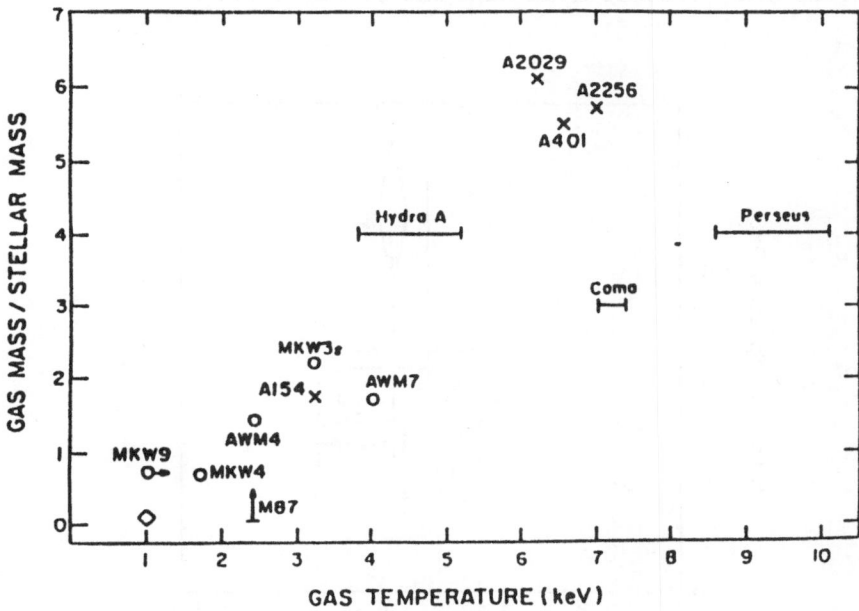

Fig. 5. Correlation of the gas to stellar mass ratio in clusters of galaxies with the observed gas temperatures from David *et al.*(1990a). The assumed galactic mass to light ratio is 8 $M_\odot/L_\odot$.

The high amount of iron in the intracluster gas requires that a large fraction of the intracluster plasma comes from galatic winds and the cluster galaxies must have lost up to 40% of their previous mass in early star burst and wind phases (David *et al.*, 1990b,c; White III, 1991). The model calculations show that the supernova explosions responsible for the iron production and the galactic winds can also contribute significantly to the heating of the intracluster plasma.

Assuming that the dark matter and the gas in a cluster follow approximately an isothermal sphere model one can also deduce an effective gas temperature or a β value from a comparison of the scale hights of galaxies and gas (Cavaliere and Fusco-Femiano, 1976; Jones and Forman, 1984). One usually obtains values of β that scatter around 0.6. That this value is lower than the spectroscopically determined β can have two reasons. In some cases the clusters are not well relaxed and the observed galaxy velocity dispersion is higher than for a thermalized model cluster, but this does not apply for all the cases. The dark matter density may also decrease faster with radius than the isothermal sphere model causing a larger scale hight for the nearly isothermal gas. Fig. 6 shows a scatter diagram of spectroscopic versus scale height β parameters determined for a sample of clusters observed with EXOSAT from Edge and Stewart (1991).

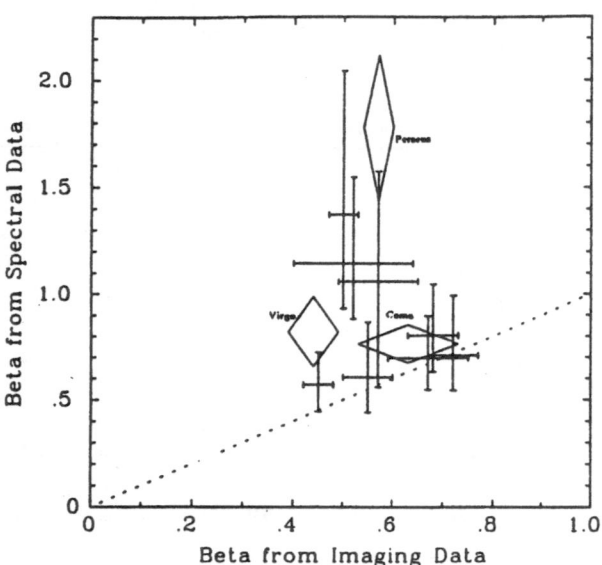

Fig. 6. Scatter plot of spectroscopic versus scale height β parameters for a sample of clusters of galaxies observed by EXOSAT from Edge and Stewart (1991).

In summary the relatively small spread in the observed β values shows that the X-ray temperature of the plasma can be used as a quite good mass indicator. This conclusion is supported by the relatively good correlation between the X-ray temperature and the galaxy velocity dispersion. This is shown in Fig. 7 for the EXOSAT cluster sample from Edge and Stewart (1991). There is also a good correlation of the X-ray luminosity with the temperature and velocity dispersion as can be seen in Fig. 7. Henry and Arnaud (1991) find the following relation for the bolometric luminosity

$$kT = 2.9 \left(h_{50}^2 L_{44}[bol] \right)^{0.265 \pm 0.035} \tag{27}$$

where $L_{44}[bol]$ is the bolometric luminosity in units of 10^{44} erg s^{-1}. The good correlation gives some hope that L_x can also be used to obtain a rough mass estimate for a cluster. This is quite advantageous since the X-ray luminosity can be determined for a much larger sample of clusters than the X-ray temperature.

If one investigates for comparison the correlation of the X-ray luminosity with the richness class parameter of Abell one finds that the correspon-

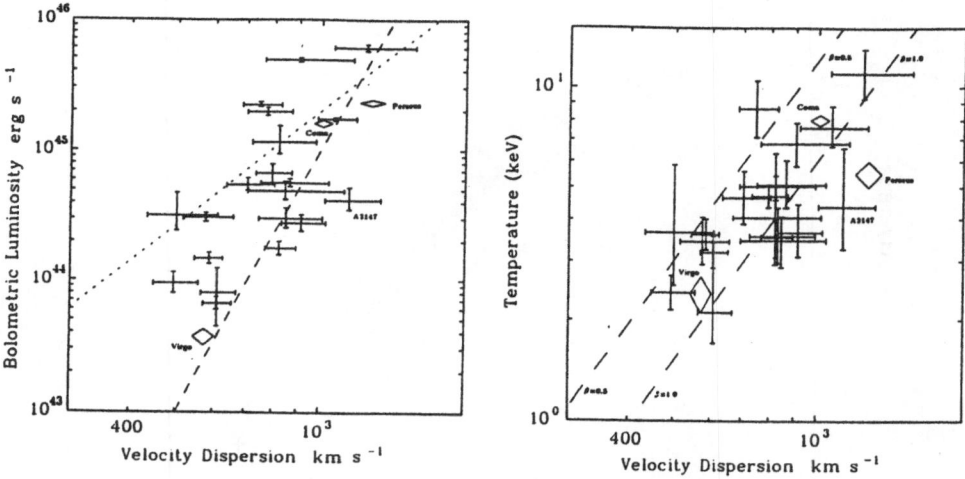

Fig. 7. Correlation of the observed gas temperature and the X-ray luminosities of clusters of galaxies with the galactic velocity dispersion for the sample of clusters observed with EXOSAT and analysed by Edge and Stewart (1991). There is a quite good correlation between all three parameters.

dence is very poor. This is shown in Fig. 8 which gives the cumulative X-ray luminosity function for clusters of different richness classes observed by EINSTEIN from Giacconi and Burg (1990). The broad, largely overlapping luminosity functions imply that the correlation is poor. Thus cluster samples selected by an optical richness criterion should be a very mixed class of objects. The selection threshold in terms of the cluster mass is a very broad function which implies that the statistical properties will be very smeared out. This is the point where there is much hope to obtain more homogeneously selected samples using the X-ray properties of clusters.

If we assume that the dark matter is distributed like in a King model and the gas follows the dark matter, and if we further assume that the gas has reached its virial temperature we can calculate the expected X-ray luminosity of a cluster. That is we assume

$$L_x(bol) \propto n_e^2 \ R^3 \ T^{1/2} \ \propto \ g^2 \ MR^3 \left(\frac{M}{R}\right)^{1/2} \tag{28}$$

where n_e is the electron density, R the characteristic radius, M the total cluster mass, and g the gas to total mass ratio. To evaluate the constants

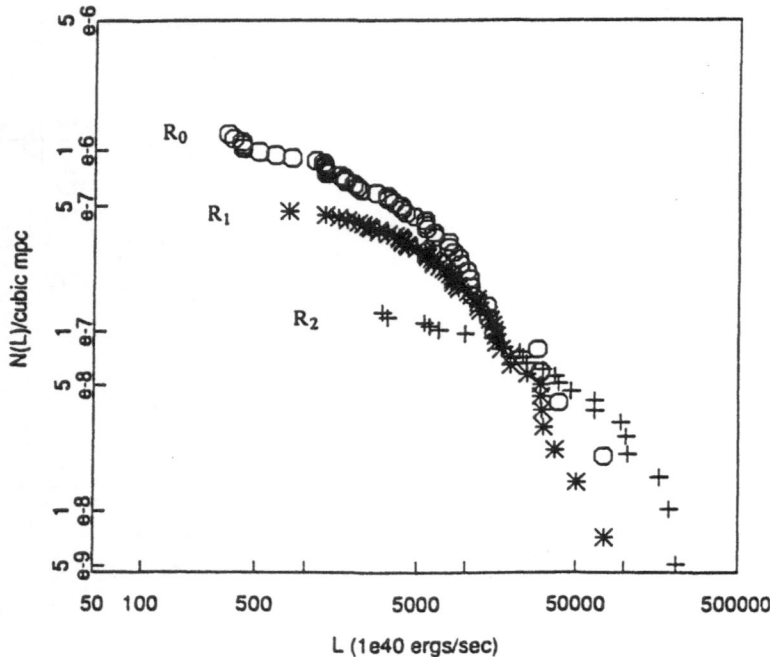

Fig. 8. Cumulative X-ray luminosity functions of a sample of clusters of galaxies observed with EINSTEIN sorted by richness class from Giacconi and Burg (1990). The broad luminosity functions show that there is no tight correlation of the X-ray luminosity with the Abell richness parameters.

involved we have also to make an assumption about the ratio of the turn around radius to the present core radius of the cluster. We have chosen a value of $r_{turn}/r_c \sim 20$ (at $z = z_f$) suggested by results of N-body simulations and obtain the following result

$$L_{44} \sim 35 \ g^2 \ M_{15}^{4/3} \ (1 + z_f)^{3.5} \tag{29}$$

where L_{44} is the bolometric luminosity in 10^{44} erg s^{-1} and M_{15} is the cluster mass in units of 10^{15} M$_\odot$. Equation (29) now allows via the relation $n(L_x) = n(M(L_x))\frac{dM}{dL_x}$ to construct the X-ray luminosity function of clusters of galaxies in the frame of the Press-Schechter theory.

6 THE EVOLUTION OF THE X-RAY LUMINOSITY FUNCTION OF CLUSTERS

The best statistical sample of clusters of galaxies with known distances and X-ray luminosities comes from the EINSTEIN Medium Sensitivity Survey (EMSS, Gioia *et al.*, 1990b). 93 clusters of galaxies were discovered among 835 X-ray sources found serendipitously by the EINSTEIN satellite. The cluster sample which covers a redshift range up to $z = 0.58$ is described in detail in Gioia *et al.*(1990a) and Henry *et al.*(1991). The data set shows a significant negative evolution of the X-ray luminosity function with redshift as given in Fig. 9. A similar evolutionary effect of the X-ray luminosity function of clusters was also found in the smaller statistical sample of the X-ray brightest clusters by Edge *et al.*(1990) from HEAO-1 and EXOSAT observations.

Henry *et al.*(1991) have analysed the EMSS results in the light of the Press-Schechter formalism. To link the X-ray luminosity to the mass of the cluster they used the mass-temperature relation found from Evrard's (1990) N-body calculations, eqn. (24) and the observed temperature luminosity relation of Henry and Arnaud (1991), eqn. (27). In addition they assumed that the central gas density has a mild dependence on the mass, $n_e \propto M^{1/4}$ and obtain

$$L_{44} = 4.5 \quad M_{15}^{11/6} \; (1 + z_f)^{3.5} \tag{30}$$

This is identical with the above given derivation of the X-ray luminosity-mass relation if we set in eqn. (29)

$$g^2 = 0.13 \quad M_{15}^{1/2} \tag{31}$$

which is well within the limits of the observed parameters as discussed above.

Henry *et al.* have fitted the theoretically predicted X-ray luminosity functions for different redshifts to their data and they could very well reproduce the evolutionary trend of the luminosity function. They obtained the best fit for the following parameters of the primordial power spectrum (if the power law form of eqn. (25) is assumed for the spectrum)

$$n = -2.1 \begin{bmatrix} +0.27 \\ -0.15 \end{bmatrix} \quad k_o = 0.0145 \begin{bmatrix} +0.004 \\ -0.0065 \end{bmatrix} \; h_{50} \; \mathrm{Mpc}^{-1} \tag{32}$$

which is within the error limits of the previous determination of these parameters by Henry and Arnaud (1991) using the observed temperature function.

We can now compare these results with the form of the primordial power spectrum predicted by the standard CDM model as shown in Fig. 1. While

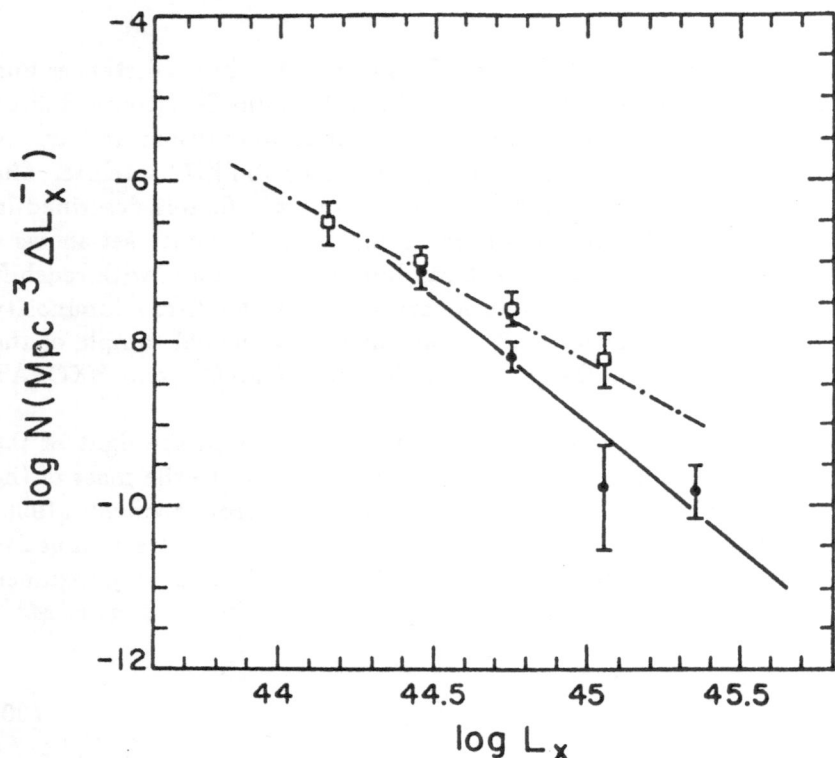

Fig. 9. X-ray luminosity functions for the clusters of galaxies found in the EIN-STEIN Medium Sensitivity Survey taken from Gioia *et al.*(1990a). Shown are the luminosity functions for the clusters in the redshift intervals $0.14 \leq z \leq 0.20$ (open squares) and $0.3 \leq z \leq 0.6$ (filled dots). The straight lines are maximum likelihood fits to the data.

the observations suggest a slope of $P(k) \propto k^{-2}$ the theory predicts a slope closer to $P(k) \propto k^{-1}$ in the interesting range around 10 Mpc. One suggestion to rectify the discrepancy is to introduce a cosmological constant that corresponds to a flat Universe with $\Omega \sim 0.2$. This was suggested by Efstathiou *et al.*(1990) to explain the new results in the galaxy-galaxy correlation function in the data from the APM survey and the IRAS survey. The same modification would help here.

7 THE CLUSTER MASS FUNCTION DETERMINED FROM PEAK STATISTICS

In the previous chapter we assumed that non-linear clumps could be identified as overdensities in the linear smoothed density field. This is a crude approximation since it does not take into account that objects of almost the same mass can have different properties and different evolution histories. A second shortcoming is the multiplication of eqn. (21) by a factor of two without justification coming from the model itself. Thus in this section we like to review an ansatz for a more rigorous solution of the problem of proto-cluster statistics. Here we will only present a brief outline of the formalism, however, and will not derive any results for specific examples.

Instead of using the overdense regions we could further restrict the class of forming objects if we consider for example only maxima above a certain threshold height in the density field. This theoretical approach has been developed by Doroshkevich (1970), Peacock and Heavens (1985) and extensively in nearly all its consequences by BBKS. In this approach the differential number density of points for which any random fields $y_1(\mathbf{r}) \ldots y_m(\mathbf{r})$ take on specific values $y_{10} \ldots y_{m0}$ is given by

$$\mathcal{N}(y_{10}, \ldots, y_{m0}) = \int P(y_1 = y_{10}, \ldots, y_m = y_{m0}, \{y_{i,j}\}) |det(y_{i,j})| d\{y_{i,j}\}$$

(33)

where $y_i(\mathbf{r}) \approx y_{i0} + \sum_j y_{i,j}(\mathbf{r} - \mathbf{r_p})$ is assumed. For Gaussian random fields the joint probability distribution $P(y_1, \ldots, y_n) dy_1 \ldots y_n$ is a multivariate Gaussian which is completely specified when the first moments $< y_i >$ and the covariance matrix are known,

$$M_{ij} = < \Delta y_i \Delta y_j > \qquad \Delta y = y - < y > .$$

(34)

Considering all regions above a certain threshold ν as forming objects the differential number density of those regions is

$$\mathcal{N}(\nu) = < \delta(\frac{\Delta}{\sigma} - \nu) >$$

(35)

Integration from the critical value $\nu_* = \frac{\Delta_*}{\sigma(R_f)}$ to infinity yields the probability of eqn. (20) needed for the Press-Schechter model. This has to be modified if we want to consider the maxima of the density above a certain threshold as observable objects. In this case the differential density eqn. (35) has to be replaced by

$$\mathcal{N}(\nu) = < \delta(\frac{\Delta}{\sigma} - \nu) |\lambda_1 \lambda_2 \lambda_3| \theta(\lambda_3) \delta(\eta) >$$

(36)

Here η is the gradient of the density field and the λ_i are the eigenvalues of the second derivative tensor ordered by

$$\lambda_1 \geq \lambda_2 \geq \lambda_3 \tag{37}$$

In the limit of high peaks the corresponding integral number density falls off like $\sim \nu_*^2 e^{-\frac{\nu_*^2}{2}}$, slightly less than the integral number density of the Press-Schechter model in eqn. (20). Thus it is clear that the Press-Schechter analysis underestimates the number of the highest peaks in the filtered fluctuation field.

A major unsolved part in these calculations is the socalled "cloud-in-cloud problem". To obtain the true mass distribution of peaks above a certain threshold the results of the peak statistics for different filterings have to be related: one ought to know how many peaks found for a small filter radius are already imbedded in more extended peaks found for larger values of R_f. In the above Press-Schechter calculations this was simply represented by a differentiation of the peak number density as a function of R_f which is equivalent to the view that the larger peaks are space filled with smaller peaks which obviously leads to a lower limit for the number density of the small peaks. This was then in some way compensated for by the *ad hoc* introduction of a factor of two.

Within the peak statistic approach this problem can in principle be solved more rigorously. One can directly investigate the probability that a small peak is imbedded in a peak that appears also in the field filtered on larger scales. Presumably this can be determined to a good approximation through the conditional probability, $P(\Delta_b|\nu_s, peak)$, of finding an overdense amplitude, Δ_b in the field smoothed on large scales at the position of a peak (with amplitude ν_s) in the field filtered on smaller scale. How this probability can be calculated is outlined in BBKS. In the determination of the cluster mass function one can thus account for the embedding problem by correcting for the clusters in each mass interval the fraction that is already included in the next higher mass range in discrete steps. In addition one has to take into account that the smaller peaks may also be embedded in peaks of a much larger size for which the variation of the field is no longer correlated with the fluctuations on the small scale. This coincidences are by definition random and are probably a minor correction to the resulting mass function. Alternative suggestions how the "cloud-in-cloud problem" may be attacted can be found in Peacock and Heavens (1990) and Bond *et al.*(1990).

8 THE SPATIAL CORRELATION OF CLUSTERS OF GALAXIES

In the frame of the Press-Schechter as well as the peak statistic theory we can calculate further properties of the linear density and velocity fields in the Universe. For example one of these properties is the two-point correlation function

$$\xi(r) = \langle \Delta(\mathbf{x})\Delta(\mathbf{x}+\mathbf{r})\rangle \tag{38}$$

which is the Fourier transform of the power spectrum

$$\xi(r) = \frac{1}{(2\pi)^3} \int_0^\infty P(\mathbf{k})e^{-i\mathbf{k}\mathbf{x}}d\mathbf{k} \tag{39}$$

Thus it might be possible to test the initial spectrum of perturbations and the corresponding transfer function by comparing (39) with observations of galaxy clustering on large scales.

The two-point correlations of clusters are observed to be much stronger than the two-point correlations of galaxies. The correlation length of Abell clusters $(R \geq 1)$ is about 50 h_{50}^{-1} Mpc (Bahcall and Soneira 1983) while the correlation length of galaxies is 10 - 15 h_{50}^{-1} Mpc (Davis and Peebles 1983; de Lapparent et al., 1988). Galaxies and clusters cannot be both good tracers of the density on large scales. As suggested by Kaiser (1984) the origin of the enhancement may be essentially statistical. The correlation function of regions above ν_* can also be defined by

$$P(r|\nu_*) = n_{\nu_*}\left(1 + \xi_{\nu_*}(r)\right) \tag{40}$$

where $P(r|\nu_*)$ is the conditional probability of finding two regions at distance r both above ν_*. Using Bayes formula one can show (BBKS)

$$P(r|\nu_*) = \frac{n_{\nu_*}(r)}{n_{\nu_*}} \tag{41}$$

with

$$n_{\nu_*}(r) = \int_{\Delta_*}\int_{\Delta_*} dy_1 dy_2 \left\{ \frac{1}{2\pi}\frac{1}{\sqrt{\xi^2(0) - \xi^2(r)}} \exp[-\frac{\xi(0)y_1^2 + \xi(0)y_2^2 - 2\xi(r)y_1y_2}{2(\xi^2(0) - \xi^2(r))}] \right\} \tag{42}$$

and n_{ν_*} as given by the complementary error function in eqn. (18). Substituting this into eqn. (40) one finds in the limit of high peaks $\nu_* \gg 1$

$$\xi_{\nu_*}(r) = \left(\frac{\nu_*}{\sigma(R)}\right)^2 \xi(r) \tag{43}$$

Eqn. (43) shows that the correlation of objects in the biased field may be stronger than the correlations in the underlying density field. The correlation function of the density field can be easily calculated for the scale free power spectrum of eqn. (25) using the relation (39)

$$\xi(r) = -\frac{\Gamma(2+n)}{2\pi^2} \sin\frac{n\pi}{2} (k_0 r)^{-(3+n)} \tag{44}$$

The variance $\sigma(R)$ of the filtered field is given by eqn. (9). Using a top-hat filter (eqn. (8)) one gets for the power spectrum of eqn. (25) with $-3 < n < -1$

$$\sigma^2(R_t) = \frac{\Gamma(2+n)}{2\pi^2} \sin\frac{n\pi}{2} \frac{9}{2^n n(n-1)(n-3)} (k_0 R_t)^{-(3+n)} \tag{45}$$

So the correlation function of the smoothed density field can be written

$$\xi_{\nu_*}(r) = \left(\frac{r}{r_0^t}\right)^{-(3+n)} \tag{46}$$

with

$$r_0^t = \left((\nu_*)^2 \frac{2^n(-n)(n-1)(n-3)}{9}\right)^{\frac{1}{3+n}} R_t \tag{47}$$

For a Gaussian filter (eqn. (6)) one gets the same form of biased correlation function but with a correlation length

$$r_0^f = \left((\nu_*)^2 \frac{\Gamma(n+3)}{\Gamma\left(\frac{3+n}{2}\right)} \left(-\frac{n+3}{n+2}\right) \sin n\pi\right)^{\frac{1}{n+3}} R_f \tag{48}$$

In both cases ξ_{ν_*} depends only on the form of the power spectrum and not on its amplitude. The enhancement of the correlation depends very critically on the threshold ν_*, but also the filter radius choosen to define the observable objects. In the above approximation ξ_{ν_*} retains the radius dependence $\xi \propto r^{-(n+3)}$ independent of the filter form and threshold.

The correlation analysis of the spatial distribution of clusters in the available redshift surveys of typically around 150 clusters results in correlation functions which have a power law form of the type $\xi_{cc} \propto r^{-\gamma}$. Results by Bahcall and Soneira (1983) on nearby northern Abell clusters give values of $\gamma = 1.8$ and $r_0 = 50 h_{50}$ Mpc. Splitting the sample of Abell clusters into different richness classes Bahcall and Soneira (1983) found that the amplitude of the correlation function increases which richness implying in accordance with eqn. (43) that there is a correlation between the Abell richness parameter and the theoretical threshold parameter. More recent investigations by Huchra et al. (1990) and Postman et al. (1992) yield $\gamma = 1.5 - 1.9$ and $r_0 = 42 h_{50}$ Mpc. Dalton et al. (1992) find a value of $r_0 = 26 h_{50}$ Mpc for $\gamma = 1.9$, however, for a sample of southern clusters selected from the

APM scans of the photographic UK Schmidt Survey. The smaller correlation lengtth of the clusters in this sample may indicate they are on average less mass rich than the Abell clusters. These values have to be compared to the autocorrelation function of galaxies (e.g. de Lapparent *et al.*, 1988) with the values $\gamma = 1.8$ and $r_o = 16h_{50}$ Mpc, which may be a good representation of the correlation of the mass density field.

From a comparison of the correlation length of the density field and of the clusters one can calculate the enhancement factor, $E = \frac{\nu^2}{\sigma^2} = \frac{\nu^4}{\Delta_*}$, defined in eqn (43). From E one can in turn derive the threshold value ν for the relevant cluster sample. For the results of Huchra *et al.* and Postman *et al.* one finds that the clusters in the sample should correspond to overdensities with $\nu \geq 2.0$. One can also determine the variance of the field for the filter scale corresponding to these clusters and gets $\sigma(R_f) = 0.84$. If we use the power spectrum that has been derived from the observed temperature function of clusters in section 4 we can also calculate the filter radius, $R_f \sim 9h_{50}$ Mpc, and the corresponding mass, $M_c \sim 2 \cdot 10^{14}$ $M_\odot$. For the sample of Dalton *et al.* the same parameters take on the values $\nu = 1.6$, $\sigma(R_f) = 1.05$, $R_f \sim 7h_{50}$ Mpc, and $M_c \sim 10^{14}$ $M_\odot$.

The values of the selection threshold, ν, and the derived effective cluster masses in the above cluster samples are smaller than the values for the clusters observed in X-rays considered in section 4 and 6. This is probably a result of the poor selection criteria in optical cluster searches. While Fig. 7 demonstrates the relatively good correlation between cluster mass and X-ray luminosity, Fig. 8 implies that the correspondence of Abell richness and cluster mass is quite poor. Abell clusters should resemble more a mixture of different $(\nu_*, R_{t,f})$ classes. The enhancement effect in the correlation function may be considerably moderated for these mixed samples with a very soft threshold. This is probably the cause of the obtained low effective threshold parameters.

Thus there is the hope that clusters found in X-ray surveys may constitute a more homogeneous data set. The only X-ray selected sample of clusters for which a correlation function has been determined is the sample of Lahav *et al.*(1989) which comprises only 53 objects. The results agree within the error limits with the correlation functions determined from optical data. But the statistics of this small sample is still too poor to improve the optical results.

9 CONCLUSIONS

The previous sections provided a brief description how clusters of galaxies can be used as a tracer of the large scale structure of the Universe mainly through the distribution function of observable parameters related to the mass function of clusters and through the correlation analysis of the spatial distribution of clusters of galaxies. The theory still contains two unsatisfactory points: the "cloud-in-cloud problem" and the question which filter function best represents the observational selection criterion for clusters of galaxies. We have hardly mentioned the large efforts conducted in using N-body simulations to attack these problems because this is a very comprehensive topic in its own. A lot of interesting problems can be studied in more detail in the simulations than in the analytical theory. A major limitation of the simulations is still the dynamical range that can be covered with current computing devices. For a resolution of about 10^{12} $M_{\odot}$ per particle in a three dimensional simulation one will only find one or a few clusters of the size of Coma in the simulation box (see Frenk *et al.*, 1990). This is some disadvantage if one is mainly interested in the evolution of the richest clusters. But it makes the analytical approach even more attractive.

The input required for the theoretical analysis outlines a most interesting observational programme that may be achievable in the next few years. Especially X-ray observations by ROSAT and the upcomming Japanese Astro D mission will allow to determine cluster masses quite accurately and provide a better correlation between the X-ray luminosities and temperatures and the masses and sizes of clusters. One should also be able to observe directly the rate at which clusters evolve by major mergers and to determine the degree to which the observed clusters are really relaxed. Last but not least the X-ray surveys will provide statistical samples of thousands of X-ray selected clusters which can be used for the statistical, cosmological tests described in this paper.

Acknowledgement We like to thank J.P. Henry for discussions and letting us use preprint versions of his papers at an early stage.

References

Arnaud, M., Rothenflug, R., Boulade, V., Vigroux, L., Vangioni-Flam, E., 1991, *Astron. Astrophys.*, , submitted.

Bahcall, N.A., Soneira, R.M., 1983, *Astrophys. J.*, **270**, 20.

Bardeen, J.M., Bond, J.R., Kaiser, N., Szalay, A.S., 1986, *Astrophys. J.*, **304**, 15.

Bond, J.R., Szalay, A.S., Turner, M.S., 1982, *Phys. Rev. Lett.* **48**, 1636.

Bond, J.R., Szalay, A.S., 1983, *Astrophys. J.*, **274**, 443.

Bond, J.R., Efstathiou, G., 1984, *Astrophys. J.*, **285**, L45.

Bond, J.R., Cole, S., Efstathiou, G., Kaiser, N., 1990, in preparation.

Bower, R.G., 1991, *Mon. Not. R. astr. Soc.*, **248**, 332.

Canizares, C.R., Stewart, G.C., Fabian, A.C., 1983, *Astrophys. J.*, **272**, 449.

Cavaliere, A., Fusco-Femiano, R., 1976, *Astron. Astrophys.*, **49**, 137.

Cavaliere, A., Santangelo, P., Tarquini, G., Vittorio, N., 1986, *Astrophys. J.*, **305**, 651.

Dalton, G.B., Efstathiou, G., Maddox, S.J., Sutherland, W.J., 1992, *Mon. Not. R. astr. Soc.*, (in press).

David, L.P., Arnaud, K.A., Forman, W., Jones, C., 1990a, *Astrophys. J.*, **356**, 32.

David, L.P., Forman, W., Jones, C., 1990b, *Astrophys. J.*, **359**, 29.

David, L.P., Forman, W., Jones, C., 1990c, *Astrophys. J.*, **369**, 121.

Davis, M., Strauss, M.A., Yahil, A., 1991, *Astrophys. J.*, **372**, 394.

de Lapparent, V., Geller, M.J., Huchra, J.P., 1988, *Astrophys. J.*, **332**, 44.

Doroshkevich, A.G., 1970, *Astrophysica*, **6**, 320.

Edge, A.C., Stewart, G.C., Fabian, A.C., & Arnaud, K.A., 1990, *Mon. Not. R. astr. Soc.*, **245**, 559.

Edge, A.C., Stewart, G.C., 1991a,b, *Mon. Not. R. astr. Soc.*, **252**, 428.

Efstathiou, G., 1989, in Physics of the Early Universe eds. J.A. Peacock, A.E. Heavens, A.T. Davies, Proceedings of the 36th Scottish Universities Summer School in Physics 1989 (a NATO Advanced Study Institute), p.361.

Efstathiou, G., Silk, J., 1983, Fundamentals of Cosmic Physics, **9**, 1.

Efstathiou, G., Kaiser, N., Saunders, W., Lawrence, A., Rowan-Robinson, M., Ellis, R.S., Frenk, C.S., 1990, *Mon. Not. R. astr. Soc.*, **247**, 10P.

Efstathiou, G., Sutherland, W.J., Maddox, S.J., 1990, *Nature*, **348**, 705.

Evrard, A. E., 1986, *Astrophys. J.*, **310**, 1.

Evrard, A. E., 1987, *Astrophys. J.*, **316**, 36.

Evrard, A. E., 1989, *Astrophys. J. (Letters)*, **341**, L71

Evrard, A. E., 1990, in *Clusters of Galaxies*, ed. M. Fitchett and W. Oegerle, Cambridge Univ. Press.

Eyles, C.J., Watt, M.P., Bertram, D., Church, M.J., Ponman, T.J., Skinner, G.K., Willmore, A.P., 1991, *Astrophys. J.*, **376**,23.

Geller, M.J., Huchra, J.P., 1990, *Science*, **246**, 897.

Giacconi, R. and Burg, R., 1990, in *Clusters of Galaxies*, ed. M. Fitchett and W. Oegerle, Cambridge Univ. Press.

Gioia, I.M., Henry, J.P., Maccacaro, T., Morris, S.L., Stocke, J.T., & Wolter, A., 1990a, *Astrophys. J.*, **356**, L35.

Gioia, I.M., Maccacaro, T., Schild, R.E., Wolter, A., Stocke, J.T., Morris, S.L., & Henry, J.P., 1990b, *Astrophys. J.*, *Suppl.*, **72**, 567.

Henry, J.P., Arnaud, K.A., 1991, *Astrophys. J.*, **372**, 410.

Henry, J.P., Gioia, I.M., Maccacaro, T., Morris, S.L., Stocke, J.T., Wolter, A., 1991, preprint.

Heydon-Dubleton, N.H., Collins, C.A., & MacGillivray, H.T., 1989, *Mon. Not. R. astr. Soc.*, **238**, 379.

Huchra, J.P., Henry, J.P., Postman, M., & Geller, M.J., 1990, *Astrophys. J.*, **365**, 66.

Jones, C., Forman, W., 1984, *Astrophys. J.*, **276**, 38.

Kaiser, N., 1986, *Mon. Not. R. astr. Soc.*, **222**, 323

Kent, S.M., Sargent, W.L.W., 1983, *Astron. J.*, **88**, 697.

King, I., 1962, *Astron. J.*, **67**, 471.

Lahav, O., Edge, A.C., Fabian, A.C., Putney, A., 1989, *Mon. Not. R. astr. Soc.*, **238**, 881.

Longair, M.S., 1989, in Lecture Notes in Physics **333**, *Evolution of Galaxies*, I. Appenzeller, H.J. Habing, P. Léna (eds.), Springer Verlag, Berlin.

Lynden-Bell, D., 1967, *Mon. Not. R. astr. Soc.*, **136**, 101.

Maddox, S.J., Efstathiou, G., Sutherland, W.J., Loveday, J., 1990, *Mon. Not. R. astr. Soc.*, **242**, 43.

McVittie, G.C., 1964, *General Relativity and Cosmology*, Chapman and Hall.

Peacock, A.J., Heavens, A.F., 1985, *Mon. Not. R. astr. Soc.*, **217**, 805.

Peacock, A.J., Heavens, A.F., 1990, *Mon. Not. R. astr. Soc.*, **243**, 133.

Peebles, P.J.E., 1971, Physical Cosmology (Princeton University Press).

Peebles, P.J.E., 1980, The Large Scale Structure of the Universe (Princeton University Press).

Peebles, P.J.E., 1982, *Astrophys. J.,* , **258**, 415.

Perrenod, S.C., 1980, *Astrophys. J.,* **236**, 373.

Postman, M., Huchra, J.P., Geller, M.J., 1992, *Astrophys. J.,* (in press).

Press, W.H., Schechter, P., 1974, *Astrophys. J.,* **187**, 425.

Szalay, A., 1988, in *Large Scale Structure of the Universe*, 17th Advanced Course of the Swiss Soc. Astron. Astrophys. Ed. L. Martinet, M. Mayer (Geneva: Observatory), p.175.

Saunders, W., Frenk, C., Rowan-Robinson, M., Efstathiou, G., Lawrence, A., Kaiser, N., Ellis, R., Crawford, J., Xia, X.-Y., Parry, I., 1991, *Nature* **349**, 32.

Schindler, S., 1991, in: *Traces of Primordial Structure in the Universe* (in press), Böhringer, H. & Treumann, R.A. (eds.), MPE Report No. 227.

Schwarz, R.A., Edge, A.C., Voges, W., Böhringer, H., Ebeling, H., Briel, U.G., 1991, *Astron. Astrophys.,* , to be published.

Smoot, G.F., *et al.*, 1991, *Astrophys. J.,* **371**, L1.

Snyder, W.A., Kowalski, M.P., Cruddace, R.G., Fritz, G.G., 1990, *Astrophys. J.,* **365**, 460.

Vittorio, N., Meinhold, P., Muciaccia, P.F., Lubin, P., Silk, J., 1991, *Astrophys. J.,* **372**, L1.

Weinberg, S., 1972, *Gravitation and Cosmology*, John Wiley and Sons.

White, S.D., 1976, *Mon. Not. R. astr. Soc.*, **177**, 717.

White, S.D., 1977, *Mon. Not. R. astr. Soc.*, **179**, 33.

White, S.D., 1989, in Physics of the Early Universe eds. J.A. Peacock, A.E. Heavens, A.T. Davies, Proceedings of the 36th Scottish Universities Summer School in Physics 1989 (a NATO Advanced Study Institute), p.1.

White III, R.E., 1991, *Astrophys. J.,* **367**, 69.

Yahil, A., Strauss, M.A., Davis, M., Huchra, J.P., 1991, *Astrophys. J.,* **372**, 380.

Zel'dovich, Ya.B., 1972, *Mon. Not. R. astr. Soc.,* , **160**, 1P.

Galaxy Formation with Gravitation, Hydrodynamics and Active Star Formation

A.Klypin [1] [2], R.Kates [3], A.Khokhlov [4]

[1] Canadian Institute for Theoretical Astrophysics, Toronto M5S 1A1, Canada.
[2] Astro-Space Center, Lebedev Physical Institute, Profsojuznaja 84/32, 117810 Moscow, USSR.
[3] Institut für Theor. Phys., Universität Kiel, 2300 Kiel, Germany.
[4] Max-Planck-Institut für Astrophysik, W-8046 Garching, Germany.

Abstract: Models for large-scale structure should in principle be able to predict the initial conditions for galaxy formation, including a reliable estimate of the gas temperature. These applications require taking into account not only gravitational clustering, but also the thermal history of the ordinary matter and in principle hydrodynamical effects as well. As a first step, useful in its own right, we have simulated the thermal history of gas and the formation of large-scale structure in a cold dark matter (CDM) scenario, using a high-resolution 3D particle-mesh algorithm for the kinematics (i.e., without explict computation of pressure forces). The results imply that gas present in superclusters should have a characteristic temperature of approximately 10^6 K. This is lower than might have been naively expected from a knowledge of peculiar velocities of matter relative to the background, which are $\approx$ 1000 km/sec. The most favorable range for detecting the diffuse radiation from this "warm" gas is about $100 - 300$ eV. We obtain estimates of the flux and the angular distribution of counts that would be contributed by these sources to ROSAT or other soft-X-ray data. For application to galaxy formation, we couple the kinematics of dark matter and the thermal evolution of the baryons to a hydrodynamic model for a multicomponent medium (here hot gas, cool gas, and stars, in addition to the dark matter). We have developed a code which uses a flux-corrected transport (FCT) algorithm to describe the gas dynamics. We report here the results of test computations which indicate that steep gradients and shocks are handled well, despite complications arising from the cosmological context, such as the amplification of small density peaks due to cooling and gravitation. This feature is particularly important for application to the problem of galaxy formation.

1 Introduction

Life would be much simpler if the universe consisted only of dark matter. In fact, the existence of a small amount of visible matter would not complicate things if its density peaks simply traced the dark-matter distribution – perhaps amplified by some biasing factor. This would simplify the process of relating n-body simulations of collisionless particles to observational results, which involve the baryonic component.

However, the formation and evolution of galaxies and the intergalactic medium is a very complicated (and thus not yet completely understood) process, and in particular the baryonic "component" is an active medium: In regions where gravitational clustering has taken place, soon after formation of the first stars, a multiphase medium rapidly evolves, with numerous and complicated links and feedback loops between different phases and between different length scales. For one thing, supernovae eject enormous amounts of energy into the surrounding gas. This process could either stimulate or inhibit star formation. In a cold, dense cloud, it could stimulate star formation by increasing the pressure at the periphery and thus causing collapse in the interior. On the other hand, propagation of the shock heats the gas and eventually turns off star formation. Even the fate of gas in dense, cold clouds is uncertain: clouds nearest to the supernova could be heated and evaporated before producing any visible star population. Supernovae also enrich the gas with metals, which drastically affects cooling rates and the conditions for star formation.

Chemodynamic models of galaxy evolution (e.g. Hensler and Burkert, 1990, Matteucci 1991) show that morphological properties of galaxies are predetermined by their initial conditions (density of baryons and dark matter, angular momentum, gas temperature, etc.) – assuming that interactions between galaxies were insignificant. Observations imply a correlation between morphology and local environment of galaxies. At the very least, models of large-scale structure ought to provide enough information to constrain the parameter space of initial conditions for simulations of galactic evolution, ultimately linking galactic morphology to cosmology.

Thermal instabilities add still more complications: In the course of cooling, the gas may become unstable with respect to the formation of clouds of cool, dense gas embedded in a hotter (but more diffuse) medium. For primordial abundances, instability occurs (assuming pressure equilibrium) if the gas is hotter than $T_{\mathrm{gas}} \approx 10^5 K$, which is quite typical in a cosmological setting. The temperature in these cool clouds drops to $\approx 10^4 K$, until the cool gas becomes neutral and stops losing energy. Now, if the heat gain due to conductivity is greater than the loss due to radiation, the cloud will not survive, but clouds larger than a critical size will grow. When the size of the cloud approaches the Jeans limit, it starts to form stars with all the ensuing complications. Eventually, the cloud presumably stops growing as a result of reheating from the surrounding material due to stellar winds, supernova explosions, and emission of young bright stars. Subsequent cooling of the hot gas could form new clouds.

A number of phenomena can couple length scales whose ratio exceeds the dynamical range of any numerical calculation. In dense regions, collisions of clouds are frequent and may reheat them at the cost of kinetic energy. So, energy is not only transfered from small to large scales (by supernovae), but also from large to small scales (by cloud collisions). Moreover, in a medium that has been enriched with heavy elements, additional thermal instabilities can arise leading to cooling below 10^4 K, and thus the "clouds" considered here will in fact have substructure.

2 A simplified picture of heating and cooling in cosmology

Valuable information about biased galaxy formation and about soft X-ray emission from superclusters can be obtained from simulations of large-scale structure (without direct simulation of hydrodynamics) by following the thermal history of the gas while imagining that, on a sufficiently coarse scale (how coarse depending on the local sound velocity), baryons are transported with the dark matter. This assumption has some justification, since both components respond to the same gravitational field, and since coarse-graining of collisionless matter resembles the dynamics of a gas whose temperature corresponds to the local velocity dispersion. The practical advantage of this picture is that the thermal history of the gas reduces to integration of one or more ordinary differential equations along *known* trajectories (those of the dark matter).

X-ray emission from superclusters can be treated with reasonable accuracy in this simplified picture: For one thing, outside of clusters, the neglect of hydrodynamics introduces errors roughly comparable to the grid resolution. For another, at least outside of galaxies, assignment of a single "temperature" to the gas (which actually is thought to consist of various components) does not lead to large errors in the emission.

The procedure as described in Kates, et. al (1990) and Klypin and Kates (1991) (hereafter Papers I and II) is briefly summarized as follows: One assumes that the local gas density is proportional to the local density of dark matter, i.e., $\rho_{gas} \equiv \rho \Omega_b$, where $\Omega_b = 0.1$ is the background fraction of baryons. This assumption is definitely valid before the formation of the first shocks, when the medium is still cold and fluctuations are small (we put $T_{gas} = 0$ at $z_{start} = 25$). As perturbations grow, eventually the first objects start to collapse, producing caustics in the dark matter and shocks in the gas. As simple pancake models show (Bond *et al.* 1984, Shapiro & Struck-Marcell 1985), shocks occur close to caustics. At a shock, the temperature acquired by gas particles with velocity $\vec{v}$ is given by

$$kT \simeq \mu_m m_H (\vec{v} - \vec{U})^2 / 3, \tag{1}$$

where $\vec{U}$ is the local velocity, m_H is the mass of hydrogen, and μ_m is the molecular weight per particle. This estimate is relatively insensitive to small errors in the position of the shock. When a particle crosses a shock and is assigned a temperature, we start to integrate the energy equation along the trajectory of the particle:

$$\frac{dT}{dt} = (\gamma - 1) \left[\frac{T}{n_H} \frac{dn_H}{dt} - \frac{\mu_m}{\mu_H} \frac{1}{k n_H} (\Lambda_{\text{rad}} + \Lambda_{\text{Comp}}) \right], \tag{2}$$

where n_H is the number density of hydrogen atoms, μ_H is the molecular weight per hydrogen atom, $\rho_{\text{gas}} \equiv \rho \Omega_b = n_H \mu_H m_H$. Here, Λ_{rad} represents radiative losses in the hot plasma with assumed primordial abundances (see Paper II), and Λ_{Comp} is the cooling rate due to Compton scattering.

When material cools below 10^4 K, stars will form, producing luminous matter. By assigning the label "cooled" to the particles with $T < 10^4$ K, we have a measure for the amount of visible matter. However, we know that the efficiency for the conversion of gas to stars is low. Thus, there should be *some* mechanism (presumably supernova explosions or heating by UV photons) preventing conversion of the gas to luminous matter in the CDM model: otherwise, all the gas would simply collapse, forming stars and ending up in globular clusters long before the formation of large objects.

To crudely simulate this effect, we "reheat" to the temperature $T_{\text{reheat}} = 5 \cdot 10^5$ those particles which have cooled to below 10^4 K with a probability of 85% and assign to the remaining 15% the label "cooled." Some of the reheated particles later will attain high temperatures, provided they enter collapsing regions of high density. Other particles can simply cool again and turn into "visible" matter. The probabilities of the reheating process (85/15 percent) were fitted to obtain a reasonable fraction of cooled gas: 1/3 of all the gas in collapsed objects (mainly superclusters) was still "hot" at $z = 0$. As long as the probability to became "cold" is small, the results are not very sensitive to the particular choice of parameters. We tried a 90/10-percent combination with essentially the same results.

It should be emphasized that *some* mechanism preventing conversion of the gas to luminous matter is *inevitable* for the CDM model: otherwise, all the gas would simply collapse, forming stars and ending up in globular clusters long before the formation of large objects.

The dark matter spends a large fraction of its time at about the same radius as the gas, because they originally had the same kinetic energy. Indeed, results of a 1D-pancake test presented in Paper I show good agreement of our model with hydrodynamical simulations (20 per cent of gas cooled, position of cooling front and shock wave). Proper inclusion of hydrodynamics and the effects of thermal instability would be expected to make a significant difference in at least two situations: First, when gas starts to cool efficiently – this happens inside dense regions and/or if the temperature becomes too low ($T < 2 \cdot 10^5$). Second, when gas undergoes secondary shocking – here, this means collapse to objects with masses larger than a galactic mass. (The smallest scale resolved corresponded to a mass of $\approx 10^{10} M_\odot$). Thus, the approach described above does not properly treat the internal regions of galaxies and clusters, but it may give a reasonable approximation for the behavior of gas which moved from voids and was trapped in the potential wells of superclusters and filaments.

3 Applications to biasing and diffuse X-ray emission

We recall that the mean density of stars (and other components of the baryonic matter) in some comoving cell depends in principle on the entire past history of the cell and not just on its density at some particular epoch. As discussed in Papers I and II, comparison of the density distribution of "cold" material – which is related to the density of visible matter – with the dark-matter density thus yields information useful for testing cooling as a physical mechanism for "biased galaxy formation."

We start from a "standard" CDM spectrum (Bardeen *et al.* 1986) normalized such that $b = 1.7$, where b is the ratio of $\delta M/M$ fluctuations in dark matter to $\delta N/N$ in galaxy counts for a randomly placed sphere of radius 16Mpc ($h = 0.5$). This value of b corresponds to a 3-d rms velocity of 760 km/sec for dark-matter particles with respect to the rest frame of the microwave background, as extrapolated by linear theory to $z = 0$ (see Paper II).

The Zeldovich approximation was used until $z = 25$. After this, using a particle-mesh (PM) code, we evolved the usual Euler-Poisson system describing the evolution of self-gravitating, collisionless matter. The computations were performed in a box of size 100 Mpc ($H_0 = 50 \ km/sec/Mpc$) with periodic boundary conditions. Our model comprised 128^3 particles on a 256^3 grid. The cell size was thus 0.39 Mpc, and the mass of one particle was 3.3 $\times 10^{10} M_\odot$.

To get an idea of the structures in the model, we show a slice of thickness 4 Mpc in Fig.1. Cooled particles (Fig. 1a) represent luminous matter. Although not all cooled particles are associated with galaxies, they trace the distribution of galaxies more closely than dark matter. The concentration of particles in the upper left corner corresponds to one of the two large clusters of galaxies which formed in the simulation. Fig. 1b shows the distribution of warm gas in this slice. Warm gas traces the more massive filaments visible in Fig.1a. (Here the term "filaments" refers either to true filaments or to walls, since in a thin slice it is difficult to distinguish between them.) These filaments are no longer evident if we consider hot gas with temperature $T > 10^7$ K (Fig. 1c), where mainly clusters and large groups are seen.

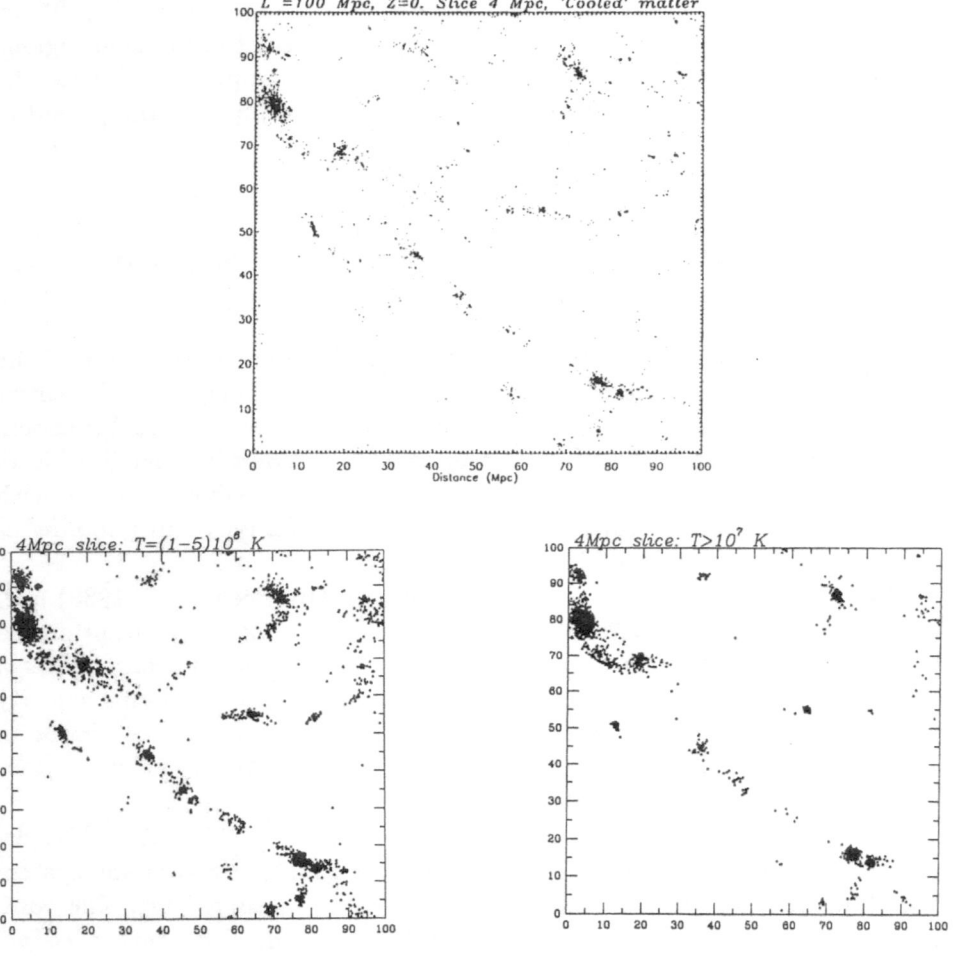

Fig. 1. Distribution of different components of matter in a slice of thickness 4 Mpc for the CDM model with the biasing parameter $b = 1.7$. The model had 128^3 particles moving in 256^3 grid. a) Cooled particles (1/12 of particles shown). b) Particles with temperature $10^6 < T < 5 \times 10^6$. c) Particles with $T > 10^7$ K.

The warm gas in superclusters, here seen as filaments, contributes to the diffuse X-ray background. Since future X-ray satellites will offer increased sensitivity in the energy window below 500 eV, comparison of the measurements with theoretical predictions of the expected flux and spatial distribution could provide a new test of different models for large-scale structure and give clues for the composition of the intergalactic medium in superclusters.

As discussed in Paper II, the CDM model gives evidence of internal velocities in superclusters of about 250 km/sec, which is consistent with observations. This would correspond to a temperature $T_{gas} \approx 3 \times 10^6$ K, if one assumes that the velocity dispersion of gas particles equals that of galaxies. A fiducial, a priori estimate of the flux from a typical supercluster would correspond to about 0.2 counts in a cell in a time 1000 sec in the ROSAT PSPC instrument. This compares to a noise level (fluctuations from the galactic background, etc.) of 30 counts in the same cell and time. (The *total* background is much larger.) Of course, an accurate estimate requires direct simulation of the thermal history of the gas in the context of a scenario for the formation of large-scale structure.

We simulated the angular distribution on the sky of the X-ray flux in the energy range 150-300 eV, appropriate for ROSAT, as shown in Fig.2. We consider a cone of diameter 60 degrees with vertex in the middle of one face of a cube consisting of 2^3 simulation boxes. The flux seen by this observer is computed using the known distribution of temperature T and hydrogen density n_H from the simulation. Contributions from particles with $5 \times 10^5 < T < 10^7$ K were included. Sources closer than 10 Mpc from the observer were excluded.

Figure 2a shows that the predicted flux density on the sky exceeds the fiducial level by a factor of about 15 for 5% of the cells. For these high-flux cells, the expected number of counts is still about a factor of 10 below the noise level. Note however that they have the appearance of a band extending across the figure, and there are large regions of at least 10 degrees on a side containing only a few high-flux cells. (A qualitatively similar pattern was found in a different projection.)

An additional improvement by a factor $\approx 10 - 20$ might be achievable if the most favorable (high-flux) cells were located along some *known* supercluster. Therefore, our simulation predicts that it is not unreasonable to hope that soft X-rays due to hot gas in superclusters may be detectable at the limits of sensitivity of ROSAT. Our results show that there are rather strong arguments in favor of searching for X-ray emission from superclusters, not in the high-energy band, but at low energies $(100 - 300 \text{ eV})$ and that there is some hope of observing superclusters in X-rays in the very near future.

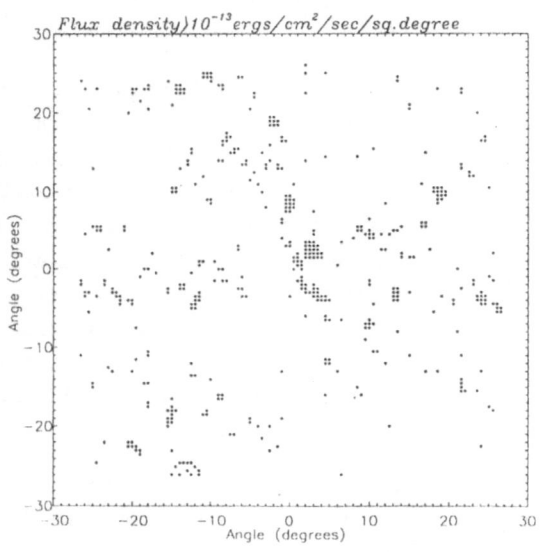

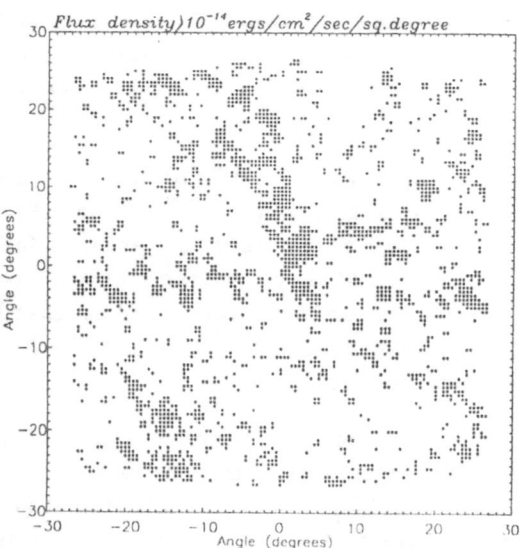

Fig. 2. Simulated angular distribution of X-ray flux density on the sky (region of size $(30 degrees)^2$ in the energy window 150 - 300 eV, binned in cells of size $(30 arcmin)^2$. a) Cells with flux exceeding $10^{-13} erg/sec/cm^2/deg^2$ (about 5% of total area). b) Cells with flux exceeding $10^{-14} erg/sec/cm^2/deg^2$ (about 17% of of total area). A flux of $10^{-13} erg/sec/cm^2/deg^2$ corresponds to about 3 counts in 1000 sec of observation for the ROSAT PSPC detector in a 30 arc minute cell. The expected background fluctuation level in a cell of this width would be 30 counts.

4 Gravitation, Cooling, and Hydrodynamics:

For problems requiring an accurate description of gas dynamics, there does not seem to be any alternative at present to direct hydrodynamical simulations such as those of Evrard (1988), Cen et al. (1990) or Katz & Gunn (1991). We now describe a new code which has advantages because it incorporates more realistic physics of star formation and provides high resolution.

Four phases are present in our model: 1) The dark matter (labeled by a subscript d) in form of weakly interacting collisionless particles (neutrinos, axions or whatever) is the main contribution to the mean density of the universe. We describe the baryonic component of the universe as a medium consisting of three interacting phases: 2) hot gas (labeled by subscript h, $T_h < 10^4$), 3) gas in the form of cold dense clouds (subscript c, $T_c < 10^4$) resulting from cooling of the hot gas, and 4) "stars" (subscript *), formed inside cold clouds . Thus, the total density $\rho(\vec{r})$ is the sum of four components:

$$\rho = \rho_{dm} + \rho_h + \rho_c + \rho_*. \tag{3}$$

At present, a rather crude approximation for thermal instability and growth of cold clouds is used. Because the typical time-step ($\Delta t = (10^7 - 10^8)$ yr) in the simulations is longer than the characteristic time-scales of star-formation effects inside clouds and the time scale of cloud evolution and because of the complexity of the star formation, we can only estimate some basic parameters of the process, such as the total energy loss due to cooling of the hot gas, the total mass converted to the cold phase, and so on. However, one important new feature of the code is the presence of multiple phases. This picture is consistent with recent work on models of galaxy evolution (Hensler and Burkert, 1990) and appears to be more realistic in some respects than the usual treatment of the gas component as a homogeneous, one-phase medium.

We start our description of the links between different phases by considering the effects of cooling of the hot gas. First, even in the presence of thermal instability, the time-scale for cooling (= energy loss) can be determined from a knowledge of the mean parameters of the hot gas. Locally, $dE = -\Lambda(\rho_h, T_h)dt$, where Λ is the rate of cooling of a hot plasma. However, the rate of energy loss expressed in terms of our numerical estimates of ρ_h and T_h, which represent averages over a cell, will be higher by some factor C. This factor takes into account the existence of small-scale density inhomogeneities and the effects of nonlinear thermal instability as well as thermal conductivity at the interfaces between cold and hot gas. We take the estimate of the enhanced cooling rate $C = 10$ used by McKee and Ostriker (1977) in their theory of the interstellar medium.

Second, we suppose that cooling results in the formation of new clouds in pressure equilibrium with the hot gas, while the entropy of the hot gas remains unchanged. That is, the energy was actually lost by the gas, which was hot and became cold. (The cold clouds are so much denser than the hot gas that their filling factor is very small.) This gives the rate of growth of mass in cold clouds. If ϵ_h and ϵ_c are the internal thermal energies per unit mass of hot and cold gas, then the change of energy of the system including the Compton cooling is

$$d(\rho_h \epsilon_h + \rho_c \epsilon_c) = -(C\Lambda + \Lambda_{Comp})dt, \qquad (4)$$

where $\Lambda_{Comp} = 7 \cdot 10^{-36} n_H T_e a^{-4}$. We take $\epsilon_c = $ constant, which means that the cold gas cannot further cool. For an ideal gas, these assumptions imply

$$\frac{d\rho_h}{dt} = -\frac{d\rho_c}{dt} = -\frac{C \cdot \Lambda + \Lambda_{Comp}}{\gamma \cdot \epsilon_h - \epsilon_c}, \qquad (5)$$

where γ is the ratio of specific heats.

The star formation, which is assumed to occur only in cold clouds, leads to a decrease in the (average) density of the cold component given by $d\rho_c/\rho_c = -dt/t_*$, where we take a constant star formation time-scale $t_* \approx 10^8$ yr . Real time-scale for star formation will depend not only on this parameter t_*, but mainly on the rate of the conversion of the hot gas to the cold phase. Thus, an *effective* time-scale for the star formation depends on the evolution of the gas. The parameter t_* defines the shortest time-scale. We assume $t_* \approx 10^8$ yr because i) even massive stars live for $t_{MS} = 3 \cdot 10^7 [M_*/10M_\odot]^{-1.6}$yr $\approx (1-3) \cdot 10^7$yr before they explode as supernovae, ii) the star-formation goes in different clouds and it is reasonable to suggest that the production of stars inside different clouds in one simulation cell is not synchronous.

The life-time of massive stars is approximately 10^7yr , thence most of stars with M_* larger than $(10-20)M_\odot$ will explode as supernovae during one time-step. Stars in the mass interval from $(5-7)M_\odot$ to $10M_\odot$ will explode on a longer time-scale, but they produce less energy, and at present the effect of these stars is neglected. Hence, due to the explosion of massive stars, the rate of growth of mass in the form of long-living stars is decreased by some factor β: $d\rho_* = (1-\beta) * \rho_c(dt/t_*)$. For the Salpeter initial mass function $dN_*/d(\log M) \propto M^{-1.35}$, one has $\beta = 0.12$ for masses from $0.1M_\odot$ to $100M_\odot$. These estimates are sensitive to both the slope of the IMF and the low-mass limit. With the slope -1.6, one obtains $\beta = 0.05$, but the IMF is probably less steep at the low-mass end anyway (Miller& Scalo 1979). So, if we were to set the lower mass limit to, say $0.5M_\odot$, then again we would have a larger value: $\beta = 0.13$. Conservatively, we adopted the value of β given by the Salpeter IMF.

The principal source of uncertainty is that this low fraction of mass in massive stars relies on the *present* IMF, whereas we would like to apply it to very early epochs of galaxy formation. The main difference is in the abundance of heavy elements. It is possible that in a hydrogen-helium plasma, only supermassive stars were formed. At the same time, enrichment of the medium could proceed very fast, changing the chemical composition of gas and, as the result, the IMF. In view of these uncertainties, we start by assuming β constant in time, while keeping in mind that other assumptions are permitted and might lead to different results.

Evaporation of cold clouds is an important effect of supernovae on the interstellar medium (McKee & Ostriker 1977, Lada 1985). We incorporated this effect by supposing that the total mass of cold gas heated and transferred back to the hot gas phase is a factor A higher than that of the supernova itself. Realistic evaporation is a much more complicated process. The rate of evaporation could depend on, among other things, the energy of the supernovae, the cloud spectrum, and the ambient density. McKee and Ostriker (1977) give an estimate of the evaporated mass which scales with the energy of the supernova E and the gas density as $E^{6/5} \cdot n_{\rm h}^{-4/5}$. However, at present we do not include the dependence on the gas density, because here $n_{\rm h}$ refers to the local density, which is relevant for the propagation of the supernova shock, whereas our $n_{\rm h}$ is the mean density of the hot gas.

Assuming that the energy input due to supernovae is proportional to the total mass of supernovae, we obtain for the rate of mass exchange due to evaporation $(d\rho/dt)_{\rm evap} = A \cdot (\beta \rho_{\rm c})/t_*$. The parameter A is directly related to the efficiency of star formation. Large A implies that a small number of supernovae will heat the surrounding cold gas, thus, stopping star formation. We know that in our galaxy, star formation, which mainly goes in giant molecular clouds, is very inefficient process. With a GMC lifetime of order 10^7 (Blitz & Shu 1980, Collins & Silk 1980) star formation must be very inefficient to have some gas left. Observations suggest that only a few per cent of the gas is converted to stars (mainly low mass stars) within the life-time of the cloud. This would imply the factor A to be of the order of a hundred.

However, these arguments are again based on the present situation in our galaxy; the problem is how to extrapolate back in time. It seems likely that the efficiency was lower in the past, but it is very difficult to make any judgement because the whole process of star formation is very nonlinear (sizes of cold clouds could for example be larger, resulting in less efficient evaporation). Consequently, we have again chosen the simplest model with constant A, which nevertheless includes the main effects and *together* with the other effects (energy input due to supernovae and radiative cooling) gives a rather complicated nonlinear picture of star formation. The value of the parameter A is restricted by the obvious condition that the energy of a supernova explosion be larger than the energy required for the evaporation of cold clouds. (In practice, one expects it it to be much larger.) This restriction leads to $\epsilon_{\rm SN} \gg A \cdot (\epsilon_{\rm h} - \epsilon_{\rm c})$. For supernova energy and mass given by $\langle E_{\rm SN} \rangle = 10^{51}$erg, $\langle M_{SN} \rangle = 22 M_\odot$ (Salpeter IMF), and for hot gas temperature $T_{\rm h} = 10^6$, we obtain the restriction $A \ll 250$. Based on these considerations, we choose $A = 20$, which allows for such occurrences as "young, active galaxies" and the transport of metal-rich, hot gas from galaxies to the intergalactic medium. This value of the parameter A is a compromise between the necessity of allowing for really active regions of star formation (low values of A) and low efficiency of conversion of gas to stars inside cold clouds (high A).

Because of the frequent exchange of mass between the hot and cold gas phases, it is reasonable to consider the hot gas and cold clouds as *one* fluid with rather complicated chemical reactions going on within it. Thus, we follow the motion of the *hot* component and integrate the change of the total density of the gas (index *gas*):

$$\rho_{gas} = \rho_c + \rho_h. \tag{6}$$

The cold and hot gas are assumed to be in pressure equilibrium. So, $P_{gas} = P_c = P_h \equiv (\gamma - 1)u_h$, $u_h = \rho_h \epsilon_h$, where u and ϵ are internal energies per unit volume and per unit mass correspondingly.

Expressed in expanding coordinates (a is the expansion parameter), the system of equations to be solved numerically thus comprises:

(1) equations for the dark matter component (collisionless particles):

$$d\mathbf{p}/dt = -\nabla\phi, \qquad d\mathbf{x}/dt = \mathbf{p}/a^2, \tag{7}$$

where $\mathbf{p} = a^2\dot{\mathbf{x}}$ and $\mathbf{x}$ are the momentum and the position of a dark matter particle, respectively (Peebles 1980).

(2) equations for the "stars" (collisionless particles) with a source term due to the production of new stars in cold clouds:

$$d\mathbf{p}_*/dt = -\nabla\phi, \qquad d\mathbf{x}_*/dt = \mathbf{p}_*/a^2, \tag{8}$$

$$\frac{\partial\rho_*}{\partial t} + 3\left(\frac{\dot{a}}{a}\right)\rho_* + a^{-1}\nabla\rho_*\vec{v}_* = \frac{(1-\beta)\rho_c}{t_*}, \tag{9}$$

(3) the continuity equation for the gas:

$$\frac{\partial\rho_{gas}}{\partial t} + 3\left(\frac{\dot{a}}{a}\right)\rho_{gas} + a^{-1}\nabla\rho_{gas}\vec{v}_{gas} = -\frac{(1-\beta)\rho_c}{t_*}, \tag{10}$$

(4) the system of hydrodynamical equations for the "hot" component:

$$\frac{\partial\rho_h}{\partial t} + 3\left(\frac{\dot{a}}{a}\right)\rho_h + a^{-1}\nabla\rho_h\vec{v}_h = \frac{\beta A\rho_c}{t_*} - \frac{C\Lambda(\rho_h, T_h) + \Lambda_{Comp}}{\gamma\epsilon_h - \epsilon_c},$$

$$\frac{\partial\vec{v}_h}{\partial t} + a^{-1}(\vec{v}_h \cdot \nabla)\vec{v}_h + \left(\frac{\dot{a}}{a}\right)\vec{v}_h = -(\rho_h a)^{-1}\nabla P_h - a^{-1}\nabla\phi, \tag{11}$$

$$\frac{\partial E_h}{\partial t} + 2\left(\frac{\dot{a}}{a}\right) E_h + a^{-1}\nabla\left((E_h + P_h)\vec{v}_h\right) =$$

$$-\frac{\gamma\epsilon_h}{\gamma\epsilon_h - \epsilon_c}[C \cdot \Lambda(\rho_h, T_h) + \Lambda_{\text{Comp}}] + \frac{\beta\rho_c}{t_*}[\epsilon_{\text{SN}} - A(\epsilon_h - \epsilon_c)], \quad (12)$$

where E_h is the total (thermal plus kinetic) energy of the hot gas per unit comoving volume ($E = \rho\epsilon + \rho v^2/2$), $kT_h = (\gamma-1)m_H\mu_H\epsilon_h$, m_H is the mass of the hydrogen atom, μ_H is the molecular weight per hydrogen atom, k is Boltzmann's constant, and $P_h = (\gamma-1)\rho_h\epsilon_h$

(5) the Poisson equation:

$$\Delta\phi = 4\pi G a^2(\rho_{\text{dm}} + \rho_* + \rho_{\text{gas}} - \rho_b), \quad (13)$$

where ρ_b is the mean density of the Universe.

The collisionless dark matter (dm) and "star" ($*$) components satisfy their continuity equations automatically by virtue of the particle-mesh method.

Some details concerning cooling rates should now be mentioned. The system of equations is solved in two different regimes: If the temperature of a volume element satisfies $T_h < 2 \cdot 10^5 K$ and if ρ_c is zero, then we suppose that the element has not yet participated in star formation. The temperature is then too low for thermal instability to be effective (Field 1965, Fall & Rees 1985). In this case the constant C is set to unity, and primordial chemical composition is assumed. If either the temperature is higher or ρ_c is not zero, then solar abundance is assumed with the cooling rate given by Raymond, Cox and Smith (1976), and the parameter A is set to $A = 20$. Compton cooling is always included. We plan to make the cooling rates more realistic by adding to the model the density of heavy elements produced by supernovae.

The time-scale of the cooling and the star-formation processes usually is shorter then the dynamical time-scale. If equal time-step is assumed for all grid cells, this difference could lead to a very short time-step, which will be defined by a tiny fraction of cells with fast cooling or active star-formation. In our code the time step for advection of the gas and motion of the dark matter was equal for all zones, but cooling and star-formation precesses were integrated with a variable time-step for each zone.

5 Numerical techniques

5.1 Equations in expanding rescaled coordinates

We are interested in gas dynamics in an expanding universe in the presence of the gravitational field generated by the gas particles themselves as well as by dark matter particles and "stars". This gravitational field is described by a potential ϕ satisfying

$$\nabla^2 \phi = 4\pi G a^2 (\rho - \rho_b)$$

where $\rho = \rho_{dm} + \rho_{gas} + \rho_*$. The density ρ_{dm} and peculiar velocity field $\vec{v}_{dm}$ of dark matter satisfy continuity and Euler equations. The gas dynamical equations consist of continuity, Euler, and energy transport equations as well as the equation of state, which is $P_h = (\gamma - 1)u_h$, where u_h is the internal energy per unit volume of the hot gas.

In order to apply the FCT method for gas dynamics in a cosmological framework, we seek transformations from the above physical quantities to rescaled quantities satisfying equations that have the "usual" form of equations of gas dynamics (i.e., in the absence of an expanding background). We adapted the transformation used by Shandarin (1980). First, we change to "expanding" variables. Auxiliary variables (denoted by a tilde) are related to the physical variables as follows. Density: $\tilde{\rho} = a^3 \rho$; peculiar velocity: $\tilde{v} = a\vec{v}$; pressure $\tilde{P} = a^5 P$; internal energy per unit volume $\tilde{u} = a^5 u$; internal energy per unit mass $\tilde{\epsilon} = a^2 \epsilon$. Second, we introduce a new time variable

$$b(t) = \frac{2}{H_0} \left(1 - a^{-1/2}(t) \right), \tag{14}$$

where H_0 is the Hubble constant now. For convenience, $b(t)$ was chosen to vanish at redshift zero ($a = 1$). Thus, b is negative for $a < 1$. (In a nonflat background, the formula for b would be changed; the definition of b would be $da = \dot{a}a^2 db$.)

In terms of the new rescaled variables, the Poisson equation and the Euler equation take the form:

$$\nabla^2 \tilde{\phi} = 4\pi G a^{-1} (\tilde{\rho}_{gas} + \tilde{\rho}_{dm} + \tilde{\rho}_* - \tilde{\rho}_b), \tag{15}$$

$$\frac{\partial \tilde{v}_h}{\partial b} + (\vec{\tilde{v}}_h \cdot \nabla)\vec{\tilde{v}}_h = -\tilde{\rho}_h^{-1} \vec{\nabla} \tilde{P}_h - a^2 \vec{\nabla} \tilde{\phi}. \tag{16}$$

The left-hand side of the continuity equation is transformed back to the usual form (i.e., no terms containing a), but the terms on the right-hand side are multiplied by a factor a^5. The same thing happens to the energy equation: the left-hand side takes the usual form, while the terms on the right-hand side are multiplied by a power of the expansion parameter (a^7). In rescaled variables, the energy equation and continuity equations for the hot gas, for the gas component, and for "stars" take the form:

$$\frac{\partial \tilde{E}_\mathrm{h}}{\partial b} + \nabla\left((\tilde{E}_\mathrm{h} + \tilde{P}_\mathrm{h})\tilde{\vec{v}}_\mathrm{h}\right) = -\frac{\gamma\tilde{\epsilon}_\mathrm{h}}{\gamma\tilde{\epsilon}_\mathrm{h} - \tilde{\epsilon}_\mathrm{c}}[a\frac{C\tilde{\rho}_\mathrm{h}^2 L(T_\mathrm{h})}{(\mu m_H)^2} + D_\mathrm{Comp}\tilde{\rho}_\mathrm{h} T_\mathrm{h}]+$$

$$+ a^2\frac{\beta\tilde{\rho}_\mathrm{c}}{t_*}(\tilde{\epsilon}_\mathrm{SN} - A(\tilde{\epsilon}_\mathrm{h} - \tilde{\epsilon}_\mathrm{c})) \qquad (17)$$

$$\frac{\partial \tilde{\rho}_\mathrm{h}}{\partial b} + \nabla\tilde{\rho}_\mathrm{h}\tilde{\vec{v}}_\mathrm{h} = \quad a^2\frac{\beta A\rho_\mathrm{c}}{t_*} - \frac{aC\tilde{\rho}_\mathrm{h}^2 L(T_\mathrm{h})/(\mu m_H)^2 + D_\mathrm{Comp}\tilde{\rho}_\mathrm{h} T_\mathrm{h}}{\gamma\tilde{\epsilon}_\mathrm{h} - \tilde{\epsilon}_\mathrm{c}}$$

$$\frac{\partial \tilde{\rho}_\mathrm{gas}}{\partial b} + \nabla\tilde{\rho}_\mathrm{gas}\tilde{\vec{v}}_\mathrm{h} = -a^2\frac{(1-\beta)\tilde{\rho}_\mathrm{c}}{t_*} \qquad (18)$$

$$\frac{\partial \tilde{\rho}_*}{\partial b} + \nabla\tilde{\rho}_*\tilde{\vec{v}}_* = \quad a^2\frac{(1-\beta)\tilde{\rho}_\mathrm{c}}{t_*}$$

Here we have defined $A(\rho_\mathrm{h}, T_\mathrm{h}) = (\rho_\mathrm{h}/\mu m_H)^2 L(T_\mathrm{h})$, where μ_H is the molecular weight and m_H is the mass of the hydrogen atom, and $D_\mathrm{Comp} = 7\cdot 10^{-36}/(\mu_H m_H)$.

5.2 Solution of the hydrodynamical equations

To calculate the flow of the ordinary (collisional) matter, described here as a multi-component medium, we apply a hydrodynamical code based on the Flux-Corrected-Transport (FCT) technique (Boris 1971, Boris & Book 1973,1976). One of the main advantages of the eulirian code is that it does not involve an artificial viscosity, but still treats shocks very accurately (for example, it does not generate postshock fluctuations and shocks are only one zone thick). The method is very fast and highly vectorizable. More specifically, in our code the low phase error algorithm is implemented in which phase errors in convection are reduced on the uniform grid to fourth order (Boris & Book 1976, Oran & Boris 1986). This algorithm is applied to gas-dynamic equations in one dimension. At each time step gas-dynamical equations are first integrated by FCT for a half-step to evaluate time-centered fluxes. Then the FCT is applied to a full timestep.

Multiple dimensions are treated through the directional timestep splitting, whereas local processes (heating, cooling, star formation, etc.), which involve various components of collisional matter, as well as self-gravitation and gravitational interaction with dark matter, are treated using process timestep splitting (Oran & Boris 1986). Composition variables (such as the concentration of heavy elements or, as introduced below, the density of hot gas) are advected through the mesh by means of the same FCT algorithm using time centered velocities from hydrodynamical step. In multiple dimensions, our code has overall second-order accuracy for regions where flow is continuous and provides a sharp, non-oscillating solution near flow discontinuities. Shock waves are spread over only one zone, as compared with 3–4 zones in the code used by Cen et al. (1990) or 3–4 point spacings in the SPH code (Hernquist & Katz 1989) . The flow chart of our code is shown in Fig.3. To avoid excessive temperature fluctuations at shocks, when estimating the temperature from known the total energy, the velocity and the density, the gas density is smoothed over seven nearest nods. Note that this smoothing was done *only* for temperature estimates.

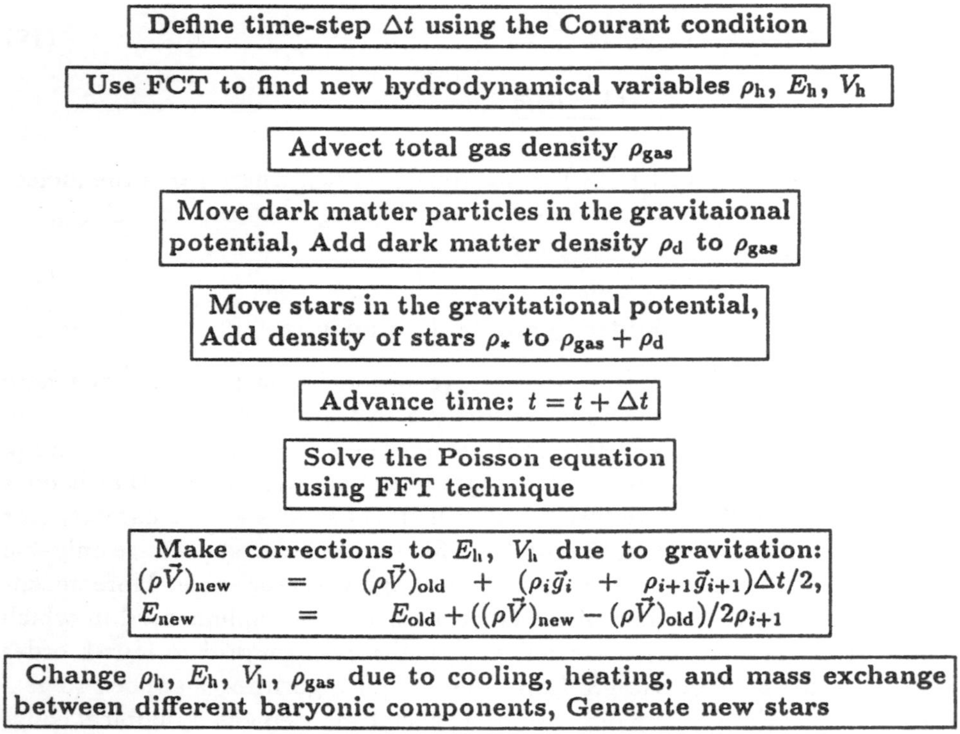

Define time-step Δt using the Courant condition

Use FCT to find new hydrodynamical variables ρ_h, E_h, V_h

Advect total gas density ρ_{gas}

Move dark matter particles in the gravitaional potential, Add dark matter density ρ_d to ρ_{gas}

Move stars in the gravitational potential, Add density of stars ρ_* to $\rho_{gas} + \rho_d$

Advance time: $t = t + \Delta t$

Solve the Poisson equation using FFT technique

Make corrections to E_h, V_h due to gravitation:
$(\rho \vec{V})_{new} = (\rho \vec{V})_{old} + (\rho_i \vec{g}_i + \rho_{i+1} \vec{g}_{i+1}) \Delta t/2,$
$E_{new} = E_{old} + ((\rho \vec{V})_{new} - (\rho \vec{V})_{old})/2\rho_{i+1}$

Change ρ_h, E_h, V_h, ρ_{gas} due to cooling, heating, and mass exchange between different baryonic components, Generate new stars

---

Fig. 3. The flow chart for the code.

The ability of a code to propagate contact and composition discontinuities without diffusing them out is crucial for problems in which simple advection prevails. Many codes which well reproduce shock waves fail to advect properly contact discontinuities and other shockless features of a flow. To trace initial fluctuations from the onset of the non-linear stage to the formation of well developed structures (clumps, etc.), great care must be taken, since numerical diffusion of a code may lead to a degradation of initial fluctuations on short spatial scales. (The ability to keep discontinuities sharp is also crucial for calculations of multi-component flows in which numerical mixing of various components has to be minimized.)

Note that the small-scale behavior of the flow is particularly important for the intended applications to galaxy formation. The non-oscillating property is especially useful, because spurious numerical fluctuations could lead to growth of spurious thermal fluctuations. The ability to keep discontinuities sharp could be of great use in distinguishing between individual galaxies within a cluster.

For a model with 128^3 dark matter particles and 128^3 grid points it takes 10-15 hours of CRAY-2 cpu time and 128Mb of the memory to run the code.

6 Tests of the new code

We carried out many tests to check the accuracy and efficiency of the code. Fig.4 shows the density and pressure for the Sod (1978) problem for the decay of a discontinuity with the parameters $\rho_L = 1$, $P_L = 1$, $u_L = 0$, $\rho_R = 0.125$, $P_L = 0.1$, $u_L = 0$. All the important features of the flow (the rarefaction wave on the left, the contact discontinuity in the middle, the shock on the right) are well represented. The contact discontinuity is spread over four zones. Fluctuations of physical parameters in the vicinity of the shock and contact discontinuities are absent. The shock is spread over only one cell.

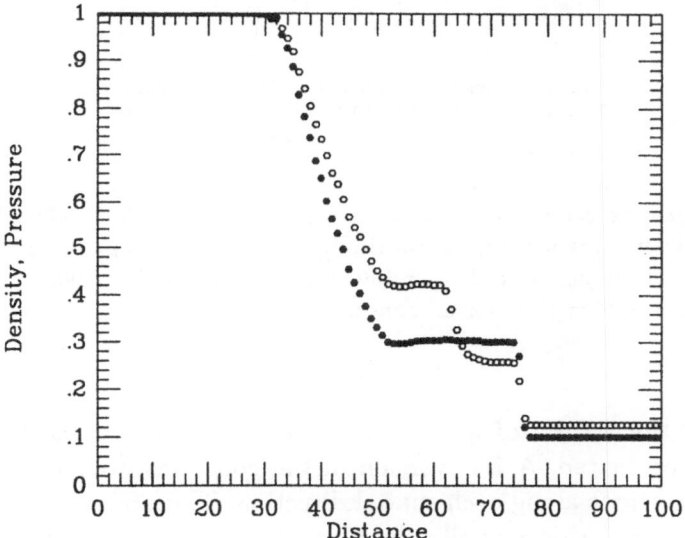

Fig. 4. The density (open circles) and the pressure (filled circles) for one-dimensional shock tube.

The ability of a code to propagate contact and composition discontinuities without diffusing them out is very important. Many codes which well reproduce shock waves fail to advect properly contact discontinuities and other shockless features of a flow. The results of a simple test of a passive advection are presented for our code in Fig.5. In this test a matter is advected with a constant velocity through the uniform mesh. Initially there is a sharp discontinuity between two regions containing different species: $\rho_{\text{left}} = 1$, $P_{\text{left}} = 100$, $\rho_{\text{right}} = 0.1$, $P_{\text{right}} = 100$, $\gamma = 1.4$. The initial velocity was $V = 1000$ for all zones. Initially the discontinuity was located at zone number ten.

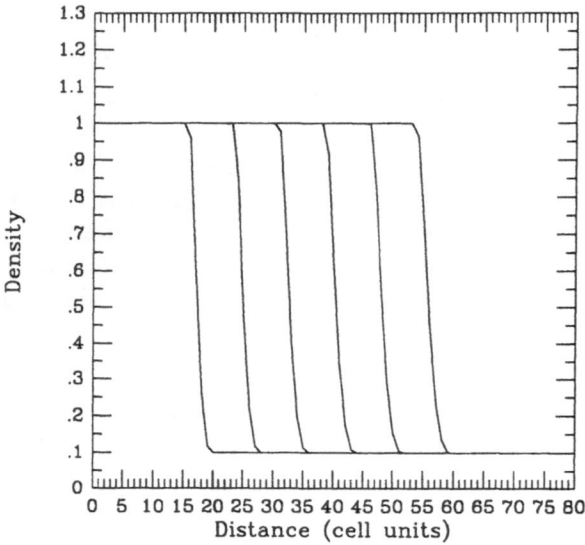

Fig. 5. The passive advection of contact discontinuity initially placed at zone number ten. There was a sharp discontinuity between two regions, which are in pressure equilibrium: $\rho_{\text{left}} = 1$, $P_{\text{left}} = 100$, $\rho_{\text{right}} = 0.1$, $P_{\text{right}} = 100$, $\gamma = 1.4$. The initial velocity was $V = 1000$ for all zones.

In Fig.6 the results of a 2-D, strong explosion in an ideal gas with $\gamma = 5/3$ are presented. A 3-D version of the code was used with periodic boundary conditions along each spatial direction. The explosion energy was initially deposited inside a small cylinder with a radius 4 cells coaxial to the z-axis. The pressure inside the cylinder initially was $P_{\text{inside}} = 10^3$, which was much larger than the pressure outside: $P_{\text{outside}} = 0.1$. The density was equal to $\rho = 1$ in both regions. After a certain time elapses, collisions of cylindrical blast waves occur at the borders of the computational domain. Reflected shocks then appear and propagate back toward the center of the explosion. Conditions for Mach reflection are not fulfilled, and regular reflection takes place.

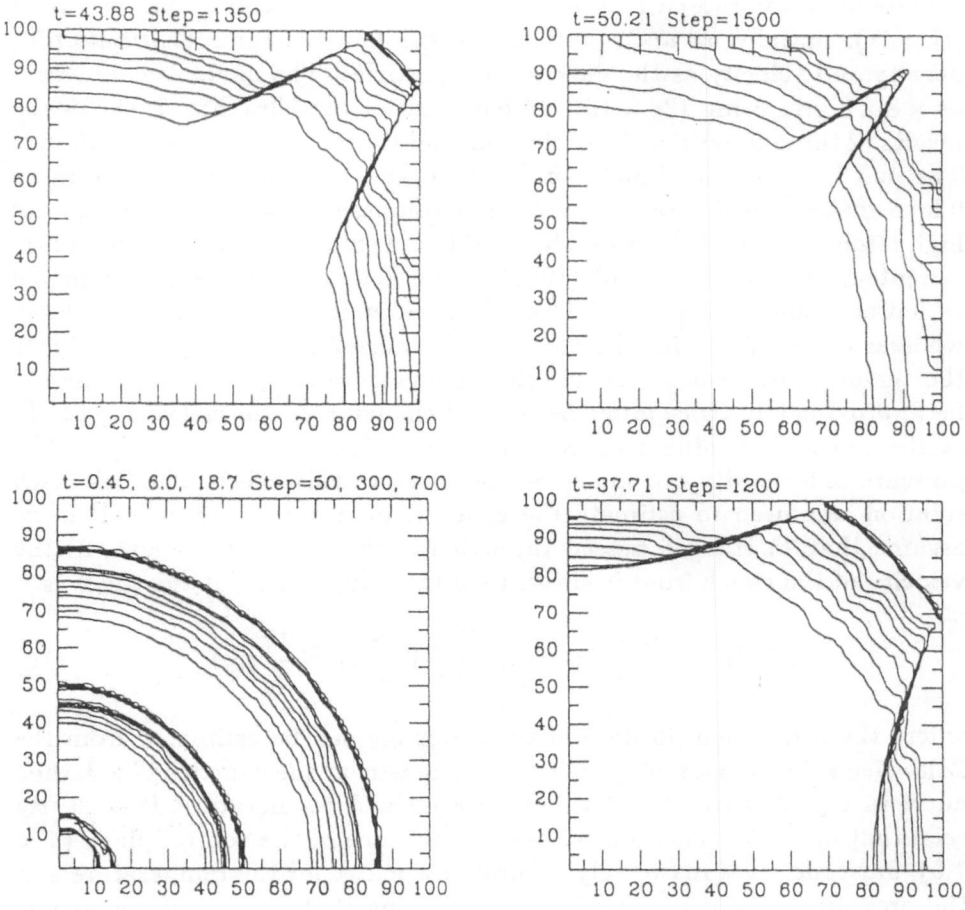

Fig. 6. The evolution of the density for a strong explosion in an ideal gas with $\gamma = 5/3$. A 3-D version of the code on the $200 \times 200 \times 1$ grid was used with periodic boundary conditions along each spatial direction. The explosion energy was initially deposited inside a small cylinder with a radius 4 cells coaxial to the z-axis. Only the upper right corner of the grid is shown. The contour levels cover the range of densities from 1 to 3 with an increment 0.25.

The evolution of a one-dimensional one-wave fluctuation was used to test our code in a quite realistic situation. The initial conditions were set in accordance with the Zeldovich solution: $x = q - (a/a_0)(L/2\pi)\sin(2\pi q/L)$, where x and q are eulirian and lagrangian coordinates $(x, q = 0 - L)$. The following parameters were used. The wavelength of the fluctuation was $L = 15\text{Mpc}$ (the Hubble constant $H_0 = 50\text{km/sec/Mpc}$, the expansion factor at the moment of collapse was $a_0 = 0.25$ (so, the redshift $z = 3$), the mean density of baryons $\Omega_b = 0.05$. The cooling rate of a gas with the primordial composition was chosen and no amplification of the cooling rate was used ($C = 1$). Initially, at $a_{start} = 0.22$, the gas had the same distribution of density and velocity as the dark matter particles. The model was ran on a $64 \times 8 \times 8$ grid using $128 \times 16 \times 16$ dark matter particles. Fig. 7 shows the results of the simulations. Note that only small part of the system is shown. The nod just at central plane of the pancake was assigned the coordinate 0.5 to make it visible on the plots. For comparison, we present results of high resolution simulations for almost the same model made by Bond (1991) using technique of Bond et al. (1984). The difference with the Bond's model is that he assumed only a small level of ionization for the gas before shocking, whereas our model is for the case of fully ionized gas. This mainly affects the temperature near the shock: the temperature in Bond's model should be approximately twice lower because of changing of the molecular weight at the shock. This difference in temperatures is clearly seen in Fig.7. The pressure and density are much less sensitive to this effect. The Zeldovich solution was used to estimate the shock temperature for our case. If we assume that 1) the gas passed through the shock does not move, 2) the velocity of the shock front is small, then the temperature at the shock is

$$kT_{sh} = \frac{(\gamma - 1)}{2\gamma} \mu_H m_H (\frac{H_0 L}{2\pi})^2 \frac{a}{a_0^2} \sin^2(\frac{2\pi}{L} q_{sh}), \qquad (19)$$

where the lagrangian position of the shock q_{sh} can be estimated from the Zeldovich solution assuming that the shock is near the caustic. The dashed curve in Fig. 7 shows this estimate of the shock temperature. It is clearly seen that once the shock moves one cell size out of the central plane ($z < 1.5$), our code gives reasonably accurate estimates for the temperature and the pressure of the gas. It was really surprising that even for the moments when the shock is still inside one cell, the code gave estimates, which are accurate within a factor of two.

We made a test for the evolution of initially one-dimensional fluctuation with the same initial parameters as for Fig. 7, but now with all effects of cooling, heating, and star formation being included. So, now $C = 10$, the cooling is assumed to produce cold clouds if $T_h > 2 \cdot 10^5$, the time scale for star formation is $t_* = 10^8 yr$, $A = 20$. When the shock forms, the cooling rate is so high that the temperature does not jump to millions degrees, as it was before. Moreover it is mainly defined by the energy input due to supernova explosions. Because of fast cooling, the pressure was much lower and the shock for a very long time did not propagate from the central plane of the pancake. Fig. 8 shows the evolution of hot gas temperature, density of hot gas and density of cold clouds in one of the zones in the central plane. The evolution of total mass in stars in the model is also present (in these units the total mass of baryons in the system is 204.8). At the moment of shock formation, which was at $a = 0.22$, the density of the hot gas started

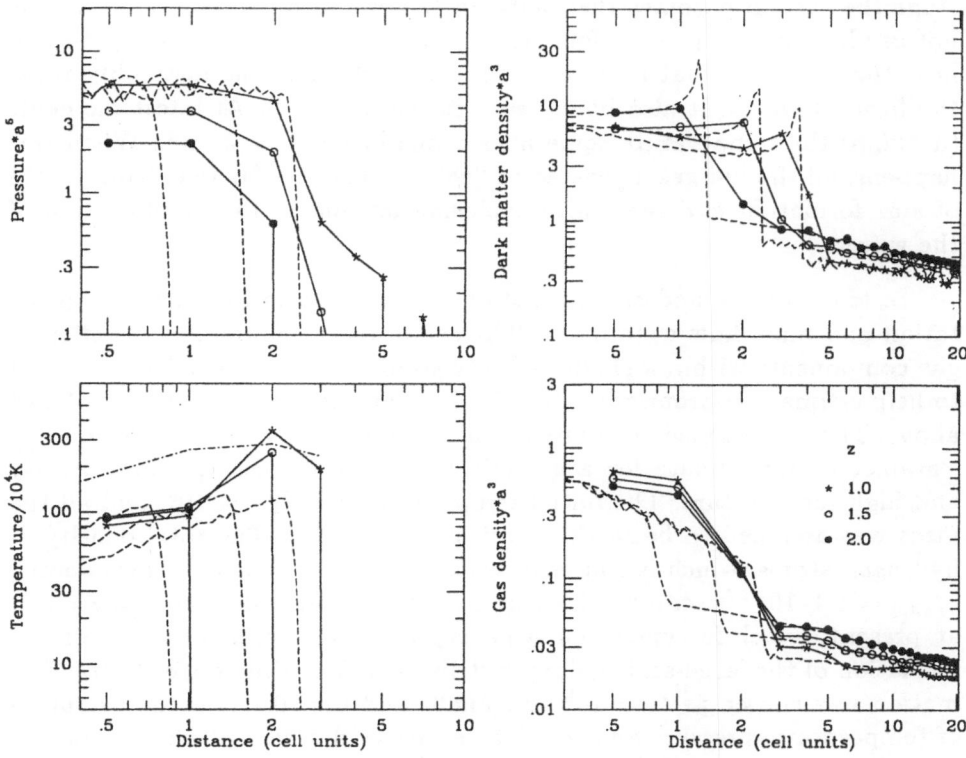

Fig. 7. The evolution of a one-dimensional sine-wave fluctuation with the wavelength $L = 15$Mpc . One cell unit is equal to 0.234Mpc . The plane of symmetry is at zero coordinate. The pancake collapsed at redshift $z = 3$. For comparison results of Lagrangian 1D simulations of Bond (1991) are shown as dashed curves. Because of a difference in initial ionization state, the temperature in Bond's model was a factor of two lower than our estimates. The dot-dashed curve present rough estimates of the temperature at shock using Zeldovich approximation. Primordial chemical composition is assumed. No corrections to the cooling rates due to substructures and thermal instability were made.

to decrease due to the formation of the cold phase (the dashed curve in the middle panel). Stars were formed and a fraction of them exploded as supernovae, which energy was only enough to reduce the rate of temperature decrease. During a short period of time from $a = 0.22$ to $a = 0.27 - 0.30$ more than a half of stars were produced. At this period the burning of stars proceeded in a steady regime with the cooling term being almost equal to the heating one. We do observe a kind of a wave in the temperature and densities, but this is explained by a sudden start of the process. Later, at $a > 0.3 - 0.35$, the temperature drops below $2 \cdot 10^5$, and the time-scale of the cooling becomes shorter than the time-scale of star formation. At this stage the system becomes thermally unstable: a small enhancement in the hot gas has time to grow before stars, produced in the cold phase, start to heat the gas. Note that because of the feedback effect the instability takes the form of waves. It also barely seen in density itself. At latest moments ($a \approx 0.5$) the temperature sometimes drops below $T = 2 \cdot 10^5$. When this happens, all the hot gas is treated as "cold", which results in enhanced rate of star formation *and* supernova explosion leading to the amplification of the waves.

Tests of cooling and star formation without taking into account gravitation and mass flow were made. The evolution of parameters of different gas components within a single cell was studied, and the results were used to help estimate appropriate values for the parameters A and C introduced above. The dependence of results on the parameter A was tested as follows. Parameters appropriate for a galactic halo were chosen, i.e., low density and high temperature. The initial temperature was $T_{h,0} = 10^6$, and all the mass was assumed to be in the "hot" phase initially. The total density of hydrogen atoms, which is constant in time, was $6 \cdot 10^{-4} \mathrm{cm}^{-3}$; equivalently, $\rho_{gas,0} = 1.4 \cdot 10^{-27} g \cdot \mathrm{cm}^{-3}$. This corresponds to an overdensity $\Delta\rho/\rho = 300$ at present, which is typical for a galaxy in numerical simulations of the formation of the large-scale galaxy distribution. The time-scale for star formation was chosen as $t_* = 10^8$ yr . Fig.9 and Fig.10 show the evolution of temperature, pressure and relative density of different phases for models with differing values of the parameter A: $A = 1$ (no evaporation), $A = 20$ ($\approx$ 10 per cent of supernovae energy goes to evaporation) and $A = 100$ ($\approx$ 50 per cent goes to evaporation). The hot gas initially cools with a characteristic cooling-time $(2-3)10^7$ yr . At $t \approx 10^7$ yr , the first stars form and begin to produce supernovae. The supernovae evaporate cold clouds, which is evident in the case $A = 20$ as a rapid drop in the density of the cold gas and as a jump in the density of the hot gas. Temperature and pressure of the hot gas also rise at $10^7 < t < 10^8$ for models with $A = 1$ and $A = 20$, but for $A = 100$ temperature and pressure remain essentially constant over this period, because too much energy goes to cloud evaporation. Note that a high rate of cloud evaporation (high A) ultimately produces relatively fast star formation, because the temperature of the hot gas is lower and it cools faster. The time-scale t_* for star formation was taken as 10^8 yr for these models, but the true time-scale of growth of mass in the "star" component was longer as a result of the nonlinear feedback processes: The star-formation e-folding time was $4 \cdot 10^8$ yr for $A = 100$, 10^9 yr for $A = 20$ and $3 \cdot 10^9$ yr for $A = 1$. The time-scale for star formation can be shorter if the density of the gas is increased, as seen from the dashed curves in Fig.11 and Fig.12, for which the gas density is ten times greater: $6 \cdot 10^{-3} cm^{-3}$ (here we take $A = 20$). Since the cooling time is shorter (10^7 yr), a larger fraction of the gas was in the cold phase when the first supernovae started to explode. As

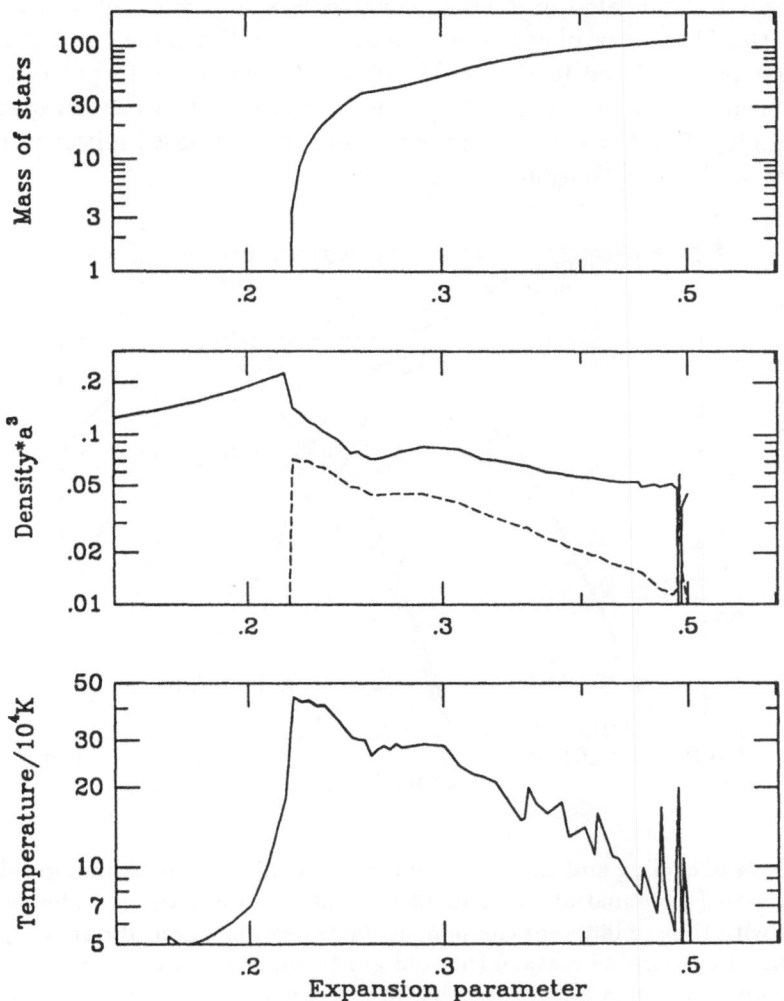

Fig. 8. The evolution of the temperature (the low panel) and the density (the middle panel; the full curve is for the hot gas, the dashed curve is for the cold gas) at the central plane of a sine-wave fluctuation with the wavelength $L = 15$Mpc . The upper panel shows the total mas of stars in the system. Effects of enhanced cooling ($C = 10$) and star formation are included.

a consequence, the time-scale for star formation was $\approx 2 \cdot 10^8$ yr , or 5 times shorter than in the lower-density case. The dot-dashed curves in Fig.11and Fig.12represent a model with a very low initial temperature ($T_{h,0} = 5 \cdot 10^4$). All the other parameters are the same as for the model designated in Fig. 9by the solid curves ($A = 20, \rho_{gas,0} = 1.4 \cdot 10^{-27} g \cdot cm^{-3}$). The model initially cools without any thermal instability and with a cooling rate corresponding to primordial chemical abundances. Nevertheless, at $t = 2 \cdot 10^7$, the temperature of the "hot" gas phase drops below the 10^4K barrier, and therefore all the gas is transfered to the "cold" phase, which starts to produce stars and supernovae. At this moment the model switches to solar composition. Interestingly, after $t = 10^8$ yr , the model becomes almost indistinguishable from that with an initially high temperature.

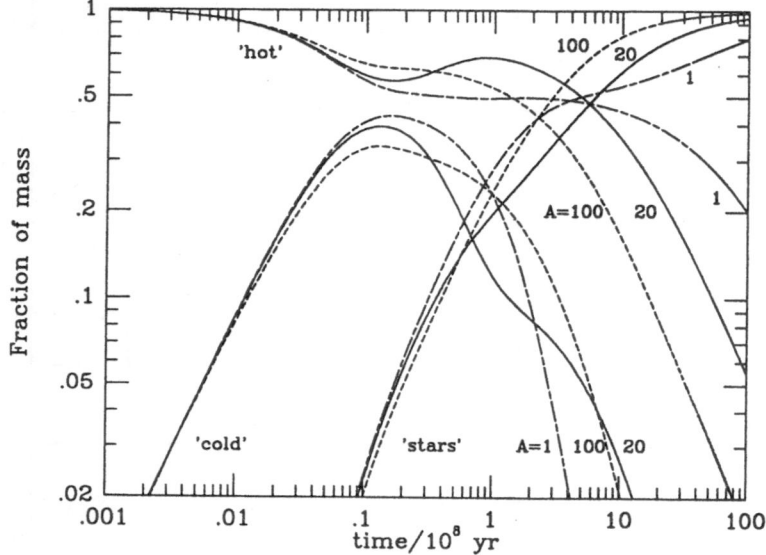

Fig. 9. Tests of cooling and star formation without taking into account gravitation and mass flow (the equations 17 and 18 without advection terms). Shown is the evolution with time of different components on the evaporation efficiency A, which is the ratio of evaporated mass of the cold gas to the mass of the supernova itself. Note that in case, when a large fraction of the supernova energy goes to heating and evaporation of the cold clouds ($A = 100$), the temperature of the hot gas remains almost constant and the star-formation is fast, while a low evaporation ($A = 1$) leads to a very small star-formation rate.

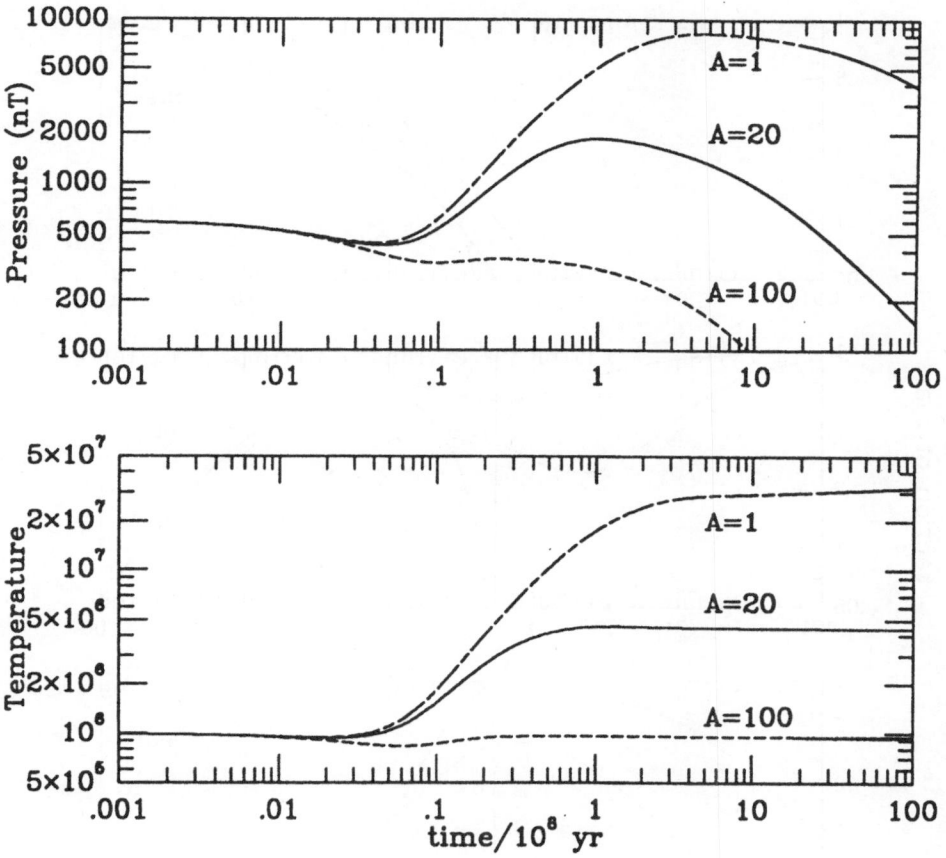

Fig. 10. The same as Fig.9, but the temperature and the pressure of the hot gas are shown.

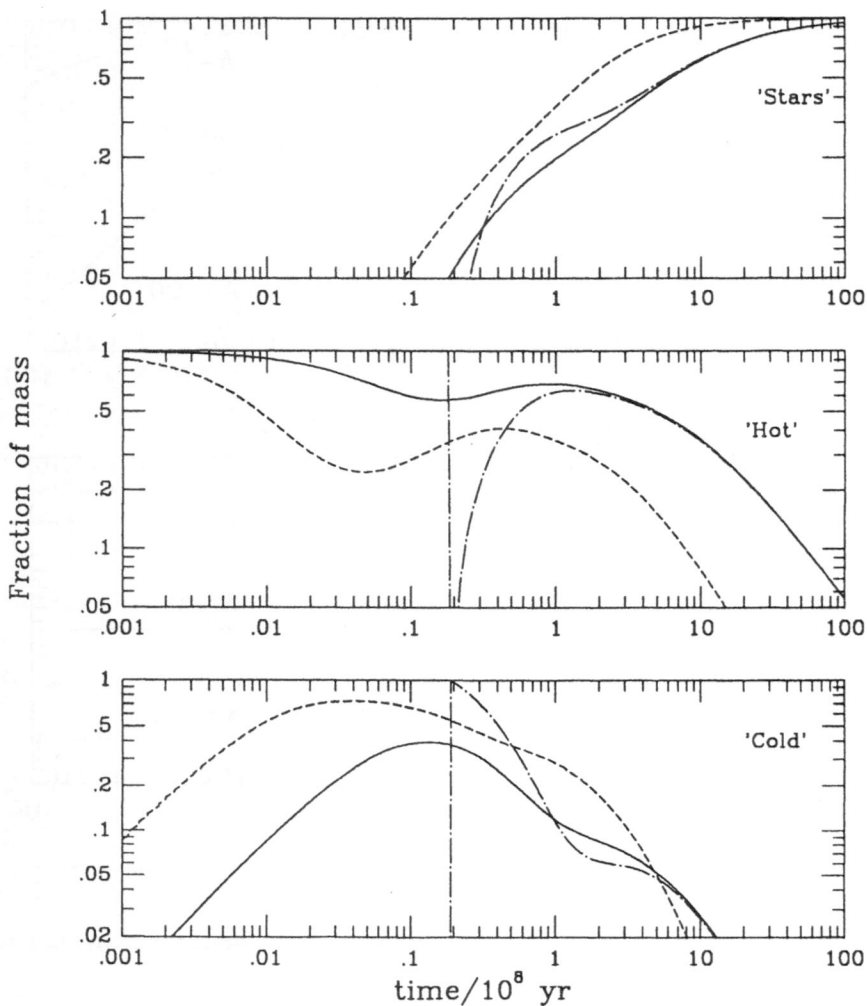

Fig. 11. The dashed curves are the same as Fig.9, but for $A = 20$ and for ten times higher gas density . The full curves are the same as full curves in Fig. 9. The dot-dashed curves present results for initially low temperature, the rest of parameters being as for full curves.

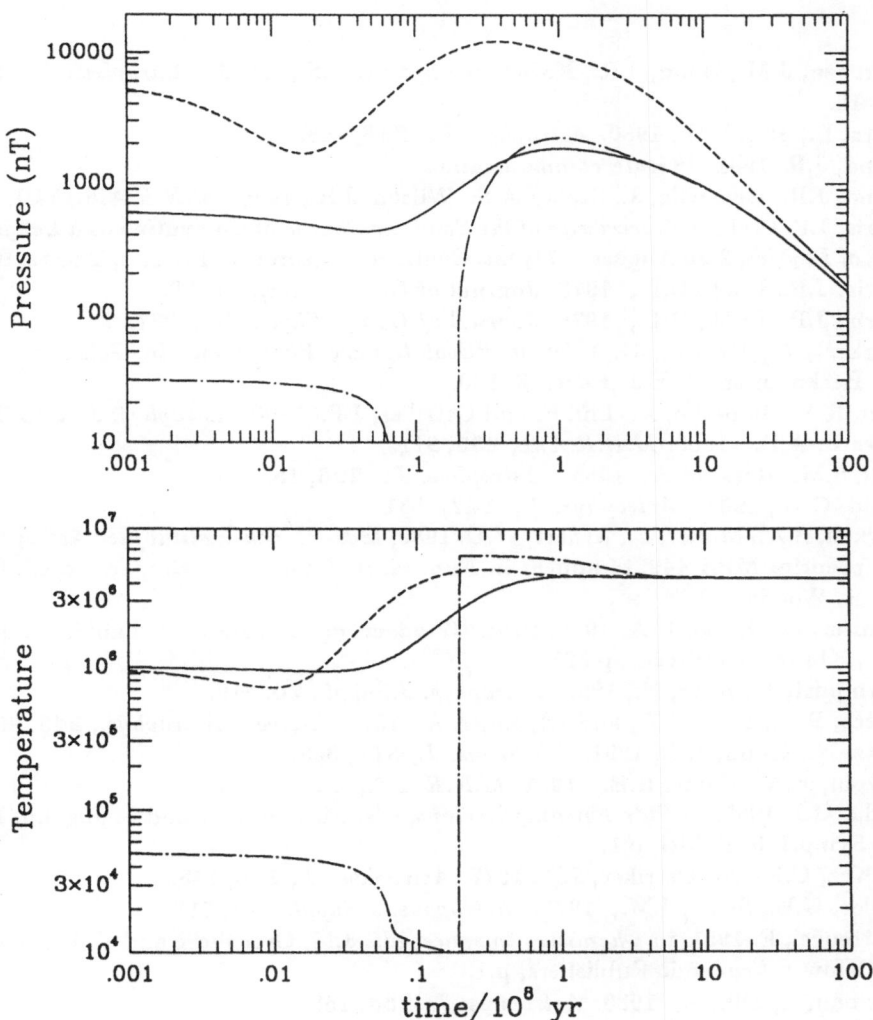

Fig. 12. The same as Fig.11, but the temperature and the pressure of the hot gas are shown.

References

Bardeen, J.M., Bond, J.R., Kaiser, N., Szalay, A.S., 1986. *Astrophys. J.*, **304**, 15.

Blitz, L., Shu, F.H., 1980. *Astrophys. J.*, **238**, 148.

Bond, J.R. 1991. *Private communication.*

Bond, J.R., Centrella, J., Szalay, A.S., Wilson, J.R., 1984. *M.N.R.A.S.*, **210**, 515.

Boris, J.P. 1971, in *Proceedings of the Seminar Course of Computing as a Language of Physics*, 2-20 August 1971, Int. Centre for Theoretical Physics, Trieste, Italy.

Boris, J.P., Book, D.L., 1973. *Journal of Comp. Phys.*, **11**, 38.

Boris, J.P., Book, D.L., 1976. *Journal of Comp. Phys.*, **20**, 397.

Burkert, A., Hensler, G. 1989, in *"Evolutionary Phenomena in Galaxies"*, J.E. Beckman and B.E.J. Pagel, S. 230.

Cen, R.Y. ,Jameson, A., Liu, F. and Ostriker, J.P., 1990. *Astrophys. J.*, **362**, L41.

Evrard, A.E., 1988. *M.N.R.A.S.*, **235**, 911.

Fall, S.M., Rees, M.J., 1985. *Astrophys. J.*, **298**, 18.

Field, G.B., 1965. *Astrophys. J.*, **142**, 531.

Fryxell, B.A., Muller, E., Arnett, W.D. 1989, *Max-Planck-Institut fuer Astrophysic* preprint MPA-449, to appear in *Numerical Methods in Astrophysics*, ed. Paul R. Woodward.

Hensler, G., Burkert, A. 1990, Proc.*"Windows on Galaxies,"* G. Fabbiano, *et al.* , Kluwer, Dordrecht, p.321.

Hernquist, L., Katz, N., 1989. *Astrophys. J.Suppl.*, **70**, 419.

Kates, R.E., Kotok, N., and Klypin, A.A., 1990. *Astron. Astrophys.*, **243**, 295.

Katz, N., Gunn, J.E., 1991. *Astrophys. J.*, **377**, 365.

Klypin, A.A., Kates, R.E., 1991. *M.N.R.A.S.*, **251**, 41p.

Lada, C.J. 1985, in *Star Forming Regions*, eds. M. Peimbert and J. Jugaku, IAU Symp.115, Reidel, p.1.

McKee, C.F. and Ostriker, J.P., 1977. *Astrophys. J.*, **218**, 148.

Miller, G.E., Scalo, J.M., 1979. *Astrophys. J.Suppl.*, **41**, 513.

Matteucci, F. 1991, in *Chemistry in space*, eds. J.M. Greenberg and V. Pirronello, Kluwer Academic Publishers, p.1.

Norman,C., Silk, J., 1980. *Astrophys. J.*, **238**, 158

Oran, E.S., Boris, J.P. 1986, *Numerical Simulation of Reactive Flow*, Elsevier, New York.

Peebles, P.J.E. 1980, *"The large-Scale Structure of the Universe"*, Princeton University Press, Princeton.

Raymond, J.C., Cox, D.P., Smith, B.W., 1976. *Astrophys. J.*, **204**, 290.

Shapiro, P., Struck-Marcell, C., 1985. *Astrophys. J.Suppl.*, **57**, 205.

Shandarin, S.F., 1980. *Astrophysica*, **16**, 439.

Sod, G.A., 1978. *J. Comp. Phys.*, **21**, 1.

Secondary Anisotropies in the CMB

J.L Sanz, E. Martínez-González

Depto. Física Moderna, Univ. Cantabria, Santander, Spain
Instituto de Estudios Avanzados en Física Moderna y Biología
Molecular, CSIC-Univ. de Cantabria, Santander, Spain

Abstract: We review the relevant physical mechanisms that can alter, or even erase, the primary anisotropies in the CMB present at the decoupling surface in the standard scenario. Special emphasis is put on the non-linear gravitational effect which provides the minimal anisotropy in the CMB generated by the present distribution of galaxies.

1 Introduction

For an $\Omega = 1$ universe (linear density perturbations -growing mode- for vanishing pressure) three types of intrinsic primordial anisotropies arise at the recombination time usually called primary anisotropies: a) photon fluctuations on the last scattering surface, b) Doppler effect due to the infall into the gravitational potential wells on the last scattering surface and c) gravitational potential fluctuations at recombination (the Sachs-Wolfe effect)

$$(\frac{\Delta T}{T})_{PA} = \frac{1}{4}\delta_{\gamma r} - \vec{n} \cdot \vec{v}_r + \frac{1}{3}\phi_r, \tag{1}$$

where

$$\vec{v}_r = -\frac{1}{3}(1 + z_r)^{-1/2}\vec{\nabla}\phi_r, \quad \Delta\phi = 6\delta(\vec{x}), \quad \delta = \delta_r(1 + z_r) \ . \tag{2}$$

δ_r is the density fluctuation at recombination. The scale factor is normalized at the present time ($a_o = 1$) and we choose units such that $c = 8\pi G = 1$ and the horizon distance at the present time is $d_{Ho} = 3t_o = 1$.

For an $\Omega < 1$ universe appears a fourth term, related to curvature, as an integrated effect along the geodesic path followed by the photon (Traschen and Eardley 1986, Sanz 1991).

However, one can ask to what extend this primary anisotropies are really preserved from the decoupling to the present time (freely propagation) or they are modified (or erased) due to some physical processes taken place during that period. Here, we shall consider four mechanisms giving extra anisotropy between recombination and the present time, that we will call secondary anisotropy: a) the differential gravitational redshifts and blueshifts of the growing non-linear density fluctuations (producing a time-dependent potential) on the photons propagating to the observer gives an integrated gravitational effect (NGE) (Martínez-González et al. 1990). b) cold gas inhomogeneously distributed between recombination and the present time generates anisotropy associated to non-linear flows (Ostriker and Vishniac 1986). c) hot gas associated to a clumpy distribution of structures (such as clusters of galaxies) also generates secondary anisotropy (Zeldovich and Sunyaev 1969). d) Finally, the fourth mechanism is dust present in the early universe (produced in the process of structure formation), this also produces anisotropy in the CMB, specially in the Wien region (Bond et al. 1991).

2 The non-linear gravitational effect (NGE)

Density fluctuations grow linearly while they are small giving rise to a static gravitational field in comoving coordinates. Their effect on the microwave temperature fluctuations is simply given, except for a factor 1/3, by the difference of the gravitational potential at recombination and at the observer (Sachs-Wolfe effect). The linear regime is the generic evolution of the density fluctuations at high redshifts; at the present time this regime is only followed by the large scale fluctuations, with scales greater than several tens of Megaparsecs. However below these scales the density contrast evolves faster than in the linear regime generating a non-static gravitational potential.

The effect of non-linear density fluctuations on the microwave temperature for a flat universe is given by an integrated effect which, up to a factor 2, represents the work performed by the microwave photons in their trip from the decoupling time to the present against the non-static gravitational potential φ (Martínez-González et al. 1990):

$$\left(\frac{\Delta T}{T}\right)_{NGE} = 2\int_r^o dt \frac{\partial \varphi}{\partial t}(t, \vec{x}) \quad , \quad \nabla^2\varphi = 6a^{-1}\Delta(t, \vec{x}) \quad , \tag{3}$$

The line integral is along the geodesic of the photon, i.e.

$$\vec{x}(t, \vec{n}) = (1 - a^{1/2})\vec{n} \quad , \tag{4}$$

with $\vec{n}$ the unit vector in the direction of observation and Δ represents the density fluctuation. Previous estimates of the effect of an isolated lump (void), represented by a Swiss-cheese model (thin shell approx.), were based on general relativity and obtained only after lengthy calculations (see Dyer 1976, Kaiser 1982, Nottale 1984, Thompson and Vishniac 1987), except for the qualitative approach used by Rees and Sciama (1968). With expresion (3) we are able to obtain $\Delta T/T$ for both, lumps and voids, via a straightforward calculation (see Martínez-González and Sanz 1990).

Below we summarize the effect produced by single structures, cosmic voids and great attractors, present in our local universe. Later on we will study the second order effect due to a statistical distribution of matter and we will give results for the cold dark matter (CDM) model.

2.1 Single Structures

A recent review of the effect on linear and non-linear structures on $\Delta T/T$ is given in Martínez-González (1992). Here we will concentrate on the later case.

Voids of matter in the universe are normally asumed, for simplicity, to be empty spherical regions surrounded by a thin shell of high density matter which compensates the interior vacuum (the total mass fluctuation is zero). The evolution of the void in a flat universe is given by the simple law $R \propto t^{\gamma}$ with $\gamma \simeq 0.13$ either for collisionless gas with adiabatic compression or collisional gas with rapid cooling (Berstchinger 1985). Thompson and Vishniac (1987) were the first to use this model to calculate the effect for several void distributions obtaining tipical values of $< 10^{-6}$. For the largest void seen in the universe up to now, Bootes, of $6000 km/s$ in size the maximum temperature fluctuation is for the photons passing through the center and it only reaches $\Delta T/T(0) \simeq -3 \times 10^{-7}$ (Scaramella 1989; Martínez-González and Sanz 1990). The angular scale of the coldspot amounts to $\simeq 15°$. Therefore we see that the effect of voids is below the expectec limits of sensibility of the present experiments.

Massive concentrations of matter can be, idealistically, represented by a concentric homogeneous internal sphere of matter surrounded by an empty shell (Swiss-cheese model). An estimation of the $\Delta T/T$ profile based on time delay was done by Rees and Sciama (1968), further studies using the framework of general relativity were done by Dyer (1976) and Nottale (1984). We have recently studied the effect of the Great Attractor and the Shapley concentration and the temperature fluctuations produced by these massive structures reach $\sim 10^{-6}$ and $\sim 10^{-5}$ respectively, being the angular scale of the coldspot of $\simeq 30°$ and $20°$ respectively (Martínez-González and Sanz 1990). A more realistic calculation has been carried out by van Kampen and Martínez-González (1991) using n-body simulations finding a similar result for the Great Attractor but asuming a smaller mass for it (the inhomogeneities present in the simulation can give rise to bigger temperature fluctuations than in the homogeneous spherically symmetric case).

For the sheet-like concentration of galaxies found in the CfA redshift survey, known commonly by the Great Wall, Atrio-Barandela and Kashlinsky (1991) have calculated the non-linear effect on $\Delta T/T$ by modelizing it with an oblate spheroid. The result for photons passing through the center is $10^{-6} \leq \Delta T/T \leq (afew) \times 10^{-5}$ depending on the physical characteristics of the spheroid an on Ω.

Summarizing, lumps produce $\Delta T/T$ fluctuations which are bigger than the ones due to voids being the effect in the range $10^{-6} - 10^{-5}$ for the most prominent overdense structures.

2.2 Statistical Distribution of Matter

We now study the temperature fluctuation produced by a non-localized distribution of matter asuming an Einstein-de Sitter universe as background. Perturbation theory up to 2nd order gives the following expression for the rms density fluctuations (Peebles 1980):

$$\Delta(t, \vec{x}) = a\delta + a^2 \left[\frac{5}{7}\delta^2 + \frac{1}{6}\vec{\nabla}\delta \cdot \vec{\nabla}\phi + \frac{1}{126}\phi_{,ij}\phi^{,ij}\right] \quad , \quad \nabla^2\phi = 6\delta \quad (5)$$

where $\delta(\vec{x}) \equiv \delta_r(\vec{x})(1+z_r)$ is the density fluctuation evolved linearly. Therefore, the second order effect, as given by equation (3), amounts to

$$\left(\frac{\Delta T}{T}\right)_{NGE} = 2\int_r^o da\Psi(\vec{x}(a)) \quad ,$$

$$\nabla^2\Psi(\vec{x}) = \frac{\partial}{\partial a}\nabla^2\varphi = 6\frac{\partial}{\partial a}\left(\frac{\Delta}{a}\right) = \frac{30}{7}\delta^2 + \vec{\nabla}\delta \cdot \vec{\nabla}\phi + \frac{1}{21}\phi_{,ij}\phi^{,ij} \quad . \quad (6)$$

On the other hand, the basic function to be calculated for any experimental set-up is the correlation function $C(\alpha, \sigma)$, where α represents the angle on the sky between two antenna beams and σ is the beamwidth associated with a Gaussian response. In the case of interest ($\sigma \ll 1$),

$$C(\alpha, \sigma) = \frac{1}{4\pi}\sum_l (2l + 1)a_l^2 P_l(\cos \alpha) \exp(-\sigma^2(l + \frac{1}{2})^2) \quad , \quad (7)$$

where P_l are the Legendre polynomials and a_l are the multipole components, given by the following expression for the Sachs-Wolfe effect (SW), Doppler effect of the last scatterers (D) and second order gravitational effect (S) (Martínez-González, Sanz and Silk 1992):

$$SW : \quad a_l^2 = \frac{8}{\pi}\int dk k^{-2}P(k)j_l^2(k) \quad , \quad (8)$$

$$D : \quad a_l^2 = \frac{8}{\pi}(1 + z_r)^{-1}\int dk P(k)[j_l'(k)]^2 \quad , \quad (9)$$

$$S: \quad a_l^2 = \frac{32 \times 36}{\pi} \int dk \, k^{-2} P_{(2)}(k) R_l^2(k) \quad . \tag{10}$$

Here, j_l is the Bessel function of fractional order, j_l' is the first derivative, $R_l(k) \equiv \int_0^1 dy (1-y) j_l(ky)$, and $P(k)$ is the power spectrum related to the density fluctuations, i.e. $\langle \delta_k \delta_{k'}^* \rangle = P(k) \delta^3(\vec{k} - \vec{k}')$. Moreover, the function $P_{(2)}(k)$ is the power spectrum associated with the 2nd order density perturbation $\delta_2 \equiv \frac{5}{7} \delta^2 + \frac{1}{6} \vec{\nabla} \delta \cdot \vec{\nabla} \phi + \frac{1}{126} \phi_{,ij} \phi^{,ij}$ and is related to the power spectrum P_{Ψ} of the time derivative of the potential Ψ by $P_{\Psi} = \frac{36}{k^4} P_{(2)}(k)$. The second order perturbation power spectrum $P_{(2)}(k)$ is given in terms of the first order power spectrum $P(k)$ by the equation (Goroff et al 1986, Suto and Sasaki 1991)

$$P_{(2)}(k) =$$

$$\frac{k^3}{98(2\pi)^2} \int_0^\infty dr P(kr) \int_{-1}^1 dx P(k(r^2 + 1 - 2rx)^{1/2}) \left(\frac{3r + 7x - 10rx^2}{r^2 + 1 - 2rx} \right)^2 \quad . \tag{11}$$

It is interesting to study the behaviour of the second order power spectrum for large scales, $P_{(2)}(k) \propto k^4$. This result, independent of the linear power spectrum, is a result of a cancelation of the three terms contributing to the second order density fluctuation and implies very litle power on large scales. Due to this cancelation the whole second order effect is one or two orders of magnitud smaller than the effect of each of the three terms considered separatly! (Sanz and Martínez-González 1991).

We have applied the previous formalism to calculate the predicted amplitudes of the $\delta T/T$ correlation function. The power spectrum of matter density perturbations was taken to be that of the CDM model with $h = 0.5$ and $\Omega = 1$ (we use the fit given in Bardeen et al. 1986). The spectrum was normalized in the standard fashion to yield mass fluctuations of order unity within a sphere of radius $800\,\mathrm{km\,s^{-1}}$. The results of our calculations are presented in figures 1 and 2.

In figure 1, we plot the square root of the temperature correlation function $\sqrt{|C(\alpha, \sigma)|}$ for a beamwidth of Gaussian dispersion $\sigma = 3°$, appropriate to the beam used in the COBE experiment (Smoot et al. 1991). The second order gravitational effect (solid line) attains an amplitude $\delta T/T \sim 10^{-7}$ at zero lag. The dashed line shows the Sachs-Wolfe effect for the same spectrum. The nonlinear correlation posses a less remarked quadrupole–like behaviour than the classical Sachs-Wolfe case. We show in figure 2 the predicted $\delta T/T(\alpha) = \sqrt{2[C(0, 3°) - C(\alpha, 3°)]}$ for single subtraction experiments as a function of beam-throw α. The second order gravitational effect (solid line) reaches 10^{-7} on almost all angular scales.

It is of interest to point out that had we included only one of the terms in the second order perturbation expression (5) (δ_2 term), we would have overestimated $\delta T/T$ by some two orders of magnitude. This is due to the significant cancellation of the three terms contributing to δ_2 at large scales

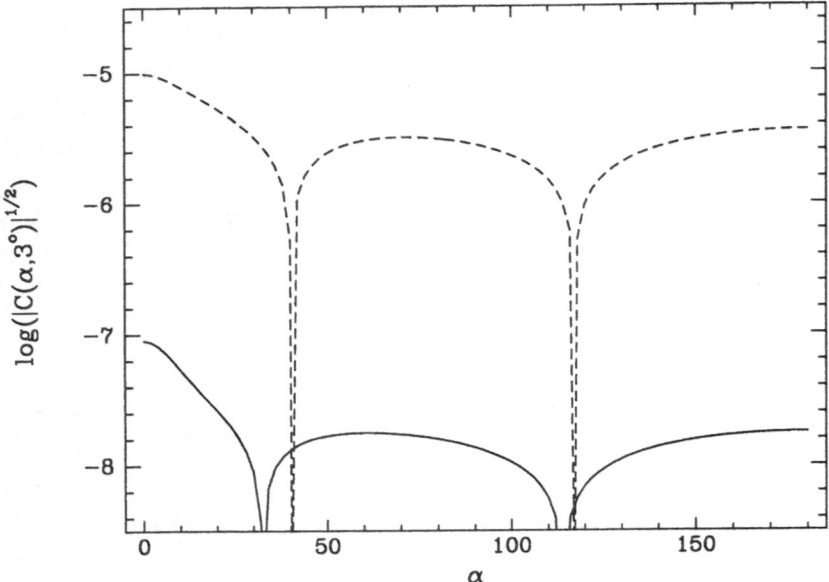

Fig. 1. $\sqrt{|C(\alpha, 3°)|}$ for the nonlinear gravitational effect (solid line) and for the Sachs-Wolfe effect (dashed line) versus throw angle α. Notice that the correlation function becomes anticorrelated at about 32° for the former case and 40° for the latter case.

as explained before. Also we wish to emphasize that about 90% of the anisotropy produced by the second order gravitational effect is generated at a redshift $z < 20$ and therefore can not be affected by any reionization of the medium taking place during the history of the universe.

Finally, for a hot dark matter (HDM) model Anninos et al. (1991) have obtained the linear and non-linear temperature anisotropy at intermediate angular scales using n-body simulations and including also the effect of Thompson scattering. These authors find that the non-linear effects do not alter the temperature maps significantly and that the amplitudes for the fluctuations are $\Delta T/T \simeq 10^{-5}$, consistent with present observational limits.

2.3 Minimal Anisotropies

From the results of the last section we may conclude that, up to the second order perturbation theory, the NGE is negligible compared to the linear SWE. However, there is a kind of scenarios where the last scale density perturbations are seeded after the epoch of matter-radiation decoupling by a late-time cosmological phase transition (LTPT) (Hill et al. 1989) and for which the situation can be quite different. In fact, Martínez-González et al.

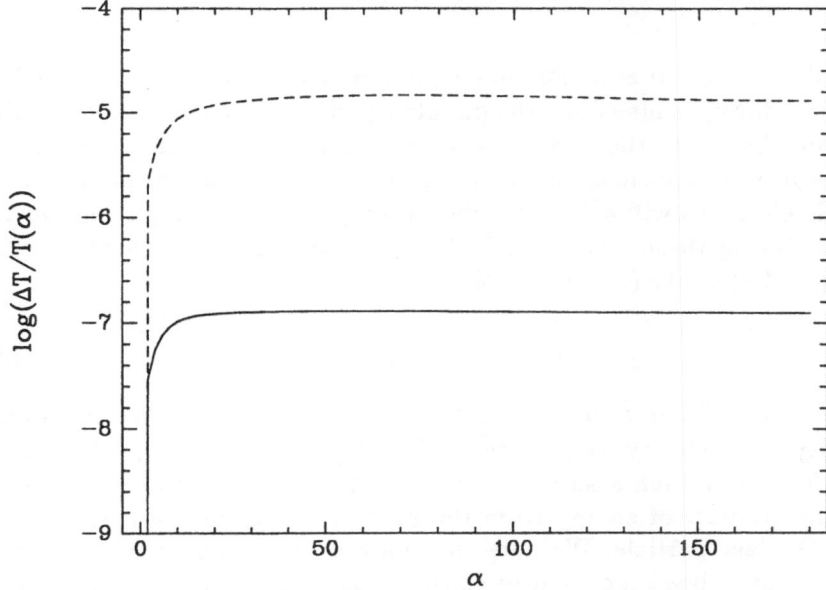

Fig. 2. Rms $\delta T/T$ predictions for single subtraction experiment anisotropy measurements as functions of beam-throw α. The lines correspond to the same cases as figure 1.

(1992) have shown that one can choose $\phi(t_i, \vec{x}) = 0$ as the initial condition for the potential at the phase transition time ($z_i < 1000$), implying a zero Sachs-Wolfe and Doppler effects at the surface of generation of the density perturbation seeds. Additionally, the gravitational potential at later times is of the retarded time and thus each point is gravitationally influenced by only that region which is inside its light cone. This result is to be contrasted with the standard scenario where each point is gravitationally influenced by the rest of the universe, even those points which are outside its particle horizon (the potential is instantaneous!). This fact makes the LTPT models specially attractive for the generation of temperature fluctuations in the CMB. Therefore in LTPT models the NGE becomes the dominant gravitational effect on the microwave photons. Moreover, because this effect is not affected by any reionization process the amplitude of the temperature fluctuations resulting from the present distribution of galaxies, through the NGE, constitutes the minimal anisotropy on the CMB. In particular, from the power spectrum derived from an analysis of the QDOT IRAS galaxy redshift survey (Kaiser 1991), the minimal anisotropy of the CMB amounts to $\Delta T/T \sim 10^{-6}$ (Martínez-González et al. 1992).

3 The Ostriker-Vishniac effect (O-V)

The standard scenario assumes that recombination take place at $z_r \sim 10^3$ but one can think, as plausible, the possibility of an early reionization of the universe maybe due to the imput of energy associated to the first generation of galaxies or more exotic phenomena. In the case of a cold plasma, the non-relativistic electrons will affect the photon propagation through Thompson scattering during these "dark ages". A photon traversing the medium will lose energy at the rate (Barbosa 1982)

$$\frac{dE}{dt} = -n_e < \sigma E >= -\sigma_T n_e (1 + \vec{n} \cdot \vec{v} + v^2) E \ , \tag{3}$$

where σ_T is the Thompson cross-section and n_e, $\vec{v}$ are the electron number density and velocity, respectively, E is the comoving energy (i.e. $E = h\nu \, a(t)$). This approach assumes for any photon a single scattering and it ignores the transfer of energy from the photon to the electron in the rest frame of the last particle. We note that only up to second order terms in the velocity have been included in eq.(3). Taking into account that $E \propto \nu$ and the relation $\nu_o / \bar{\nu}_o - 1 = \Delta T / T$ because $T_o \nu_r = T_r \nu_o$ when there is conservation of photon number, the integration of eq.(3) from recombination to the present time leads to

$$\frac{\Delta T}{T} = -\sigma_T \int_r^o dt \, \bar{n}_e (\delta + \vec{n} \cdot \vec{v} + v^2 + \delta \vec{n} \cdot \vec{v}) \ , \tag{4}$$

where $\bar{n}_e$ and δ represent the mean number density and the density fluctuation, respectively. The first and the second terms in eq.(4), at the linear level, usually cancel the primary anisotropies, e.g. Efstathiou (1988) has obtained that for baryonic models $\Delta T / T \sim 10^{-6}$ with a wide range of parameters. Ostriker and Vishniac (1986) and Vishniac (1987) have shown that the second order term due to the coupling of the bulk flow of the electrons with their density fluctuation generates a secondary anisotropy that is relevant for standard models. For CDM ($\Omega = 1, \Omega_b = 0.1, h = 0.5$) and isocurvature perturbations, the O-V effect appears relevant for scales of $\sim 1'$ (we note that for the Owens Valley experiment the effect is slightly increased over models without reionization), whereas For HDM ($\Omega = 1, \Omega_b = 0.1, h = 0.5$) the level of fluctuation is of the same order as in the standard recombination case (Vishniac 1987). Bond and Efstathiou (1987) and Efstathiou (1988) have explored the parameter space (Ω, n') for isocurvature baryon models, assuming that the primordial spectrum of entropy fluctuations is $P_s(k) \propto k^{n'}$. In order to satisfy the Owens Valley upper-limit $\Delta T / T (7'.2, 46''.8) \leq 2.1 \times 10^{-5} (95\% c.)$ (Readhead et al. 1989) either $n' > -1, \Omega < 0.6$ or $n' < -0.9, \Omega > 0.6$.

The lesson to be learned from this appealing to an early epoch of reionization is that primary anisotropies are erased but secondary ones appear at the same level for most popular models of galaxy formation, but the problem is not closed because in eq.(4) appear other terms related to second order density fluctuations that must be taken into account for the full effect.

The previous formulae can also be applied to an isolate structure (i.e. a cluster of galaxies), then we obtain an anisotropy with respect to any direction pointing to the background given by

$$\frac{\Delta T}{T} = -\tau_c \vec{n} \cdot \vec{v}, \quad \tau_c = \sigma_T n_e D \ , \tag{5}$$

where τ_c, D and $\vec{v}$ are the optical depth, diameter and peculiar velocity of the cluster, respectively (Sunyaev and Zel'dovich 1980).

4 The Sunyaev-Zeldovich effect (S-Z)

The difussion of microwave photons by hot gas (e.g. associated to clusters of galaxies) leads to another secondary anisotropy. Invers Compton scattering of the CMB photons by non-relativistic electrons generates a spectral distortion of the CMB. In this case, if one takes into account multiple scattering and the possibility of induced and spontaneous emission, the basic equation for the evolution of the photon occupation number n is (Kompanneets 1956)

$$\frac{\partial n}{\partial t} = \sigma_T n_e x^{-2} \frac{\partial}{\partial x} x^4 [\frac{1}{3} < v^2 > \frac{\partial n}{\partial x} + \frac{kT}{m_e} n(1+n)] \ , \tag{6}$$

where $x = h\nu_o/kT_o$, T_o and T are the temperature of the CMB at the present time and at time t, respectively. n_e is the electron number density and $< v^2 >= 3kT_e/m_e$ for a Maxwellian distribution for the electrons. From the previous equation and the relations $I \propto n x^3$ (I is the blacbody intensity), $\Delta T/T = (1 - e^{-x})\Delta I/I$, one obtains (Zel'dovich and Sunyaev 1969)

$$\frac{\Delta T}{T} = -2y[2 - \frac{x/2}{tanh(x/2)}], \quad y = \sigma_T \int_r^o dt\, n_e(\frac{1}{3} < v^2 > -\frac{kT}{m_e}), \tag{7}$$

where y is the Thompson parameter. The effect appears as a decrement ($\Delta T/T \simeq -2y$) in the R-J region of the spectrum changing to $\Delta T/T > 0$ at $\lambda 1.25mm$.

Recently, Cen et al. (1990) have considered an hydrodynamic scheme to study the intergalactic medium in an universe dominated by cold dark matter. They obtain, in the R-J part of the spectrum, fluctuations of 5.6×10^{-7} on arcminute scales and a mean $< y >$ parameter of 5.5×10^{-7} due to the S-Z effect.

On the other hand, for an isolate cluster ($kT_e \sim 10keV$) the y parameter is given by

$$y_c = \frac{kT_e}{m_e}\tau_c \quad , \tag{8}$$

where τ_c is given by eq.(5). The typical value expected for some clusters is $\Delta T/T \simeq -4 \times 10^{-4}$, which agrees with the order of magnitude obtained from direct observations at Owens Valley, operating at $\lambda 1.5\,cm$. At least for three clusters of galaxies (0016+16, A665 and A2218) there is unequivocal detection of the S-Z effect at the levels $(\Delta T/T)_{R-J} \simeq -(5.1, 2.6, 2.6) \times 10^{-4}$, respectively (Birkinshaw and Gull 1984, Birkinshaw 1989). We comment that the existence of cooling flows in the central parts of the cluster can modify through the free-free emission process the decrement in the R-J region whereas the contribution of it in the Wien region is smaller (Schlickeiser 1991). Of course, some measurements in the Wien region are needed to stablish without ambiguity the S-Z effect. The effect of hot gas in an early-type galaxy on the CMB has been estimated by Trester and Canizares (1989) to be $(\Delta T/T)_{R-J} < 3 \times 10^{-5}$ for the fractional attenuation at the center.

Moreover, some authors have estimated the anisotropy generated by a statistical distribution of structures. Schaeffer and Silk (1988) and Cole and Kaiser (1988) have considered clusters of galaxies in a universe dominated by dark matter which confine the hot gas, obtaining $\Delta T/T \simeq 10^{-5}$ on arcmin scales for the S-Z effect. Finally, Trester and Canizares (1989) have computed the same effect due to hot gas contained in a population of early-type galaxies out to some redshift, their result is $\Delta T/T < 3 \times 10^{-8}$ on sub-arcmin scales in the R-J region.

5 The Effect of Dust

The presence of dust in the universe related to the process of formation of bound structure generate spectral distortions and anisotropies on the CMB (specially in the Wien region) due to inhomogeneities in the dust density. Bond et al. (1986) have estimated $\Delta T/T \sim 10^{-5}$ on arcmin scales at 400μ when the dust is distributed like galaxies for current galaxy-formation scenarios (CDM), and this rises up to $\sim 10^{-4}$ around the peak of the CMB. More recently, the same authors (Bond et al. 1991) have explored the parameter space for different dust models, obtaining that the Owens Valley upper-limit $\Delta T/T(7'.2, 46''.8) \leq 2.1 \times 10^{-5}(95\%c.)$ (Readhead et al. 1989) is not violated, whereas for the COBE experiments (DIRBE, FIRAS) the theoretical prediction is in the range $\Delta T/T = 10^{-7} \div 8 \times 10^{-3}$ representing a clear challenge for models.

References

Anninos, P., Matzner, R.A., Tuluie, R. and Centrella, J. 1991, *Ap. J.*, **382**, 71.

Atrio-Barandela, F. and Kashlinski, A. 1991, *preprint.*

Barbosa, D.D. 1982, *Ap. J.*, **254**, 301.

Bardeen, J.M., Bond, J.R., Kaiser, N. and Szalay, A.S. 1986, *Ap. J.*, **304**, 15.

Berstchinger, E. 1985, *Ap. J. Supplement Series*, **58**, 1.

Birkinshaw, M. 1990, "Measurement of the Suyaev-Zel'dovich effect" in Physical Cosmology, ed. Tran Thanh Van, J., Frontieres.

Birkinshaw, M. and Gull, S.F. 1984, *Nature*, **309**, 34.

Bond, J.R., Carr, B.J. and Hogan, C.J. 1986, *Ap. J.*, **306**, 428.

Bond, J.R., Carr, B.J. and Hogan, C.J. 1991, *Ap. J.*, **367**, 420.

Bond, J.R. and Efstathiou, G. 1987, *Mon. Not. R. ast. Soc.*, **226**, 655.

Cen, R.Y., Jamesom, A., Liu, F. and Ostriker, J.P. 1990, *Ap. J. Lett.*, **362**, L41.

Cole, S. and Kaiser, N. 1988, *Mon. Not. R. ast. Soc.*, **233** 637.

Dyer, C.C. 1976, *Mon. Not. R. ast. Soc.*, **175**, 429.

Efstathiou, G. 1989, Proc. of the Vatican Conference "Large Scale Motions in the Universe".

Hill, C., Schramm, D.N. and Fry, J.N. 1989, *Comments on Nuclear and Particle Physics*, **19**, 25.

Kaiser, N. 1982, *Mon. Not. R. ast. Soc.*, **198**, 1033.

Kaiser, N. 1991, A. I. P. Conference Proceedings 222: *After the first three minutes.* Eds. S. S. Holt, C. L. Bennett and V. Trimble.

van Kampen, E. and Martínez-González, E. 1991, Proc. of the IInd Rencontres de Blois "Physical Cosmology". Ed. A. Blanchard et al. Pergamon Press, p. 522.

Kompaneets, A.S. 1956, *Zh. Eksp. Teor. Fiz.*, **31**, 876 [*Sov. Phys. JETP*, bf 4, 730 (1957)].

Martínez-González, E. 1992, Proc. of the NATO ASI "Infrared and Submillimeter Sky after COBE", Les Houches, Kluwer (in press).

Martínez-González, E. & Sanz, J.L. 1990, *Mon. Not. R. ast. Soc.*, **247**, 473.

Martínez-González, E., Sanz, J.L. and Silk, J. 1990, *Ap. J. Lett.***355**, L5.

Martínez-González, E., Sanz, J.L. and Silk, J. 1992, submitted to *Phys. Rev. Lett.*

Nottale, L. 1984, *Mon. Not. R. ast. Soc.*, **206**, 713.

Ostriker, J.P. and Vishniac, E.T. 1986, *Ap. J.*, **306**, L51.

Peebles, P.J.E. *The Large Scale Structure of the Universe*, (Princeton, Princeton University Press, 1980).

Goroff, M.H., Grinstein, B., Rey, S.-J. and Wise, M.B. 1986, *Ap. J.*, **311**, 6.

Suto, Y. and Sasaki, M. 1991, *Phys. Rev. Lett.*, **66**, 265.

Smoot, G.F. et al. 1991, *Ap. J. Lett.*, **371**, L1.

Readhead, A.C.S., Lawrence, C.R., Myers, S.T., Sargent, W.L.W., Hardebeck, H.E. and Moffet, A.T. 1989, *Ap. J.*, bf 346, 566.

Rees, M.J. & Sciama, D.W. 1968, *Nature*, **517**, 611.

Sachs, R.K. & Wolfe, A.M. 1967, *Ap. J.*, **147**, 73.

Sanz, J.L. 1991, Proc. of the 2nd. I.A.C. Winter School "Observational and Physical Cosmology", Tenerife. Ed. F. Sánchez et al., p. 145.

Sanz, J.L. and Martínez-González, E. 1991, Proc. of the NATO ASI "Observational Tests of Cosmological Inflation", Durham. Ed. T. Shanks et al., p. 47.

Scaramella, R. 1989, *Ap. J.*, **346**, 607.

Schaeffer, R. and Silk, J. (1988), *Ap. J.*, **333**, 509.

Schlickeiser, R. 1991, *Astrom. Astrophys.*, **248**, L23.

Sunyaev, R. and Zel'dovich, Ya. 1980, *Mon. Not. R. ast. Soc.*, **190**, 413.

Thompson, K.L. & Vishniac, E.T. 1987, *Ap. J.*, **313**, 517.

Traschen, J. and Eardley, D.M. 1986, *Phys. Rev.*, **34**, 1665.

Trester, J.J. and Canizares, C.R. 1989, *Ap. J.*, bf 347, 605.

Vishniac, E.T. 1987, *Ap. J.*, **322**, 597.

Zel'dovich, Ya.B. and Sunyaev, R. 1969, *Astrophys. and Space Sci.*, 4, 301.

The Relikt Missions:
Results and Prospects to Detect
the Microwave Background Anisotropy

A.A. Klypin [1] [2], I.A. Strukov [3], D.P. Skulachev [3]

[1]Canadian Institute for Theoretical Astrophysics, Toronto M5S 1A1, Canada.

[2]Astro-Space Center, Lebedev Physical Institute, Profsojuznaja 84/32, 117810 Moscow, USSR.

[3]Space Research Institute, Profsojuznaja 84/32, 117810 Moscow, USSR.

Abstract: We review the soviet space program "Relikt" to measure anisotropies in the cosmic microwave background (CMB) at mm-wavelengths. The first experiment - "Relikt-1" - was successively carried out in 1983-84 yrs. It produced a radio brightness map of the sky at 8 mm with $5°.8$ angular resolution. The map was used to get the dipole component of the CMB, to measure flux from galactic plane and to set constraints on large-scale angular fluctuations. For the Harrison-Zeldovich spectrum of primordial fluctuations the upper limit on the quadrupole component of the spectrum of angular fluctuations was found to be $(\Delta T/T)_{quad} < 1.6 \cdot 10^{-5}$ at 95 per cent confidence level. The model-independent upper limit on the quadrupole was $(\Delta T/T)_{quad} < 3.0 \cdot 10^{-5}$. The angular correlation function of fluctuations was constrained at the level: $\langle \Delta T \cdot \Delta T(\theta) \rangle \leq 5 \cdot 10^{-9}$ for angles θ from 20° to 160°. Results of the "Relikt-1" experiment (Klypin et al. 1987) are close to that of the COBE differential microwave radiometer (DMR) (Smoot et al. 1991). In mid 1993 another soviet satellite, carrying the "Relikt-2" experiment will be launched. The "Relikt-2" will have much better sensitivity and more channels than the COBE DMR. The main strategy of "Relikt-2" to large extend will follow that of "Relikt-1". The galactic emission (synchrotron, thermal and dust emissions) probably will be the main obstacle to detect fluctuations of the CMB. Here we discuss expected level of this emission and methods to remove this contamination.

1 Introduction

In the years since the discovery of the microwave background radiation (Penzias & Wilson 1965) many experiments have been carried out to detect any deviations from isotropic planckian radiation coming from outer space. In spite of many efforts, the only found deviations are the dipole anisotropy due to the motion of the observer relative to the CMB (Henry 1971, Fixsen *et al.* 1983, Lubin *et al.* 1985) and fluctuations of the CMB in the direction of rich galaxy clusters due to the Sunyaev-Zeldovich effect (Sunyaev & Zeldovich 1970, Birkinsaw *et al.* 1984). The lack of any observed cosmological anisotropy on scales from seconds of arc to 90 degrees and stringent upper limits on the fluctuations essncially restrict the choice of possible cosmological models. In other words, even without having detected cosmological fluctuations, experiments on $\Delta T/T$ measurements played a very important role in contemporary cosmology. Measurements of large-scale anisotropies have its advantages and disadvantages when compared with intermediate (degrees) or small (arcminutes) angular scale experiments. Because atmosphere effects are not so severe at small and intermediate angular scales, it is possible to carry out these experiments using onground radio-telescopes or balloons. At large scales it becomes very difficult to get rid of atmosphere fluctuations. Instead, one can use rather small and easy-to-operate horns, not dishes for large-scale experiments. As a result, equipment for large-scale mm-wavelength measurements is small, compact and light. This is the reason why there is a tendency to use satellites for these experiments (i.e Relikt-1, COBE and Relikt-2). From the theoretical point of view the large-scale experiments are very important because we *must* have large-scale $\Delta T/T$ fluctuations. It is possible to suggest that there was a secondary ionization of the Universe, which essentially reduced fluctuations on scales less than, say, one degree. But this does not affect large-scales: we know that the Universe is transparent at present and that the only mechanism which matters is the gravitation itself (the Sach-Wolfe effect). Another important feature of large-scale measurements is that by measuring these fluctuations we are probing very long-scale density fluctuations. The fluctuations (scales about the size of present horizon) were formed during the inflation and were not changed by complicated processes at recombination or at the moment when dark matter started to dominate in the Universe. So, by measuring large-scale angular anisotropies we confront with very fundamental ideas about the Universe as a whole.

The inflation theory in its simplest and most attractive version predicts specific spectrum of density fluctuations, so called the Harrison-Zeldovich spectrum $(\delta\rho/\rho)^2_k \propto k^1$. The spectrum results in $\Delta T/T$ spectrum of angular fluctuations of the form $(\Delta T/T)^2_{lm} \propto 1/l(l+1)$. The knowledge of shape of the spectrum can be used in two ways. First, if no cosmological fluctuations are detected (as it it is now), it helps to get more stringent constraints on possible fluctuations because one can test many harmonics in data analysis, not only the quadrupole, say. Second, if cosmological fluctuations were detected, we would be able to compare its angular spectrum with that predicted by the inflation theory.

With rising sensitivity of receivers, next and very difficult problem comes – contamination due to galactic sources. Because the effective temperature of these sources is different at different wavelengths and the CMB temperature does not depend on frequency, one can hope to subtract or reduce the contamination by carrying out multifrequency experiments. Unfortunately, at present the number of channels and the sensitivity is not enough to do this subtraction in a model-independent way. In this paper we use results low-frequency measurements of Reichs (1982, 1986, 1988) and Haslam (1982) to make predictions of the level of galactic contamination (mainly due to synchrotron emission). These predictions, though not reliable enough to be used for data analysis, show the high level of expected contamination. We also use the predictions as models for testing different schemes of data reduction.

2 "Relikt-1": Instrumentation

Detailed description of the "Relikt-1" instrumentation was given by Strukov and Skulachev (1988). Here we present some basic parameters of the system. The instrument was designed using modulation scheme with superheterodyne receiver and microwave amplifier. The system operated at frequency $f_0 = 37\text{GHz}$ (wavelength 8 mm) with bandwidth $\Delta f = 0.4\text{GHz}$. The RMS noise for an integration time 1 second was 31mK, which corresponds to system noise temperature 300 K. The sampling of the signal was made twice in a second, but for subsequent data analysis every two readings were averaged.

The noise generated by system was correlated. In other words, it had some rather complicated spectrum shown at Fig. 1. When presenting the plot, we suppose that the noise was periodic with the period 120 seconds equal to one revolution of the satellite around its axis. So, $l = 1$ corresponds to the first harmonic on a big circle on the sky. One can think about the spectrum as if it were the spectrum of $\Delta T/T$ which gave the same measured spectrum as the system noise. Two points should be mentioned in relation with the system noise. First, because the noise signal of two nearest (0.5 sec) readings was highly correlated, averaging over them does not reduce the noise too much. So, the RMS noise of 1 sec readings was actually 38mK and more distant readings were slightly correlated, which means that the effective number of independent points was reduced by ≈ 20 per cent. Second, the part of the system, which suppressed the short-scale part of the system noise also and exactly in the same way suppressed any received signal. So, Fig. 1 also presents system transfer function of a signal. This function can be considered as additional smoothing of a signal and it is comparable to the antenna smoothing presented at the Fig. 1

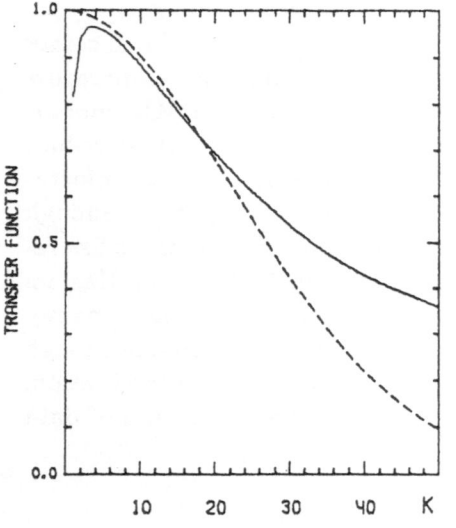

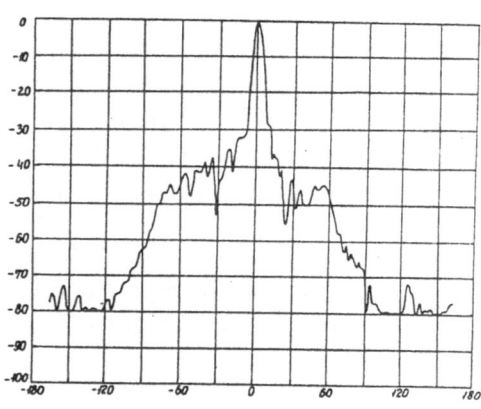

Fig. 1. The transfer function of the receiver (solid line) and the measuring antenna (dashed line).

Fig. 2. The antenna directivity (db) of the Relikt-1 experiment.

The system measured the difference between temperatures of two antennae. One of the antennae, called the reference antenna, had its optical axis parallel to the spin axis of the spacecraft, which was directed to antisolar direction on the sky. Another (measuring) antenna was directed 90° from the spin axis. Fig. 2 gives antennae derictivities. The beamwidth of the measuring antenna was 5.°8 (FWHM). A parabolic mirror was used to deflect the measuring antenna beam by 90° from the optical axis of the reference antenna. As a result, configuration of measuring antenna was changed from simple corrugated horn. This lead to high side lobes at angles 30–60 degrees at the level 10^{-4}.

In order to eliminate the contamination of data by the Moon and the Earth emission a threshold on the *expected* signal due to this sources was set: a measurement was rejected if *expected* contaminating signal was high than 0.5mK. When estimating the signal, both antennae were considered with side lobes for measuring antenna being assumed to be $\approx 10^{-4}$ for angles in the range 20–60 degrees. As a result, all observational points for which angular distance to the Moon or the Earth were smaller than 30–40 degrees were lost (actual angular distance depends on the source and on the distance to the source). In total, almost half of the measurements were lost. *Actual* level of contamination due to the Moon and the Earth emission was probably smaller than *expected* one: the level of *measured* noise was very close to that estimated in pre-flight laboratory tests. It also must be noted that antenna side lobes did reduce the efficiency of the experiment, but it does NOT reduces the reliability of the results because the upper limits on possible anisotropy of $\Delta T/T$ found by "Relikt-1" include any signal, no matter whether it is cosmological signal or the contamination. The situation would be different if the anysotropy were detected. But because no signal was detected, this "No" means no cosmological signal except the dipole and no remaining contamination.

The radiometer and the internal noise source were calibrated before the flight with the accuracy 5%. Emission of the Moon was used for the callibration during the flight. Because of large number of measurements ($6 \cdot 10^5$ every week), it was possible to estimated the level of the apparature noise during the flight with very high accuracy. The comparison of the levels found for in a week intervals gives relative calibration of the instrument. The level of week-to-week fluctuations of the apparature noise was 4 per cent. The receiver also was calibrated every four days using internal noise generator.

Because of possible drift of the null-point, an automatic balancer was added to the system, which kept the average signal output at zero level. The balancer had time constant ≈ 30 sec. If T_{meas} and T_{ref} are temperatures of measuring and reference antennae, then the value at the output of the system was $(T_{\mathrm{meas}} - T_{\mathrm{ref}}) - \langle T_{\mathrm{meas}} - T_{\mathrm{ref}} \rangle$, where averaging was made over rotational period. Because the reference antenna was pointed at the same direction during a week of observations, $T_{\mathrm{ref}} = \langle T_{\mathrm{ref}} \rangle$. So, measured value was $T_{\mathrm{meas}} - \langle T_{\mathrm{meas}} \rangle$ and no information on the temperature of the reference antenna was actually recorded. Fig. 3 demonstrates the reaction of the system (antennae not included) to different signals: a dipole, a quadrupole and a point source. The amplitude of a point source was reduced by a factor of two because of smoothing. Note small negative tail with maximum negative value about 1/20 of the input amplitude of the source.

3 "Relikt-1": observations

The instrument was carried on board the high apogee satellite "Prognoz-9", which was launched 1 July 1983. The experiment ceased operating in February 1984. The satellite was put on highly eccentric orbit with the perigee ≈ 1000 km and apogee $\approx 750,000$ km. The orbital period was 26 days and the rotational period was 120 seconds. The axis of rotation was directed towards the Sun and the reference antenna was always pointed in the antisolar direction. The optical axis of the measuring antenna was set to 90° from the rotation axis of the satellite. Thus, one revolution of the spacecraft produced a set of measurements along a big circle on the sky. A measurement was removed from the subsequent data analysis if the expected

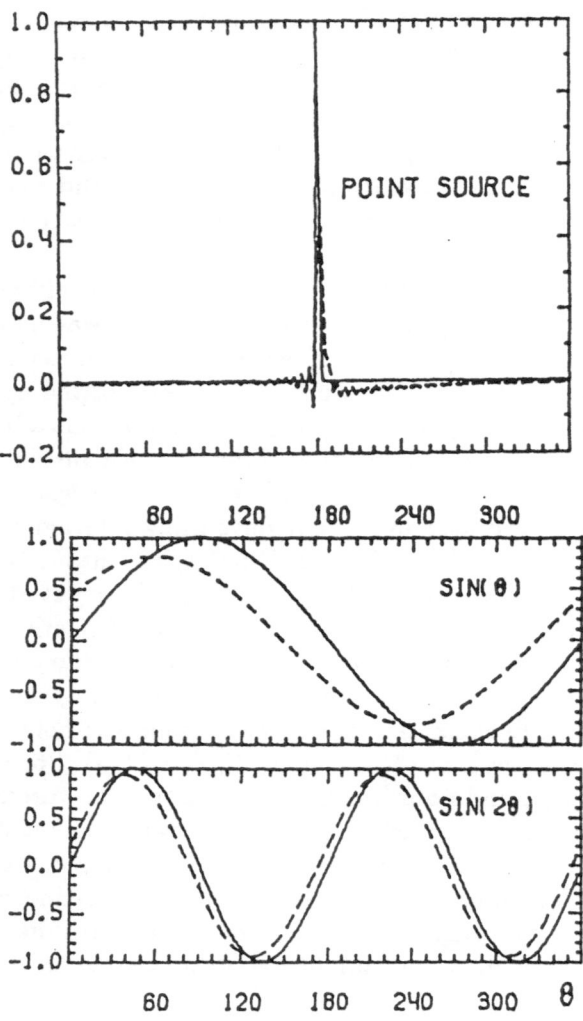

Fig. 3. The response (dashed) of the receiver to input signals of different kinds (full curves).

level of the contamination due to the Earth or the Moon was higher than 0.5 mK. Spikes above 3σ level were also eliminated.

The axis of rotation was kept unchanged for a week. This set of measurements, which refers to one big circle on the sky, is called "scan". Each 3° pixel of the big circle was observed 5040 times. The experimental data were stored and once in four days transmitted to the Earth. After a week of observations engines of the satellite were used to redirect its axis. Half a year of operation of the system gave 31 scans of data, mapping all the sky. Because the axis of satellite rotation lay in the plane of ecliptics, all scans intersected at the ecliptical poles. Scans for which the mean number of measurements per 3° pixel was too low because of contamination (less than 3000 seconds of "noncontaminated" observations) were removed leaving for data analysis 22 scans.

4 "Relikt-1": data analysis and results

4.1 Data reduction and matching scans

In order to analyse data on different scans, one need to match them and make corrections for systematic effects. In case of the "Relikt-1" experiment this matching procedure and data reduction were as follows.

– Readings of all scans were normalized to the same system gain. Let $\langle \sigma_i \rangle$ be the mean measured noise for i-th scan, which is proportional to the system gain. Then in order to correct temperatures for the variation of the gain, all measurements of the scan were multiplied by $\langle \sigma_1 \rangle / \langle \sigma_i \rangle$, which scales all measurements to the gain of the first scan.

– The artificial signal produced by the system when it crossed the galactic plane was removed. Fig. 3 presents the response of the system to a compact source. Knowing parameters of the system and having observed the emission of the galaxy, we estimated the artificial signal and subtracted it from pixels lying out off the galactic plane. The correction was small. The maximum observed signal in the plane was about 5 mK and the maximum of the 'false' signal was $-1/20$ of the input signal, so the correction was less than 0.25 mK.

– The galactic plane is a strong source at 8mm. Measured temperatures and the discussion of the subject is given by (Strukov & Skulachev 1987). For the analysis of the cosmological large-scale fluctuations this is another source of the contamination. To eliminate this contamination and to remove some fraction of systematic errors related with system response to the galaxy as discussed above, a belt of $\pm 15°$ width around the galactic plane was removed from the data set.

– The mean value of the measured temperature on each scan $\langle T_j \rangle_{i-\text{th scan}}$ was set to zero.

This technique is reasonable because the only observed signals on the sky are the galactic plane and the dipole. There is an independent test for this matching procedure. All data scans intersect at the ecliptic poles. If there is no offset from one scan to another, than after subtracting the dipole component and averaging over scans the mean temperature at the ecliptic poles should go down by a factor $\sqrt{N}_{\text{scans}}$, which was really observed. Moreover, because temperatures at the poles were lower than the typical temperature of other pixels, no additional matching, which would lower temperatures at the poles, should be applied.

After subtraction of the dipole component on the sky, a small residual systematic effect was found in the data: there remains a small SIN signal at each scan. The signal is small in comparison with the dipole and it affects only the first Fourier component, but it is large enough when compared with high order multipoles. This is the only one unexplained systematic effect of the "Relikt-1" experiment. One of suspected reasons is a drift of the parameters of the balancer. The main effect of the balancer on the first Fourier component on a scan $(\sin(\theta))$ was to rotate the phase of the component by $\approx 37°$. If the phase-shift were slightly fluctuating from scan to scan with variation about 3–5 degrees, this could account for the additional phase-shift. But the parameters of the balancer were calibrated during the flight and it seems unlikely that this is the main reason. Small nutation of the satellite axis also can add to the phase-variation. In any case, for the analysis of high order multipoles of the CMB anisotropy it was necessary to subtract from the scan data any first Fourier component of the signal. This does not affect the quadrupole and all even multipoles because these multipoles do not have a SIN component on a big circle. The main effect was for the octupole with the outcome of the lost of sensitivity of the experiment to the octupole by a factor of two.

The map of temperature of the sky was produced on a grid of $2° \times 2°$ rectangular cells, with the polar caps ($2°$ radius) treated as two circular grid cells. If T_i, N_i are the measured temperature and number of measurements (number of seconds of observations) for a pixel on a scan, then the temperature on the sky T_j and the weight W_j of j–th grid cell are

$$T_j = \frac{1}{W_j \cdot \Delta\Omega_j} \overset{\text{over data points}}{\underset{i,\ \Theta_{ij}<2\Theta_0}{\sum}} T_i \cdot N_i \cdot \omega_{ij},$$

$$W_j = \frac{1}{\Delta\Omega_j} \overset{\text{over data points}}{\underset{i,\ \Theta_{ij}<2\Theta_0}{\sum}} N_i \cdot \omega_{ij}, \tag{1}$$

$$\omega_{ij} = \frac{1}{\sum_{j',\ \Theta_{ij'}<2\Theta_0}^{\text{over cells}} \Delta\Omega_{j'} \exp(-(\Theta_{ij'}/\Theta_0)^2)} \cdot \Delta\Omega_j \exp(-(\Theta_{ij}/\Theta_0)^2).$$

Here $\Delta\Omega_j$ is the solid angle of the cell, Θ_{ij} is the angular distance between i–th data point and j–th grid cell, Θ_0 is the width of a smoothing filter necessary to get the map of temperatures. The parameter ω_{ij} is evaluated in a way which insures the conservations of number of measurements on a grid with finite cell-size. The weight W_j was set to zero for the galactic belt of $\pm 15°$ width. Fig. 4 presents the map of the sky at 8 mm with the dipole subtracted (Klypin et al. 1987).

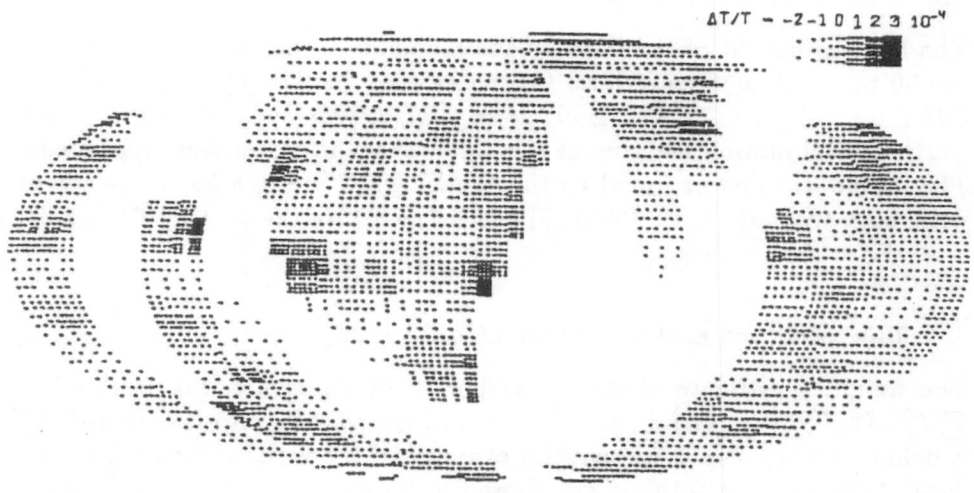

Fig. 4. The measured temperature map of the sky at 8 mm in the ecliptical coordinates.

4.2 Monte-Carlo simulations of instrumentation effects and simulations of cosmological signal

In order to take into account all main instrumental and observational details of the experiment, the Monte-Carlo method was chosen. Statistical significance of some characteristic (like amplitude of the quadrupole) was obtained from a set of many simulated maps of temperature distribution on the sky. The procedure starts with the simulation of one scan. The position of the scan on the sky and the mean number of measurements per pixel are input parameters. The later defines the noise level. The code generates a realization of the noise with the same mean level and the same *spectrum* as that of the observational scan. It subtracts the mean value of the signal and nulls off the first Fourier component on the scan. A map of temperatures is produced using the complete set of simulated scans. If necessary, a cosmological signal is added to each scan. A realization of the cosmological signal with some assumed spectrum and amplitude is generated on the map of $2° \times 2°$ resolution and then is interpolated to the position of each pixel. The spectrum of the cosmological signal is suppressed by factor $\exp(-l \cdot (l+1) \cdot \theta^2_{meas}/2)$ to account for the antenna smoothing. Here l is the order of the multipole and $\theta_{meas} = 2°.46 \cdot \pi/180$ is the gaussian beamwidth of the measuring antenna. Afterwards, the cosmological signal is processed through all instrumental and observational steps.

4.3 The dipole anisotropy

The least-square fit of the data by the dipole gave the following results at the 90 per cent confidence level (for details see Strukov et al. 1987): $T_D = 3.16 \pm 0.12$mK, $\alpha = 11^h 17^m \pm 10^m$, $\delta = -7.^\circ5 \pm 2.^\circ5$. Corrections due to the Earth orbital motion and due to the balancer transfer function were made. The dipole parameters found by the COBE DMR are in a good agreement with ours (Smoot et al. 1991): $T_D = 3.3 \pm 0.2$mK, $\alpha = 11^h 12^m \pm 12^m$, $\delta = -7.^\circ0 \pm 2^\circ$.

4.4 The variance and the correlational analyses

The weighted estimate of the dispersion σ^2 of the signal was used: $\sigma^2 = \sum T_j^2 \cdot W_j / \sum W_j$, where sums are taken over all cells and the weight W_j is defined by (1). Fig. 5 presents an example of the variance analysis of the data. Additional smoothing was chosen to be $\Theta_0 = 7^\circ$, which corresponds to $4^\circ.9$ for a gaussian filter. The arrow shows $\sigma = 0.181$ mK for the map of measured data. The full histogram presents $\Delta N/(N \Delta \sigma)$ probability to find a realization with the level of noise in the interval $(\sigma, \sigma + \Delta \sigma)$. The dashed histogram is for the simulated distribution of σ, in case when cosmological signal has the spectrum of angular fluctuations $\propto 1/l \cdot (l+1)$ and the amplitude of the quadrupole $(\Delta T/T)_2 = 1.6 \cdot 10^{-5}$. The distribution had 5 per cent of realizations below the observed value 0.181 mK. One hundred realizations was used to get the ensemble. Later simulations with 1000 realizations were made, which changed the estimates only slightly. Table 1 presents results for different smoothing filters and different spectra of the cosmological signal.

Table 1. Results of the variance analysis

Θ_0	Measured RMS (mK)	Noise($\pm 2\sigma$) (mK)	Upper limit for $(\Delta T/T)_2$	Spectrum C_l^2
7°	0.18	0.20 ± 0.02	$1.6 \cdot 10^{-5}$	$(2l+1)/l(l+1)$
$3^\circ.5$	0.31	0.32 ± 0.02	$1.2 \cdot 10^{-5}$	Constant
$3^\circ.5$	0.31	0.32 ± 0.02	$6.0 \cdot 10^{-6}$	$2l+1$

Fig. 6 shows the measured correlation function $\langle \Delta T(x) \cdot \Delta T(x + \theta) \rangle$ and $\pm 3\sigma$ fluctuations due to the noise. Fig. 7 presents upper limits on the correlation function of the cosmological signal. The correlation function of the cosmological signal with the spectrum $1/l \cdot (l+1)$ and the quadrupole amplitude $2 \cdot 10^{-5}$ is also shown.

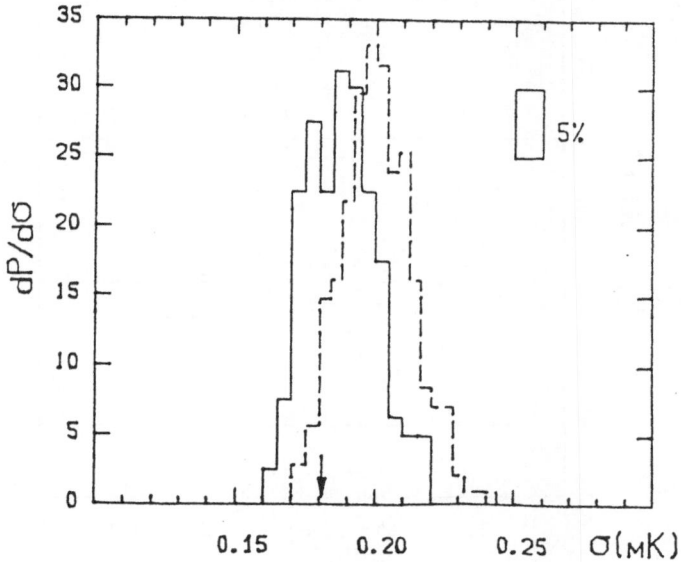

Fig. 5. The simulated distribution of the RMS noise on the sky (full histogram). The arrow shows the measured value of the RMS temperature fluctuations. The distribution of the RMS fluctuations in case when the system noise is added to the cosmological signal with the Harrison-Zeldovich spectrum $(1/l(l+1))$ and the quadrupole RMS$=1.6 \cdot 10^{-5}$ is shown as the dashed histogram.

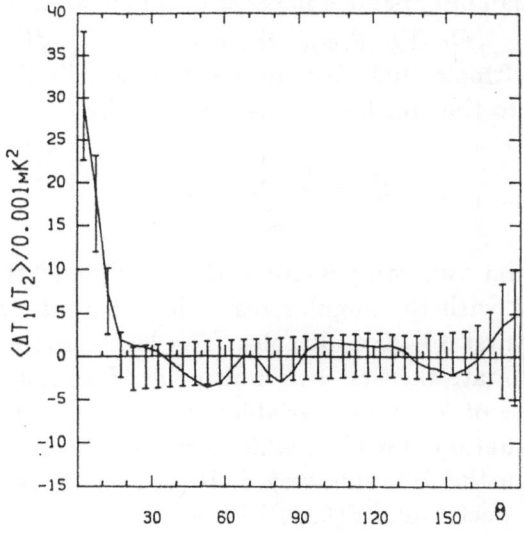

Fig. 6. The correlation function of measured fluctuations is shown as full curve. Error bars present the level of fluctuations of the noise ($\pm 2\sigma$)

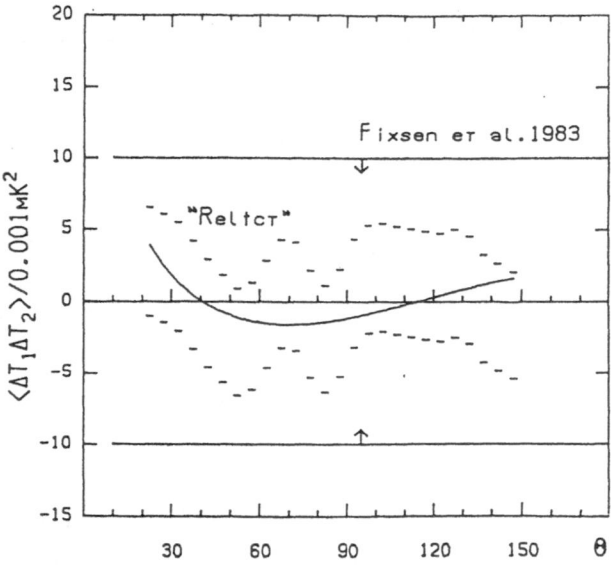

Fig. 7. The upper limits on the correlation function of a cosmological signal. The full curve shows the correlation function for a signal with the quadrupole $(\Delta T/T)_{\text{quad. rms}} = 2 \cdot 10^{-5}$ and the spectrum $1/l(l+1)$.

4.5 The multipole analysis

A function $f(\theta, \phi)$ can be expanded in series of spherical harmonics $Y_l^m(\theta, \phi)$:
$f(\theta, \phi) = \sum_l \sum_{m=-l}^{l} C_{lm} Y_l^m(\theta, \phi)$, where $C_{lm} = \int_{4\pi} f(\theta, \phi) Y_l^m(\theta, \phi) d\Omega$.
The goal of the multipole analysis is to find the contribution C_l^2 of components with equal l to the rms fluctuations on the sky:

$$C_l^2 = \frac{1}{4\pi} \sum_{m=-l}^{l} C_{lm}^2. \qquad (2)$$

Usually only the first two components – the dipole and the quadrupole – are looked for. But with the angular resolution $\approx 5°$ one can hope to get information about first 10–20 multipoles. The least-square fit to the data, usually used to estimate the first two multipoles, does not seem to be very encouraging because of too many variables to find. Instead, one can solve a system of linear equations for C_{lm}, which can be obtained from the least-square method or in the following way. If $W(\theta, \phi)$ is the weight defined by (1) and b_{lm} is the spectrum of $f(\theta, \phi)W(\theta, \phi)$:

$$b_{lm} = \int_{4\pi} f(\theta, \phi) W(\theta, \phi) d\Omega, \qquad (3).$$

then by expanding in this formula the function $f(\theta, \phi)$ in spherical harmonics, we obtain the following system of linear equations

$$\sum_{l'm'} W_{ll'}^{mm'} C_{l'm'} = b_{lm},\tag{4}$$

where $W_{ll'}^{mm'}$ is the spectrum of the product $W(\theta, \phi) Y_l^m(\theta, \phi)$:

$$W_{ll'}^{mm'} = \int_{4\pi} W(\theta, \phi) Y_l^m(\theta, \phi) Y_{l'}^{m'}(\theta, \phi) d\Omega\tag{5}.$$

The solution of the system (4) gives desirable components C_{lm}. But there are two obstacles. First, one must truncate the system at some multipole $l_{\max}$. Then one observes that on the left-hand side of the system there is a dependence on harmonics with $l \leq l_{\max}$, while the integrals on the right-hand side does depend on high order multipoles with $l > l_{\max}$. This also means that fitting the results with a dipole and a quadrupole does not give accurate estimates of these multipoles. How accurate – depends on the spectrum of the window function W. It also depends on the spectrum of the noise: it is a good idea to filter the map before doing the multipole analysis. In any case, it is important to demonstrate that high order multipoles did not infiltrate to the solution. Second, if one really wants to benefit from the multipole analysis – to get information about multipoles upto 10th order, say, then the instability of the system is the main problem: small variations of b_{lm} lead to enormous fluctuations of C_l. In case of the "Relikt-1" data, the amplification of errors was about 30-100, when multipoles up to $l_{\max} = 10 - 15$ were included to the system.

The regularization method of Tikhonov (Tikhonov & Arsenin 1977) was used to stabilize the solution. This method leads to new (stable) system of linear equations:

$$(\mathbf{D} + \alpha \mathbf{I})\mathbf{C} = \mathbf{K},\tag{6}$$

where $\mathbf{I}$ is the unit matrix, and $D_{ll'}^{mm'} = \sum_{ln} W_{ln}^{mr} W_{l'n}^{m'r}$, $K_{lm} = \sum_{l'} b_{l'}^{m'} W_{ll'}^{mm'}$. The idea of the method is as follows. The system of equations (4) $WC = b$ is almost singular, i.e. some of its eigenvalues are close to zero. This results in erratic behaviour of the solution and its large norm $\|C\|^2 = \sum_{ij} C_{ij}^2$. Because of the instrumental noise and because of systematic errors, the right-hand-side of the (4) is known with some uncertainty δ. In other words we solve not the system (4), but a disturbed system $WC - \tilde{b} = 0$, where $\|b - \tilde{b}\| \leq \delta$. Among all possible solutions of the disturbed system $WC - \tilde{b} = 0$ it is reasonable to find a solution $\tilde{C}$, which gives an approximate solution of the original system ($\|W\tilde{C} - b\| < \delta$) and has a minimum norm $\|\tilde{C}\|$. So, the problem reduces to minimizing the functional $\|C\|^2$ on a set of vectors C satisfying the condition $\|WC - \tilde{b}\| \leq \delta$. Tikhonov and Arsenin (1977) prove, that for $\delta \to 0$ the solution of the disturbed system converges to the solution of (4), which has a minimum possible norm. Using the Lagrange method, the problem reduces to finding the vector C minimizing the smoothing functional $\|WC - \bar{b}\|^2 + \alpha\|C\|^2$, where α is a small parameter determined by a condition that the discrepancy must be $\|WC - \bar{b}\| = \delta$. The minimizing the functional leads to the system of equations (6). The small parameter α was chosen so that the discrepancy in the norm of the solution was small and equal to $\|W_{ll'}^{mm'} C_{l'm'} - b_{lm}\|/\|b_{lm}\| = 0.005$. Tests showed that the solution of the stabilized system is close to the "true" solution within 20 per cent. Fig. 8 presents results for systems up to $l_{\max} = 10$ and $l_{\max} = 15$.

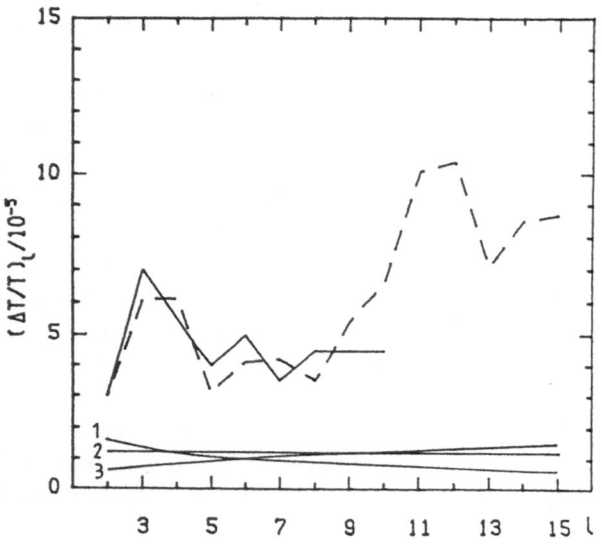

Fig. 8. Upper limits on the spectrum of a cosmological signal. The full broken curve is for the case when multipoles up to $l_{max} = 10$ are taken into account. The dashed broken curve shows the case $l_{\max} = 15$. Full labeled curves at the bottom present constraints from the variance analysis in the Table 1 (1 - for the Harrison-Zeldovich spectrum).

It is interesting to compare results of the Relikt-1 (Klypin *et al.* 1987) with results of the COBE DMR (Smoot *et al.* 1991). It appears that both experiments give essentially identical results. The rms quadrupole amplitude is less then $(\Delta T/T)_2 < 3 \times 10^{-5}$. For gaussian fluctuations the pixel-to-pixel variance of the COBE data results in upper limit on intrinsic sky fluctuations $\Delta T/T < 4 \times 10^{-5}$. This roughly corresponds to the upper limit $(\Delta T/T)_2 < 1.6 \times 10^{-5}$ of the Relikt-1 on the inflationary spectrum $\propto 1/(l \cdot (l+1))$ because the total rms signal observed with 6° antenna for the inflationary spectrum normalized to have the quadrupole 1.6×10^{-5} would be about 3.5×10^{-5}. Nevertheless COBE has great advantage over the Relikt-1 because COBE is still operating, more data are coming. Also there is a hope that it would be possible to remove some of the systematic effects, which currently restrict the sensitivity of the COBE DMR.

5 "Relikt-2": multifrequency observations

The "Relikt-2" experiment scheduled to mid 1993 will have 5 channels from 1.5 mm to 1.5 cm. The experiment will have the same basic strategy as the first one, but it will be launched to a very distant orbit and will not have problems with the Moon and the Earth emission. The antenna system is still not settled: all antennae will be corrugated horns with a bit wider ($7°$) angle as compared with "Relikt-1", but it is not clear how reference and measuring antennae will be set. Nevertheless, it seems that the main source of systematic effects will be the emission of our galaxy.

To have an impression and rough estimates of the expected emission from our galaxy at these frequencies, we present maps of predicted temperatures, which were kindly given us by W.Reich. The maps at Fig. 9 and 10 present predictions of galactic emission at 34 GHz and 59 GHz. Fig. 11 gives the map of used power-law slopes. Predictions were made by an extrapolation from the low frequency observations of Haslam *et al.* (1982) and Reich (1982) and Reich & Reich (1986, 1988). For each $0.°5$ pixel on the sky ($\delta > -40°$), the measurements at 408 MHz and 1420 MHz were used to estimate the amplitude and the slope for a power-law approximation: $T(f) = A \cdot f^{-\alpha}$, where f is the frequency. Temperature was extrapolated to high frequencies using this simple approximation. There are some obvious problems with these extrapolations. Extrapolation to frequencies, which are 30 times high, is not very reliable because even small change in the slope α can result in very different prediction. This definitely happens near the galactic plane, where the slope α becomes smaller due to free-free emission. But this region will be excluded from $\Delta T/T$ analysis anyway. Far from the galactic plane the main source of the emission is the synchrotron emission, which has a steep slope. The map of slopes, which we use, does show reasonable for the synchrotron emission values of slopes out of the galactic plane. There are two regions on the sky (one is below the plane at $-60° < l < -120°$, the other is above the plane at $l \approx -100°$) with too shallow ($\alpha \approx 2.4 - 2.5$) slopes for the synchrotron emission, but the rest shows slopes in the range typical for the synchrotron emission: $2.6 < \alpha < 3..$ If we exclude these two suspicious regions and will subtract monopole component, which cannot be observed by $\Delta T/T$ measurements, then the typical level of fluctuations at 34.5 GHz is about $0.05 - 0.1$ mK and $0.02 - 0.04$ mK at 59 GHZ, which is below the current constraints given by Relikt-1 or COBE, but not much below! So, the lesson we learn from low-frequency observations is: the galactic emission *can be* much higher than expected cosmological signal, which for the Cold-Dark-Matter model is on the level $20\mu K - 30\mu K$ depending on the normalization and the angular resolution. In other words, multifrequency observations are necessary for the detection of cosmological fluctuations below 0.1 mK.

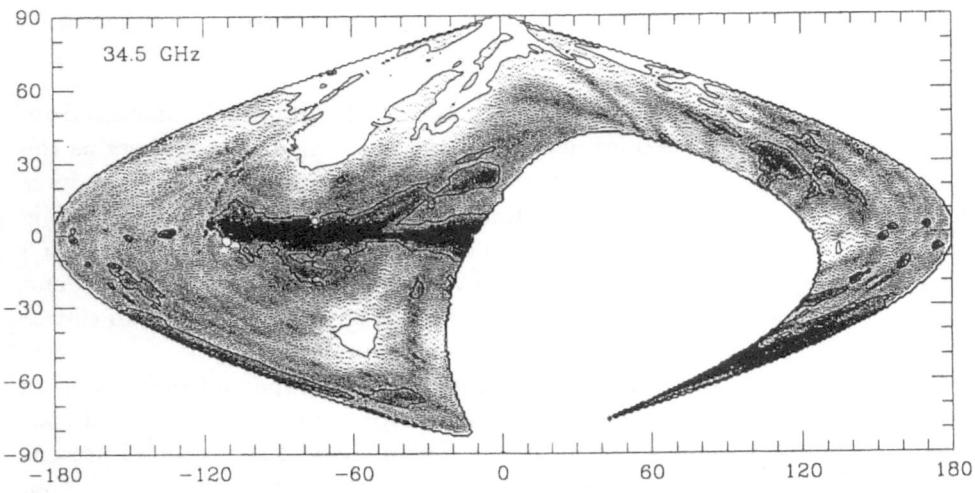

Fig. 9. The temperatures due to the galactic emission predicted at 34.5 GHz from the extrapolation of low-frequencies observations. Predictions were made by W.Reich (see Reich & Reich (1986,1988)). The level shown by the thick contour corresponds to the temperature level 0.250 mK. The thin contour is for 0.10 mK, shades cover temperature range 0.1 mK – 0.5 mK.

Sensitivities of mm-receivers of the "Relikt-2" will be much better than that of the COBE DMR. Table 2 presents frequencies and expected sensitivities of the "Relikt-2" experiment. The fifth channel at 1.6 mm has low sensitivity and probably will not be interesting for cosmological fluctuations. So, we will have measurements at four frequencies to find unknown temperature of the CMB and temperature of the galactic emission. The later must be parameterized in some way. The form of parametrization is crutial. In principal, one can think of four parameters: the temperature and the slope of the synchrotron emission, the temperature and the slope of the thermal emission.

There possibly could be dust emission, but at 9 mm, where the sensitivity is the best, the dust emission is probably negligible. Having four measurements and one parameter for the CMB, we can find at the most only three parameters for the galactic emission. So, the first idea is to fix slopes of the synchrotron and the free-free emission (say, -3 and -2) and find amplitudes. Unfortunately, this doesn't work either because the level of the noise in the temperature of the CMB appears to be much higher than the noise levels in channels. The reason is that the three functions – a constant, f^{-3} and f^{-2} are not orthogonal. For example, on a finite frequency interval

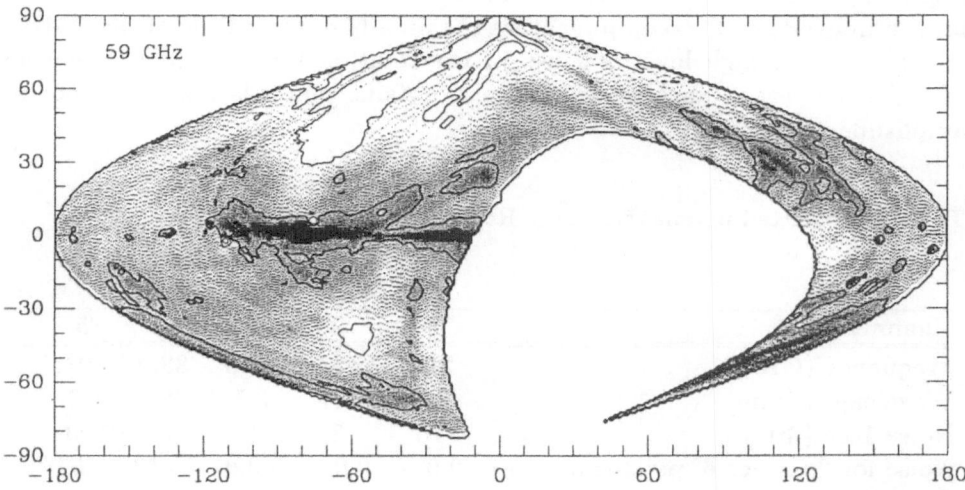

Fig. 10. The temperatures due to the galactic emission predicted at 59 GHz using extrapolation of from the low-frequency observations. The predictions were made by W.Reich. The level of 60μK is shown by the thick contour. The thin contour above and below the galactic plane corresponds to 20μK. Shades are for temperatures 20μ K – 200μ K.

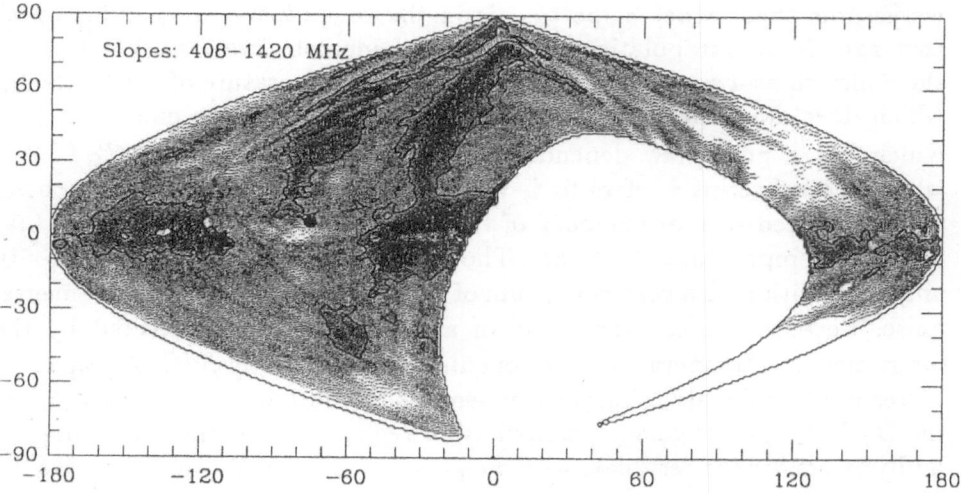

Fig. 11. The power-law slopes of temperatures used to make predictions at Fig. 9, 10. Shades span slopes from -2.4 to -3.0. The contour shows the level -2.75.

a sum of f^{-3} and a constant could mimic f^{-2} law. As the result, instead of a large amplitude of f^{-2}-component we obtain a high amplitude of the CMB, which means a high level of noise in the solution for the CMB. Because of the same reason another approximation fails: power-law with unknown amplitude and slope.

Table 2. Expected parameters of the Relikt-2 experiment

Channel	1	2	3	4	5
Frequency (GHz)	21.7	34.5	59.0	83.0	193.
Wavelength (mm)	13.8	8.7	5.1	3.6	1.6
Noise 1sec (mK)	15	3	5	10	100
Noise for 2yrs per 6° pixel (mK)	0.077	0.015	0.026	0.051	0.51
Contrib. to the noise of CMB (mK)	0.012	0.006	0.015	0.009	0
Contrib. to CMB temperature (mK)	-0.152	0.389	0.594	0.169	0

Reasonable results were obtained for a very simple power-law model with unknown amplitude and fixed slope α_{solve}. The difference between real slope α_{sky}, which varies from pixel to pixel, and suggested slope α_{solve} results in systematic errors in the amplitude of the CMB. In the following analysis we assume that "true" slope α_{sky} is in the range $2.5 < \alpha_{sky} < 3.$, which corresponds to extrapolation from low-frequency observations. We describe the emission as a sum of two components: the temperature of the CMB T_R, which does not depend on the frequency, and the galactic emission $T_{G,\,i}$, which has a power-law dependence on the frequency $T_{G,\,i} = T_G f_i^{-\alpha_{sky}}$, where the subscript i refers to i−th frequency channel and all frequencies are normalized to the frequency of the second (34.5 GHz) channel, so that T_G is the temperature at 8.7 mm. The measured temperature $T_i(j)$ at j−th angular position on a scan is the sum of $T_R(j)$, $T_{G,\,i}(j)$ and the instrumental noise. Because the constant level of a signal cannot be detected by the instrument, all temperature components will include only *fluctuating* part of corresponding signal. Using the least-square method, we find temperature of the CMB $\tilde{T}_R(j)$ and galactic contribution $\tilde{T}_G(j)$. When doing the fitting, we multiply squares of residuals with weights which are taken to be $w_i = \sigma_i^{-2}$, where σ_i is the rms noise at i−th frequency channel. Because of the noise and because of the difference between α_{sky} and α_{solve}, tilded temperatures are estimated with errors. The least-square fitting leads to a system of two linear equations, which solution is:

$$\tilde{T}_R(j) = \sum_1^4 T_i(j)\epsilon_i, \qquad (7)$$

where

$$\epsilon_i = \frac{w_i \sum_k^n w_k f_k^{-\alpha_{\text{solve}}}(f_k^{-\alpha_{\text{solve}}} - f_i^{-\alpha_{\text{solve}}})}{\sum_k^{n-1} w_i \sum_{i+1}^n w_k (f_k^{-\alpha_{\text{solve}}} - f_i^{-\alpha_{\text{solve}}})}. \tag{8}$$

Setting amplitudes of signals to zero, we obtain the rms fluctuations in estimated amplitude of the CMB due to the instrumental noise: $\langle \tilde{T}_R^2 \rangle = \sum \sigma_i^2 \epsilon_i^2$. Thus, ϵ_i and $\sigma_i |\epsilon_i|$ are the contribution of i−th channel to the temperature and the noise of estimated CMB. These contributions are presented in Table 2. The first and the third channels are the main sources of the noise. Small contribution of the fourth (83 GHz) channel to the temperature of the CMB is very encouraging because it implies that the contamination by the dust, if it would be present in the observations, will be reduced by a factor of five. Note that the total level of the noise of the CMB will be 0.0218 mK, which is low enough to *detect* fluctuations with the inflation-like spectrum having the quadrupole on the level $(2–3)10^{-6}$. Note also that here we estimate the noise in a pixel on a scan and we do not take into account overlapping of pixels on different scans. This overlapping will reduce the noise by additional factor of two. So, we expect the level of random noise in the solution of the CMB amplitude to be on the level $10\mu K$.

Thus, the main concern is the systematic effects due to wrong slope of the power-law. To demonstrate the expected level of systematic effects, we assume a very simple model of the galactic emission, which nevertheless covers practically all possible cases (in the sense of stability of the CMB solution). We suggest that the galactic emission is on the level 0.1 − 1. mK at 34.5 GHz, which is what we expect from the low-frequency extrapolation and which is already on the level of detectibility of COBE. Specifically, we assume the following dependence on the angle along a scan: $T_{G,\,i} = 0.1$ mK$/(0.1 + |\cos(\theta)|) \times f_i^{-\alpha_{\text{sky}}}$ and we exclude points within $\pm 15°$ from the galactic plane. Because our analysis does not include angular dependence (we solve the problem pixel by pixel), this particular shape is not important, but we need to assume some. Fig.12 and 13 give two examples for the model with different slopes α_{sky}. The systematic errors grow when real slope starts to deviate from assumed slope 2.75. But if real slope stays in the range 2.5 − 3.0, the systematic errors remain too small to be detected. For α_{sky} equal to 2.5 or equal to 3.0, the rms pixel-to-pixel fluctuations are equal to 0.0223 mK as compared with 0.0218 mK for $\alpha_{\text{sky}} = 2.75$. So, this small difference probably will not be detected. The only problem with this systematics is that this "false" signal, with rms fluctuations $4.5\mu K$ can have and probably has large-scale correlations. So, after proper filtering of observational data, the "signal" possibly can be detected. Now, we can argue that presented level of systematics was probably overestimated because i) we took maximal amplitude of galactic emission (just on the level of sensitivity of Relikt-1 and COBE) and ii) a large fraction of pixels above the galactic plane at these frequencies has slopes around the value 2.7 − 2.8, which greatly reduces the errors.

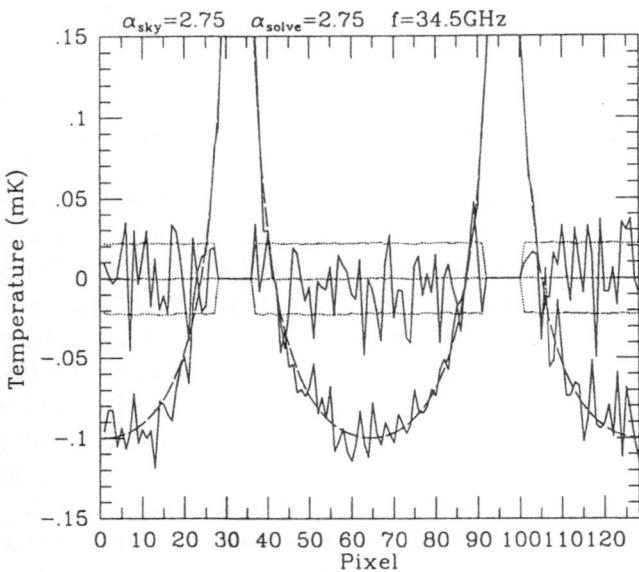

Fig. 12. An example of multifrequency analysis of a data scan crossing the galactic plane twice (at pixels ≈ 30 and ≈ 95). The model of the galactic emission at 34.5 GHz is shown as the dashed curve. "Measured signal" is the model plus a realization of the noise of the receiver (broken line fluctuating around the model). The broken line around zero level shows estimated cosmological signal, which "true" value was zero. After averaging over many realizations the mean level of the cosmological signal was found to be nearly zero (the middle dotted curve) with $\pm 1\sigma$ fluctuations shown by lower and high dotted curves. The noise of the system at 34.5GHz was 15μK and the slope of the galactic emission was properly "guessed" and is $\alpha = 2.75$

The method we presented in this paper is one of the probable approaches to handle multifrequency data. It demonstrates that i) it *is* possible to subtract the galactic contamination and to detect the cosmological signal much below the level of the galactic emission; ii) although the assumptions about the galactic emission look plausible, at present the subtraction cannot be maid in a model independent way.

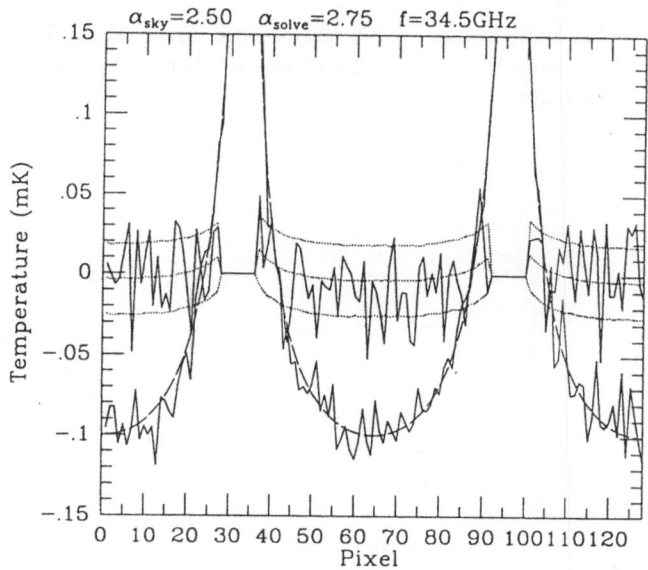

Fig. 13. The same as at Fig.12 but the real slope of the galactic emission −2.50 is very different from that the solver assumes. Note the systematic error at the level 5μK for "found" cosmological signal and the level of noise 22μK.

References

Birkinshaw, M., Gull S.F., Hardebeck, H., 1984. *Nature*, **309**, 34.

Haslam, C.G.T., Salter, C.J., Stoffjel, H., Wilson, W.E., 1982. *Astron. Astrophys.Suppl.*, **47**, 1.

Henry, P.S., 1971. *Nature*, **231**, 516.

Fixen, D.J., Cheng, E.S., Wilkinson, D.T., 1983. *Phys. Rev. Lett.*, **50**, 620.

Klypin, A.A, Sazhin, M.V, Strukov, I.A., Skulachev, D.P., 1987. *Sov. Astron. Letters*, **13**, 104.

Lubin, P.M., Villela, T.V., Epstein, G.L., Smoot, G,F., 1985. *Astrophys. J.*, **298**, L1.

Penzias, A.A., Wilson, R.W., 1965. *Astrophys. J.*, **142**, 419.

Reich, W., 1982. *Astron. Astrophys.Suppl.*, **48**, 219.

Reich, P., Reich, W., 1986. *Astron. Astrophys.Suppl.*, **63**, 205.

Reich, P., Reich, W., 1988. *Astron. Astrophys.Suppl.*, **74**, 7.

Smoot, G.F. *et al.* , 1991. *Astrophys. J. Lett.*, **371** , L1.

Strukov, I.A., Skulachev, D.P., 1984. *Sov. Astron. Letters*, **10**, 1.

Strukov, I.A., Skulachev, D.P., Boyarskii, M., Tkachev, A., 1987. *Sov. Astron. Letters*, **13**, 65.

Strukov, I.A., Skulachev, D.P., 1987. *Sov. Astron. Letters*, **13**, 191.

Strukov, I.A., Skulachev, D.P., Klypin, A.A., 1988. *Acta Astronautica* , **17**, 903.

Strukov, I.A., Skulachev, D.P., 1988. *Astrophys. and Space Phys. Review*, **6**, 147.

Sunyaev, R.A., Zeldovich, Ya.B., 1970. *Astrophys. Space Sci.*, **7**, 3.

Tikhonov, A.N., Arsenin, V.Y., 1977. *Solutions of ill-posed problems*, pp.95-102, Halsted Press, Washington.

ROSAT Observations of Clusters of Galaxies

Hans Böhringer

Max-Planck-Institut für Extraterrestrische Physik, D-8046
Garching, Germany

Abstract: During the first part of its mission the X-ray observatory ROSAT conducted the first All Sky Survey with an X-ray telescope. Of the order of 50,000 X-ray sources were detected in the survey data. Presumably thousands of clusters of galaxies are among the detected sources and several hundred clusters have already been identified. First results suggest that the final catalogue of X-ray selected clusters of galaxies will probably be quite different from the present optical catalogues. This catalogue will provide the basis for very interesting studies of the luminosity function, cluster evolution, and the spatial distribution of clusters in the Universe. In the ongoing pointed observation program of ROSAT individual clusters of galaxies are studied. Due to the higher spatial resolution and sensitivity compared to previous instruments a more detailed investigation of the dynamical state, the mass distribution, and the gas content of clusters will be possible. This is especially interesting with respect to the problem of the "missing mass" and the evolution of the intracluster medium. First results from both observing programs are presented in this paper. The cosmological relevance of the observations and the propects for the further ROSAT mission are discussed.

1 INTRODUCTION

The ROSAT X-ray Observatory which was launched on June 1st, 1990 (Trümper *et al.*, 1991) with its higher spatial resolution and increased sensitivity compared to previous X-ray telescopes is an ideal instrument for the study of clusters of galaxies. Firstly the hot intracluster gas that makes these galaxy clusters very prominent X-ray sources has its radiative emission maximum right in the soft X-ray region partly covered by the detection band of the ROSAT X-ray telescope (0.1 to 2.4 keV). And secondly clusters of galaxies are very large objects that can nicely be imaged as extended objects out to large distances provided that the exposure is sufficiently deep.

The ROSAT mission is divided in two parts. Half a year at the beginning of the mission was devoted to an All Sky Survey that lasted from August 1990 to January 1991 and was completed by scanning a small missing patch in August 1991. The survey is followed by pointed observations which will also cover detailed studies of individual clusters. There is therefore a twofold interest in research on custers of galaxies conducted with the ROSAT X-ray Observatory: statistical studies of a large sample of X-ray detected clusters and the detailed investigation of the physical properties of individual clusters.

Up to date thousands of clusters of galaxies have been catalogued notably by Abell (1958) and Abell, Corwin, and Olowin (1989) comprising 5250 objects (referred to as ACO clusters in the following). However, the selection criteria for the detection of clusters on photographic plates are poorly understood and it is even unclear how many of the objects detected as rich clusters on the plates are actually chance projections of a series of smaller galaxy groups in the line of sight. There is now evidence for these projection effects from spectroscopic studies (Huchra *et al.*, 1990), clustering analysis of ACO clusters (Sutherland, 1988), and N-body simulations of the growth of structure in the Universe (Frenk *et al.*, 1990). Therefore observations are needed that lead to a large statistical sample of galaxy clusters with well understood selection rules. This goal may be achievable with X-ray observations.

Clusters of galaxies are - except for quasars - the most powerful X-ray sources in the sky with luminosities of 10^{43} to $3 \cdot 10^{45}$ erg s^{-1}. The emission originates from hot gas that fills the gravitational potential well of the cluster close to hydrostatic equilibrium. The X-ray surface brightness is roughly proportional to the square of the galaxy density and can be even more peaked in clusters with cooling flows. Clusters of galaxies can therefore be detected in X-rays as three dimensional, gravitationally bound entities. The observations are very much biased to the dense core of the clusters and chance projections are hardly a problem. Thus the ambiguity adherent to the optical observations is practically removed in X-ray surveys. So far only a few hundred clusters have been observed in X-rays mostly with no underlying strategy for the construction of a statistically complete sample. The two presently available statistical samples of X-ray selected clusters from HEAO1/EXOSAT observations (Lahav *et al.*, 1989; Edge *et al.*, 1990) and from the EINSTEIN Medium Sensitivity Survey (Gioia *et al.*, 1990a, b) comprise of the order of 50 and 100 clusters, respectively. Nevertheless these samples have been found to be very interesting in particular in showing a trend of evolution in the X-ray luminosity function of clusters with redshift.

Thus there is a great potential for cosmologically relevant observations of clusters of galaxies for X-ray missions. Especially the first All Sky Survey with an imaging X-ray telescope by ROSAT will provide the basis to construct large statistically complete samples of X-ray selected clusters of galaxies. In addition pointed observations of individual clusters with ROSAT will help to understand the relation of observable quantities like X-ray luminosity, spatial extent, X-ray temperature and morphological features. It is especially important to understand how these observables are related to the cosmologically interesting parameters as mass and the form of the gravitational potential of the cluster. This can be learned from detailed observations and modelling (e.g. Hughes, 1989). The relation of the cluster mass to the X-ray luminosity as deduced from a well studied nearby sample of clusters will be a necessary ingredient for the cosmological work on the larger sample of clusters of galaxies for which hardly more than the X-ray luminosity will be known.

This paper provides some details about the ROSAT observatory and the observing mission in section 2; first statistical results form the ROSAT All Sky Survey with respect to clusters of galaxies are described in section 3; the first results on individual clusters are discussed in section 4, and section 5 gives a summary and an outlook on cosmological research that can be conducted with ROSAT observations of clusters of galaxies.

2 THE ROSAT MISSION

A description of the ROSAT Observatory can be found in (Trümper *et al.*, 1991). The main instrument on board of the ROSAT observatory is the X-ray telescope that consists of a fourfold nested Wolter type mirror configuration. It covers an energy range from 0.07 to 2.4keV. The second instrument is the Wide Field EUV Camera with a spectral window from 20 to 300 eV. Due to the strong absorption of the interstellar medium of the Galaxy in the EUV band the Wide Field Camera is only of limited application for extragalactic objects such as clusters and we will here concentrate on the results from the X-ray telescope.

The X-ray telescope carries two types of focal plane detectors. The Position Sensitive Proportional Counter (PSPC) has a spatial resolution better than 30 arcsec within a radius of 20 arcmin around the telescope axis. It has a limited energy resolution which is roughly about 45 % FWHM at 1 keV and the relative resolution varies roughly inversely proportional to the square root of the energy. The second type of detector, the high resolution imager (HRI), has a better spatial resolution of less than 5 arcsec near the telescope axis. This detector has no energy discrimination, however, and is about half as sensitive. Compared to the EINSTEIN IPC the ROSAT PSPC has an increased sensitivity that is roughly a factor of two to three higher (depending very much on the type of X-ray spectrum; for most cluster observations the sensitivity increase is about a factor of two).

The first observational results show that the ROSAT XRT-PSPC instrument has a much lower internal and particle background compared to

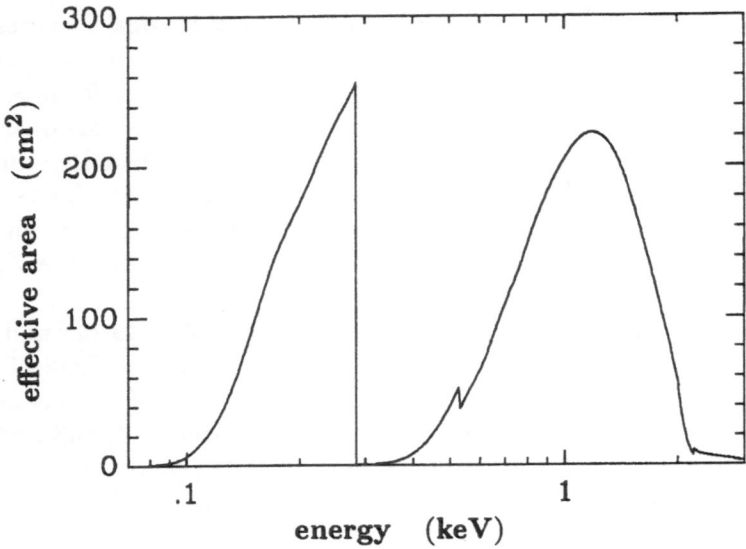

Fig. 1. PSPC X-ray spectra for optically thin thermal plasma at temperatures of $3 \cdot 10^6$ K, (a), $1 \cdot 10^7$ K, (b), $3 \cdot 10^7$ K, (c), and $1 \cdot 10^8$ K, (d).

the EINSTEIN telescope. Also the photon scattering in the telescope that produces a faint halo around bright sources could be reduced due to the extremly smooth surfaces of the ROSAT mirrors. This is especially important for the study of the faint outer parts of the X-ray halos in clusters of galaxies. It will enable the estimate of the gas and gravitational masses of clusters out to larger radii than previously possible.

The X-ray emitting hot intracluster gas in galaxy clusters is optically thin and has a temperature roughly equivalent to the virial temperature of the cluster potential well. The expected and actually observed temperatures range from 2 to 10 keV. Fig. 1 shows the spectra for hot plasma with temperatures in this range normalised to the same emission measure. One notes that in the temperature range relevant for clusters of galaxies the spectra are quite similar. The differences are becoming larger only for the lower temperatures. The ROSAT energy window is unfortunately too small to observe the exponential cutoff in the Bremsstrahlung spectrum for the high cluster temperatures. Also the absolute energy flux integrated over the ROSAT window (defined as 0.1 to 2.4 keV) is not very sensitive for the gas temperature. Therefore temperatures of the cluster gas can only be determined for spectra of very high quality (as shown for the example of Perseus below). This effect has the advantage on the other hand that the

X-ray flux is directly related to the emission measure and thus gas densities can be determined without having precise temperature information.

The small variation of the spectral shape shown in Fig. 2 can be used to determine the temperatures for spectra with very good photon statistics. There are two major features that characterize the change of the PSPC X-ray spectrum for temperatures between 10^7 K and 10^8 K which can been seen in the Figure. The temperature dependence of the gaunt factor causes a change of the slope of the free-free radiation spectrum which is reflected in the total spectrum. The second feature is the peak around 1 keV which is produced mainly by a blend of iron L-shell lines. These lines only become prominent at temperatures below about 2 keV. The height of this line emission peak depends therefore not only on the gas temperature but also on the metallicity of the gas.

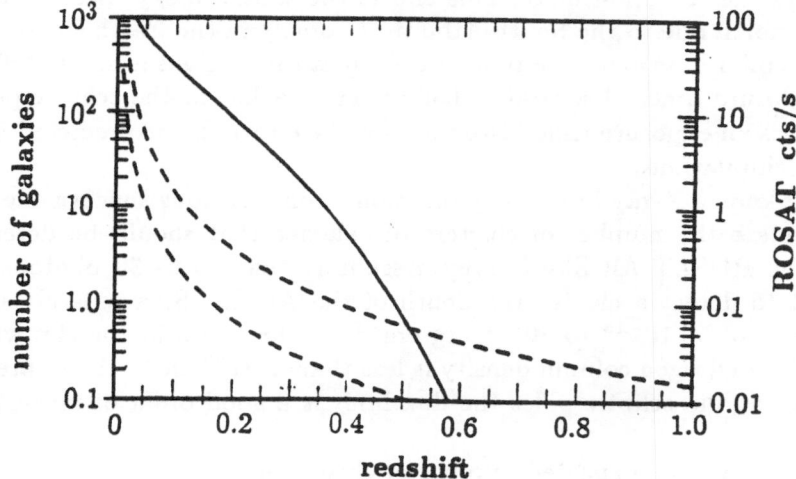

Fig. 2. Comparison of the depth of optical versus X-ray cluster surveys. The continuous line gives the number of galaxies that can be observed on an optical plate with a limiting magnitude of $m_V = 20$ for a Coma type cluster with an Abell number of galaxies of about 100. The dashed lines give the ROSAT PSPC count rates for clusters with a luminosity of 10^{44} and 10^{45} erg s^{-1} in the 2 - 10 keV band.

Compared to optical observations X-ray detections of clusters of galaxies are biased to richer and more evolved and compact clusters which can more easily be detected out to larger distances. Fig. 2 compares the dependence of the X-ray flux on distance to the number of galaxies that can be seen on a photographic plate down to $m_V = 20$ mag for a Coma type cluster (with an Abell number of galaxies equal to 100). The X-ray luminosities for the two X-ray flux curves are 10^{44} and 10^{45} erg s^{-1} (in the 2 - 10 keV band), the typical luminosity range for rich clusters. For the typical exposure time of the ROSAT survey with ~ 400 sec very luminous clusters can be seen out to $z \sim 0.5$ while a distance of $z = 1$ can be reached with deeper exposures. One also notes the relatively sharp cut off for the optical plates compared to the long tail in the X-ray flux curves.

After two month of calibration and verification observations half a year of the ROSAT mission was devoted to an All Sky Survey ended at the end of Jan. 1991. A small patch that was left open due to spacecraft and instrumentation problems was scanned in August 1991 thus completing the All Sky Survey. During this survey the sky was scanned by the ROSAT telescope in great circles in a plane perpendicular to the solar direction. Following the sun the whole sky is thus covered in half a year. Due to the overlap of the scan circles at the ecliptic poles the exposure is much greater at the poles. The instrument has to be shut down frequently when the orbit penetrates the radiation belts. This effects the southern sky more severely than the north due to the South Atlantic Anomaly in the Earth's magnetic field. The minimum exposure time in the equatorial regions is about 400 sec while exposure times of several 10 000 sec are reached at the ecliptic poles. Fig. 3 gives an exposure time histogram for the entire sky as predicted from preflight simulations.

Using known X-ray luminosity functions from previous surveys one can now estimate the number of clusters of galaxies that should be detected during the ROSAT All Sky Survey. Assuming that 15 to 30 photons are sufficient to detect a cluster the depth of the All Sky Survey is given by a flux limit of $5 \cdot 10^{-13}$ to 10^{-12} erg cm^{-2} s^{-1} for areas in the sky where interstellar hydrogen column density is less than $4 \cdot 10^{20}$ cm^{-2}. In an area of ~ 30 deg^2 at the ecliptic poles the flux limit is a good order of magnitude lower.

Fig. 4 shows the expected number of cluster detections in redshift bins for a flux limit of $5 \cdot 10^{-13}$ erg cm^{-2} s^{-1} for an area of 8 ster of the sky (excluding a 40 deg wide strip around the galactic plane) and for a flux limit of $5 \cdot 10^{-14}$ cm^{-2} s^{-1} in the area of 30 deg^2 around the ecliptic poles. For the calculations an X-ray luminosity function for the 2 - 10 keV energy band of

$$n(L_x) = 3 \cdot 10^{-7} \ exp(L_x/8.2) \ L_x^{-1.6} \tag{1}$$

has been used where L_x is in units of 10^{44} erg s^{-1} and $n(L_x)$ in units of Mpc^{-3}. Eq. (1) is a good fit to the combined results for HEAO 1 of Kowalski et al.(1984) and the EXOSAT/HEAO 1 results of Edge et al.(1990). A temperature - flux relation of $L_x = 24 \cdot (T/10^8 K)^{2.7}$ was used to convert these fluxes into fluxes for the ROSAT band. Standard cosmological paramters ($H_o = 50$ km Mpc^{-1} s^{-1}, $\Omega = 1$, and $\Lambda = 0$) were used for the calculations. The results show that about 4000 to 8000 clusters of galaxies should be

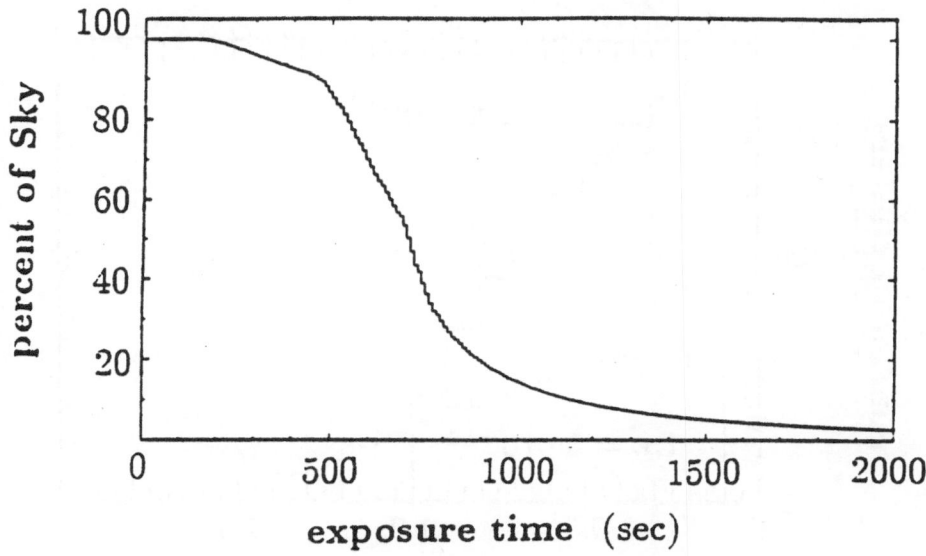

Fig. 3. Exposure time histogram for the ROSAT All Sky Survey from preflight simulations

found in the All Sky Survey for a flux limit between $5 \cdot 10^{-13}$ and 10^{-12} erg cm^{-2} s^{-1}.

The above calculations do not take any evolutionary effects of clusters of galaxies into account. If such effects are taken into account for the rich clusters as they were found in the EINSTEIN Medium Sensitivity Survey (EMSS; Gioia *et al.*, 1990 a,b) and in the HEAO 1/EXOSAT data (Edge *et al.*, 1990) the histograms have a steep cutoff at around $z \sim 0.5$ and $z \sim 0.9$, respectively. We also did not account for the effect that for nearby clusters the detect area is larger and thus for a given background flux the detection efficiency will decrease for the nearby more extended clusters. In summary most of the clusters to be detected in the survey will have redshifts below 0.3 but a few rich clusters out to $z \sim 1$ should also be found.

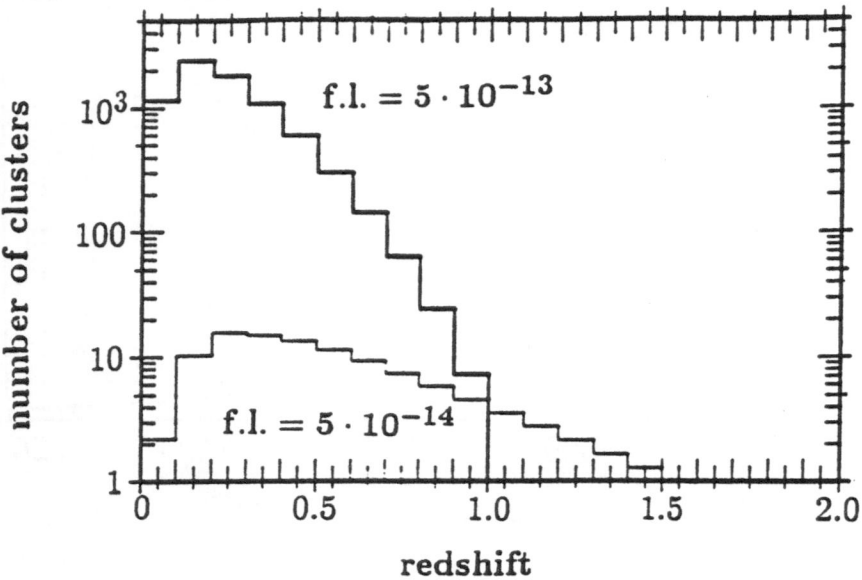

Fig. 4. Histogram of the expected number of cluster detections in the ROSAT All Sky Survey as a function of redshift for a flux limit of $5 \cdot 10^{-13}$ erg cm^{-2} s^{-1} for the entire sky and for a flux limit of $5 \cdot 10^{-14}$ erg cm^{-2} s^{-1} for an area of 30 deg^2 around the ecliptic poles.

3 FIRST STATISTICAL RESULTS FROM THE ROSAT ALL SKY SURVEY

In the first round of the standard analysis of the ROSAT All Sky Survey the data are organized in 90 two degree wide strips following great ecliptic circles. So far 85 strips have been analysed and a total of about 50,000 X-ray sources have been found. The significance of the sources was tested by means of a maximum likelihood method and only sources with a likelihood equivalent to a significance of more than about 4σ were accepted. First tests with the All Sky Survey data show that a sensitivity limit of about 10^{-12} erg s^{-1} cm^{-2} (for the energy band 0.1 to 2.4 keV) is easily reached for most of the sky for objects with cluster type spectra.

3.1 DETECTIONS OF KNOWN CLUSTERS OF GALAXIES IN THE ROSAT ALL SKY SURVEY

For an area covering 38 strips (corresponding to 45 %) of the sky a search for X-ray source counterparts of clusters of galaxies from the catalogues by Abell (1958), Abell, Corvin, and Olowin (ACO) (1989) and other publications was completed (Ebeling, 1991). A study of the detection statistics leads to the conclusion that X-ray sources with a spectral hardness ratio greater than zero (see below) found within a coincidence radius of 3.7 arcmin of the ACO cluster position have a 95% certainty to be cluster sources. The analysis also shows that more clusters are detected with larger off-sets. But for these sources the individual identification is much more uncertain. Out of a total of 2686 ACO clusters of galaxies in the study area (this includes the poor supplementary clusters listed in ACO) 479 clusters were detected at the high confidence level, while the total statistical number of detections is slightly larger than 500. Thus only about 19% of the ACO clusters appear in the X-ray survey.

The efficiency of the detection of ACO clusters with ROSAT as a function of the distance class and richness class is shown in Fig. 5. Some of the very close clusters in distance class 1 and 2 are missed by the standard analysis software because they are too extended whereas the software is geared to the detection of point sources. But they are easily found by a further inspection of the survey data. The detection rate is then high for the clusters up to distance class 3. From there on the detection rate is rapidly decreasing. The richness class 0 clusters that constitute about half of the ACO catalogue entries are less well represented while the detection rate increases towards the higher richness classes which is expected. As it has been found earlier (Bahcall, 1977) that the X-ray luminosity correlates very poorly with the Abell number (number of galaxies in the cluster within a radius of 3 Mpc and two magnitude intervals), one only expects to see a statistical trend of increasing detection rate with increasing Abell number but not a sharp richness dependend selection function. This expectation is very well reflected in the present X-ray data.

Of the order of 300 additional clusters from the Zwicky catalogue and other published compilations were found in the same study region. Extrapolation of these detection rates to the entire sky gives a number of only 1500 to 2000 catalogued clusters of galaxies that should be found in the ROSAT All Sky Survey. Compared to the expected number of cluster detections the known clusters would make up only about one third of the final X-ray selected cluster sample. Where are the missing clusters? They are probably previously unknown clusters that should be found by the optical cluster identification programme described below.

The information on the X-ray properties of the detected sources that is provided by the standard analysis are parameters for the hardness ratio of the X-ray source spectrum as well as an extent parameter and likelihood.

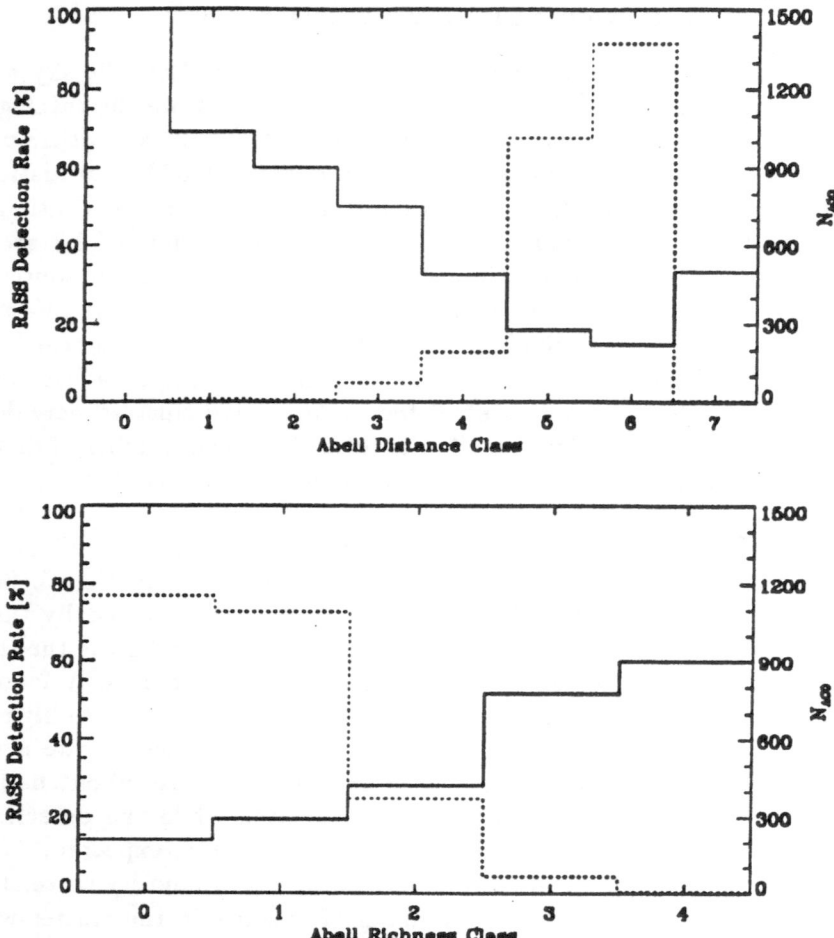

Fig. 5. Number of dectected ACO clusters of galaxies (dotted line) in an area covering 45% of the ROSAT All Sky Survey as a function of distance class and richness class. The solid lines shows the fraction of ACO clusters detected in the ROSAT Survey for each category.

The hardness ratio is given by $RH = \frac{H-S}{H+S}$, where S is the flux in the 0.1 - 0.4 keV band and H the flux in the 0.4 - 2.4 keV band. Clusters of galaxies should be found among the hardest X-ray sources in the survey. Only neutron stars and young supernova remnants which are confined to the Milky Way and nearby galaxies have harder X-ray spectra. One also expects to detect a fraction of the clusters as extended objects.

The analysed survey data show indeed that clusters of galaxies are considerably harder than other X-ray sources such that the sample of clusters can for example be enriched among the detected sources without loosing many cluster candidates by selecting only sources with a positive hardness ratio. This was applied in the X-ray source/ACO cluster correlation analysis described above. A fraction of clusters of galaxies is found with a high likelihood for an extent.

3.2 OPTICAL IDENTIFICATION OF NEWLY DETECTED CLUSTERS

The identification of ROSAT X-ray sources which do not coincide with catalogued objects is in a next step based on the correlation with optical data from scans of the Palomar and UK Schmidt Survey plates in collaboration with Space Telescope Science Institute (Baltimore), Royal Observatory (ROE, Edinburgh), and Naval Research Laboratory (NRL, Washington). The first results from the identification of clusters in the southern sky on the basis of the COSMOS/UK Schmidt data indicate that indeed 10 to 15 % of the ROSAT sources outside the galactic plane can be found to be clusters of galaxies. The identification is based on two approaches. A machine produced catalogue of cluster candidates based on the galaxy distribution in the COSMOS data was produced by ROE and NRL and is compared with the list of detected ROSAT sources. Alternatively the galaxy density is evaluated in small concentric rings about the ROSAT sources and compared to the background density. The results of the two methods are combined to produce a list of cluster candidates that is subsequently studied by optical observations and spectroscopy.

Results of the second method of galaxy counts around X-ray sources will briefly be described for a sample of 1800 All Sky Survey sources from a study area in the southern sky. Galaxies detected on the optical survey plates by COSMOS were counted in circles with radii from 3 to 10 arcmin around the X-ray source positions. In addition the cumulative probability distribution of the galaxy number counts for the same radii were derived for 1000 random positions on each photographic plate. The so obtained random probability distribution function allows to assign a cumulative probability value to each count results around a ROSAT source position. Using different random probability distributions for different optical plates takes account of the plate to plate variation of the overall galaxy density.

If there were no correlation between ROSAT sources and galaxy density enhancements a histogram of the cumulative probabilities found for the X-ray sources should give a constant line with the respective superposition of statistical noise. The actual results for counts within 3 arcmin radius from the X-ray sources are shown in Fig. 6. There is a clear excess of significant galaxy overdensities associated with ROSAT sources. A careful analysis shows that the statistical content of this sample is about 280 clusters. Using a significance threshold for the optical identification such that the sample is less than 20% contaminated by chance coincidences and selecting X-ray sources with a flux limit well above 10^{-12} erg cm^{-2} s^{-1} and with a positive

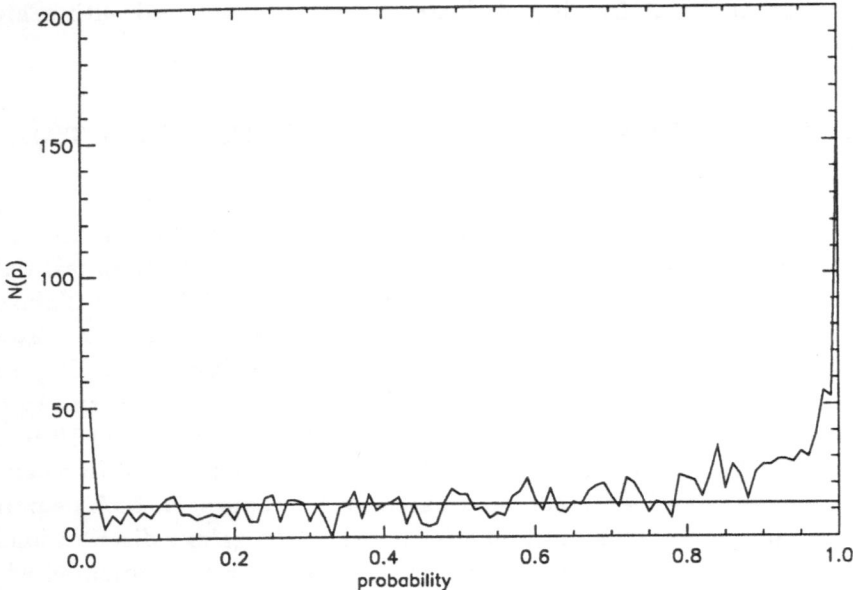

Fig. 6. Results of the number counts of galaxies in the COSMOS data within three arcmin of an X-ray source position for a sample of 1800 ROSAT sources in a study area in the southern sky. As explained in the text a histogram of cumulative probabilities assigned to each of the ROSAT sources is shown in the plot. Random coincidences would result in a constant line plot. The excess in the right part of the histogram shows that there is a significant association of ROSAT sources with galaxy overdensities which comprise 10 to 15% of the sample of X-ray sources.

hardness ratio results in a sample of higher quality containing 152 clusters. Thus this study shows that 10 to 15% of the extragalactic ROSAT sources from the ALL Sky Survey might well be clusters of galaxies.

This comes very close to the estimated number of clusters that should be found. It is also consistent with previous experience from the EMSS which is about comparable in depth to the All Sky Survey. There about 12 % of the sources detected at high galactic latitude were clusters of galaxies. Of these about 100 detections only 16 were clusters from the ACO catalogue.

4 OBSERVATIONS OF INDIVIDUAL CLUSTERS
OF GALAXIES

In this section results on observations of nearby clusters in the ROSAT All Sky Survey and from an early pointed observation on A 2256 will be reviewed. For the scaling of physical parameters we will assume a value for the Hubble constant of $H_o = 50$ km s^{-1} Mpc^{-1} throughout this article.

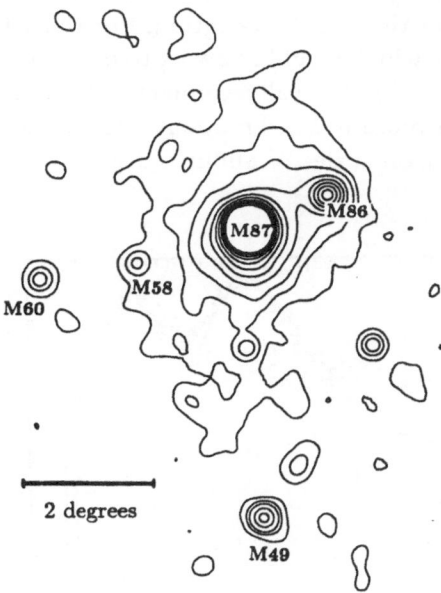

Fig. 7. Contour plot of an X-ray images of the Virgo cluster from the ROSAT All Sky Survey. Only hard photons ($E \geq 0.4$ keV) and a Gaussian filter with a width of 20 arcmin were used to construct this image. The most prominent elliptical galaxies with observed X-ray emission are labelled in the plot.

The nearest cluster of galaxies is located in the constelation Virgo and has a distance of about 23 Mpc. In optical images the Virgo cluster extends over an area of at least 10 degrees diameter. Due to its large size the cluster has so far never been imaged as a whole in X-rays. Scans of the cluster region by the GINGA collimated detector indicate that the X-ray emission is very extended (Takano *et al.*, 1989).

In the ROSAT All Sky Survey the Virgo cluster region was scanned with an average exposure time of about 460 sec. Fig. 7 shows a contour plot of the Survey data. For this plot only the hard photons have been used to enhance the signal to noise ratio and the image has been smoothed with a Gaussian filter. The most prominent features are the giant X-ray halo of M87 and the much smaller halo of M49 (NGC 4472). Besides the almost spherically symmetric halos there is also diffuse emission between the two halos that is observed clearly for the first time. The bright elliptical galaxiey M58 - and possibly M60 - which are labelled in the Figure, are also enclosed by the diffuse X-ray emission region. It is interesting that the diffuse emission is extended asymmetrically to the east between the two halos, because a similar shape is visible in the contour plots of the galaxy densities from the optical survey by Binggeli *et al.*(1987).

The lowest contour in Fig. 7 has been chosen such that it lies well above the observed variations of the X-ray background in the ROSAT Survey. A deeper look at the cluster which would allow us to determine wether the two X-ray halos of M87 and M49 are really connected by diffuse emission can only be performed when more is known about the variations of the X-ray background in the Survey on scales of about a degree.

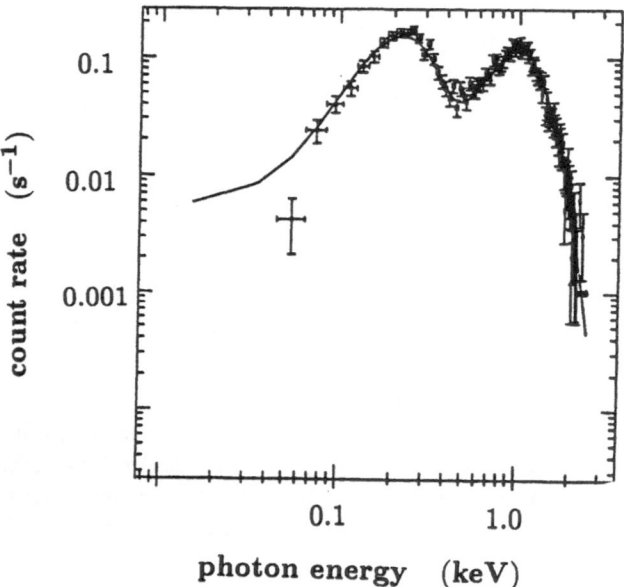

Fig. 8. ROSAT PSPC X-ray spectrum of the innermost region of the X-ray halo of M87 within a radius of 7 arcmin. A spectrum calculated by means of the Raymond and Smith code has been fitted to the data yielding a best fit at 1.86 keV which is shown in the Figure as straight line.

In the following we will concentrate on the analysis of the X-ray halo of M87. The azimuthally averaged surface brightness profile looks very much like that from the EINSTEIN IPC observations (Fabricant *et al.*, 1980), but in the present data the X-ray emission can be traced to larger radii. One can therefore determine that the gas mass contained in the halo of M87 within a radius of 0.63 Mpc (2 degrees) is about $8 \cdot 10^{12}$ $M_\odot$.

The image of the M87 X-ray halo contains more than 25,000 photons within a radius of 60 arcmin. This allows us to split the large data sample into several radial bins and to obtain a good X-ray spectrum for each of the regions separately. We have chosen here a binning with radial boundaries at 7, 15, 30, and 60 arcmin. Fig. 8 shows for example the background subtracted X-ray spectrum of the innermost bin together with the best fitting theoretical spectrum obtained with the radiation code of Raymond and Smith (1977). A region outside the 1° radius, where the cluster emission is very faint, has been used to determine the background spectrum that was subtracted from the data.

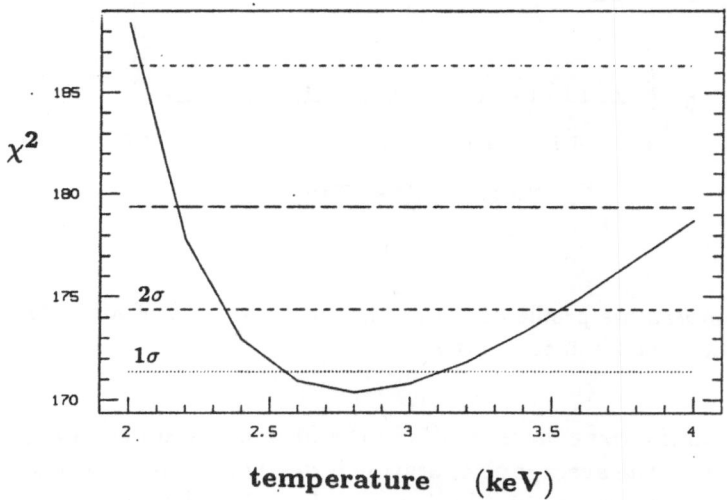

Fig. 9. χ^2 fit of theoretical spectra as a function of temperature to the X-ray spectrum of the gas in the halo of M87 in the region between 7 and 15 arcmin radius.

χ^2 fits of theoretical spectra with various temperatures to the same X-ray spectrum, where the normalization factor, the interstellar absorption column density, and the metallicity of the gas have been treated as free fitting parameters, show that the temperature in the innermost bin is quite well confined to the range 1.7 - 2.1 keV. The number of photons is lower in the outer bins while the temperature is increasing. Therefore the determination of the temperature becomes more and more difficult. Fig. 9 shows for example the χ^2 fits as a function of temperature for the second bin (7 to 15 arcmin). The error bars are asymmetric with a smaller uncertainty towards the lower temperatures, which is also a result of the fact that the variation of the spectrum is larger at lower temperatures.

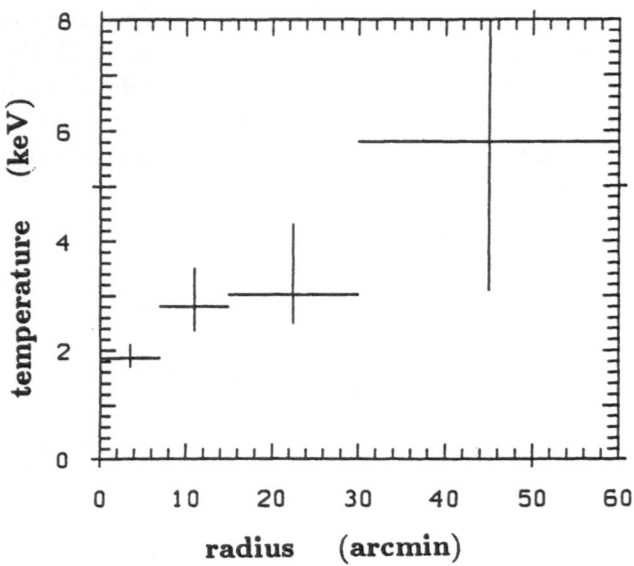

Fig. 10. Temperature profile of the gas in the X-ray halo of M87. The vertical error bars show the 2σ uncertainties.

Fig. 10 summarizes these results in the form of a temperature profile for the M87 halo. If the average temperature is determined by χ^2 fitting over the whole selected region one obtains a value of $2.4 \left[^{+0.35}_{-0.2}\right]$ keV which is in good agreement with the results of the GINGA satellite by Koyama *et al.*(1991) who found a temperature of 2.19 ($\pm$0.04) keV. Only in the innermost part of the cluster, at radii $\leq$ 15 arcmin, is a significant temperature drop observed.

A deprojection analysis of the surface brightness profile (following Fabian *et al.*, 1981) yields a cooling flow radius of about 120 kpc inside which the cooling time is shorter than the Hubble time. This cooling flow radius of $\sim$ 20 arcmin is consistent with the observed temperature drop inside a 15 arcmin radius and also its magnitude is within the limits of the predictions of a one-phase cooling flow model. The flow rate in the cooling flow that is determined from the surface brightness profile has a value of 20 ($\pm$4) $M_{\odot}$ y^{-1} in good accord with the results by Stewart *et al.* (1984).

One can also try to determine the metallicity of the gas through spectral fitting. If this is done for the coldest, innermost region one finds a value of 0.44 ($\pm$0.13) of the solar value which is in surprisingly good agreement with the GINGA results obtained by Koyama *et al.* (1991) of 0.44 ($\pm$0.018). But one has to be very careful in the interpretation of the ROSAT spectra. The gas in the cooling flow region should be expected to have a broad temperature distribution. The main spectral feature that has a dependence on the metallicity is the peak in the spectrum at 1 keV which is mainly produced by iron L-shell lines. These lines have their emission maximum at temperatures which lie mostly below the bulk temperature of the gas (see e.g. Canizares *et al.*, 1983). Therefore gas colder than about 10^7 K can give a more than proportional contribution to the 1 keV peak resulting in erroneously high metallicities. A correct analysis can only be performed through a multiphase cooling flow model where calculated spectra for the proper temperature distribution are fitted to the observations.

The brightest cluster of galaxies detected in X-rays is the Perseus cluster. In the ROSAT All Sky Survey it was observed for $\sim$ 505 sec and detected with a count rate of 33.6 cts/s, where the counts were integrated out to a radius of 42 arcmin (Schwarz *et al.*, 1991). Fig. 11 shows the X-ray image of Perseus from the All Sky Survey superposed on the POSS plate image. Also shown in Fig. 11 at the same scale is the image obtained previously with the EINSTEIN IPC (Branduardi-Raymont *et al.*, 1981) which clearly demonstrates the advantage of the "unlimited field of view" in the survey mode. The X-ray halo was detected over a diameter larger than 2.5 Mpc (where $H_o = 50$ km s^{-1} Mpc^{-1} was assumed here and in all the following calculations). The asymmetric extension of the inner contourlines to the east which was already seen in the EINSTEIN image is also visible here. More surprisingly a trace of the large prominent chain of galaxies in Perseus is also clearly seen in the X-ray image on an even larger scale.

The second maximum of the X-ray surface brightness within the western tail in the X-ray contours following the chain of galaxies, coincides with the galaxy IC 310. The spectrum of the photons in this region is hard enough to be consistent with hot cluster or galactic halo gas. The luminosity of this feature around IC 310 is about $2 - 3 \cdot 10^{42}$ erg s^{-1} in the ROSAT band. This is consistent with the luminosity of a group of galaxies(e.g. Kriss *et al.*, 1983). Therefore this feature can be interpreted as a small group of galaxies just falling onto the main cluster.

The image of Perseus contains more than a total of 16,000 photon counts. Thus the image can be subdevided into smaller regions and spectra of good statistical quality can be obtained for each area. Fitting of theoretical thermal spectra (Raymond and Smith, 1977) to the data binned in concentric rings around the central galaxy NGC 1275 result in the temperature profile shown in Fig. 12. There is a significant temperature drop in the centre of the cluster which is due to the strong cooling flow in Perseus. Indications

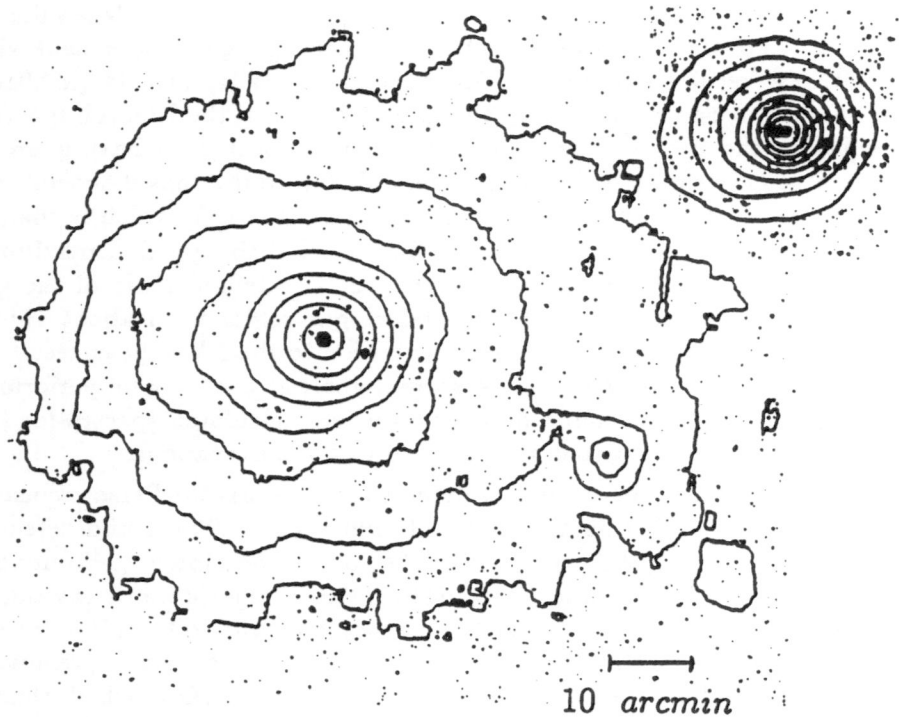

Fig. 11. X-ray image of the Perseus cluster observed in the All Sky Survey. The contour levels are logarithmically spaced. The X-ray map is superposed onto the POSS plate. The main maximum coincides with NGC 1275 while the small western maximum coincides with IC 310. The inset at the upper right corner shows the EINSTEIN IPC image from Branduardi-Raymont *et al.*.

for this temperature drop have also been infered from the collimated X-ray telescopes SPARTAN (Snyder *et al.*, 1990) and Skylab (Ponman, 1991). The temperature profile found is consistent with that derived from modelling the surface brightness distribution assumning the presence of a strong cooling flow (Fabian *et al.*, 1981).

One of the puzzles about the Perseus cluster is the fact that reasonable models reproducing the observed distribution of the X-ray emitting gas have to assume a shallower gravitational potential than is suggested by the optically observed galaxy velocity distribution (e.g. Fabian *et al.*, 1981; Kent and Sargent, 1983). Analysing the X-ray spectra for different parts of the X-ray image of Perseus one notes a temperature gradient across the cluster at radii outside the cooling flow region. While the overall temperature outside the cooling flow can be determined to be $5.27\left[^{+1.3}_{-1.2}\right]$ keV in good agreement with previous results from GINGA and EXOSAT, one finds a significantly lower temperature of $2.06\left[^{+0.71}_{-0.36}\right]$ keV in an eastern and north-eastern sector

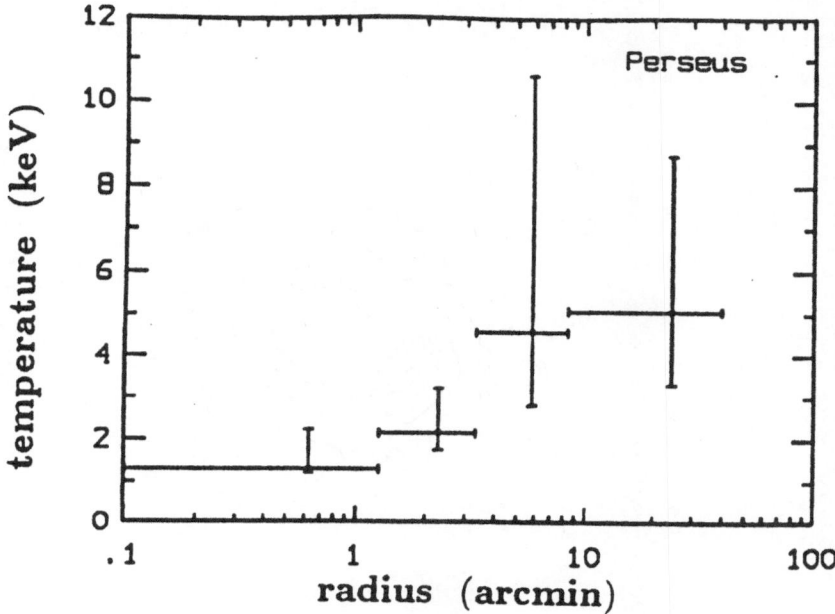

Fig. 12. Temperature profile of the X-ray emitting gas in the Perseus cluster derived from fits of thermal spectra to the ROSAT X-ray spectral data. There is a significant temperature drop in the centre of the cluster which is consistent with the presence of a strong cooling flow centred on NGC 1275.

(The errors quoted correspond to 2σ deviations). An indication of an east-west temperature gradient was also found in the SPARTAN observations by Snyder *et al.* (1990). Such a strong temperature gradient could have been caused by a recent infall of a major subcluster. s Similar to the case of A2256 presented below there could have been a recent merger of a smaller subcluster with lower gas temperature with the main body of Perseus and the small component may not be thermalized yet. Alternatively but less likely the cluster could have been heated by an infall from the western side. This implies that Perseus is a very unsettled cluster and it therefore may explain the discrepancy between the mass estimates from velocity and gas distribution data.

During the calibration phase the Abell cluster 2256 was observed for 17,323 sec. It was previously considered as a symmetric and well relaxed Coma type cluster even though a slight ellipticity was observed with EIN-STEIN (Fabricant *et al.*, 1984, 1989). The ROSAT image which is shown in Fig. 13 clearly shows two maxima in the X-ray surface brightness. Further analysis which is described in detail by Briel *et al.*(1991) shows that the cluster can be decomposed into two subcomponents.

Fig. 13. X-ray image of the cluster A 2256 taken with the ROSAT PSPC. The X-ray contours are in intervals of 2 PSPC counts per 8 × 8 arcsec pixels starting at 2 and the image has been smoothed by a Gaussian with a width of 48 arcsec.

In the 270 degree wide sector from position angle 310 to 220 the cluster appears azimutally symmetric around the central dominant galaxy which coincides with the eastern maximum. A surface brightness profile of the form used by Jones and Forman (1984) can be well fitted to this part of the cluster yielding a core radius of 480 ± 70 kpc and a scale height parameter $\beta = 0.755 \pm 0.025$. If these parameters are used to calculate the surface brightness of the symmetric component of the cluster in the remaining sector and if this result is subtracted from the original image one is left with a very compact second cluster component which is shown in Fig 14. This subcluster is clearly distorted indicating that the two cluster components are interacting.

If the sample of 87 galaxy redshifts is divided into the same two sectors the mean galaxy velocity in the undistorted sector is 17817 km s^{-1} while for the sector containing the second component the mean velocity is 16977 km s^{-1}. The probability that the two velocity distributions belong to the same distribution function is less than 5 %. A fit of two Gaussian distributions to the total velocity dispersion histrogram yields a velocity difference of about 2000 km s^{-1}, where the smaller component has the lower velocity. For a total mass of the system of $\sim 10^{15}$ M$_\odot$ this velocity corresponds to the case

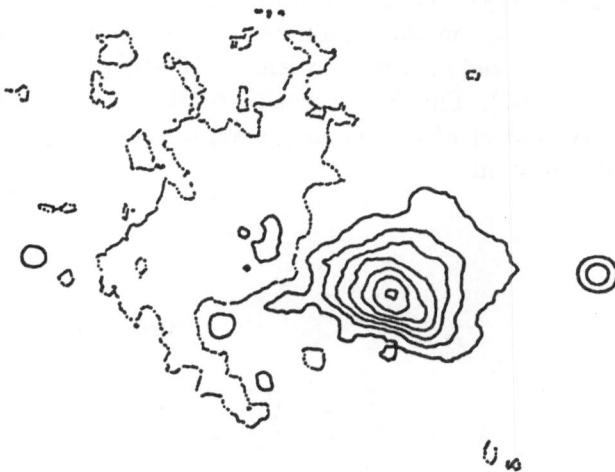

Fig. 14. X-ray image of the second component in A 2256. This contour image has been obtained by subtracting a surface brightness distribution which was fitted to the undisturbed part of A 2256 and which is symmetric around the central cD galaxy and the eastern maximum. The contour level intervals are 2 PSPC counts per 10 × 10 arcsec pixels starting at 2.

where the two subcomponents started at rest and are today approaching each other at a distance of 1 to 2 Mpc. At this distance the observed X-ray halos already overlap which is consistent with the above implications from the X-ray image that the distortion of the smaller component is due to interaction of the intracluster medium of the two cluster components.

A2256 is also one of the few strong and extended cluster radio halos discovered so far (Hanisch, 1982). The very complex radio morphology of A2256 (see Bridle and Fomalont, 1976; Bridle *et al.*, 1979) has been a puzzle to the observers. Knowing that A2256 is a merging cluster explaines many of the radio features. Three prominent head-tail radio galaxies that are probably members of the smaller clump have their radio tails pointing in the direction from which the smaller component has presumably fallen in. The very extended radio emission could well be a result of cosmic ray acceleration at the gas shock of the merging clusters. Estimates show that a fraction of a percent of the shock power is sufficient to power the synchrotron electron population of the observed radio halo. Thus the radio appearance of A2256 provides further support for the merging szenario of A2256.

The Ophiuchus cluster is another very bright low galactic latitude cluster with a redshift of $z = 0.028$. Even though its soft X-ray flux is highly absorbed by an interstellar column density of $\sim 2 \cdot 10^{21}$ cm^{-2} it is detected in the ROSAT All Sky Survey with a high count rate of ~ 7 cts/s. Fig. 15 shows the X-ray image on the Ophiuchus cluster from the All Sky Survey. Ophiuchus was observed previously with the EINSTEIN HRI and EXOSAT (Arnaud *et al.*, 1987). The X-ray source to the north of the cluster was identified by Arnaud *et al.*as a 11 mag star, probably a pre main sequence star or RS CVn system.

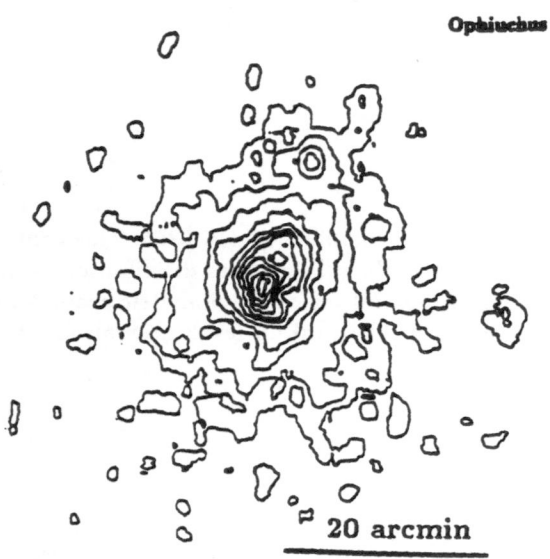

Ophiuchus

20 arcmin

Fig. 15. X-ray image of the Ophiuchus cluster from the ROSAT All Sky Survey. The emission of the cluster is centred on the bright central galaxy. A second X-ray source within the cluster north of the cluster centre was previously identified with a 11 mag star.

For these bright clusters of galaxies in the ROSAT survey the surface brightness profile can be determined out to larger radii than previously possible. In the X-ray image of Ophiuchus the X-ray luminosity of the cluster can be traced out to about 1.6 Mpc (35 arcmin). The model fits give a quite well defined value for $\beta = 0.67$ and a core radius of $r_c = 5.1$ arcmin. This will allow a more precise determination of the mass profile, the gas scale height parameter, and the gas content of the cluster.

AWM 7 was found as a group of galaxies around the prominent cD galaxy NGC 1129 (Albert *et al.*, 1977). This object was observed for about

420 sec in the ROSAT All Sky Survey and detected with a count rate of $\sim$ 7.8 cts/s (inside a cluster radius of 30 arcmin, corresponding to $\sim$ 0.9 Mpc). The cD-group has a galactic latitude below 20° and thus did not enter the ACO catalogue, but it is at least comparable to a richness class 0 Abell cluster. Fig. 16 shows the X-ray image of AWM7 from the All Sky Survey. The central cD galaxy NGC 1129 is found to have a small offset to the X-ray centre determined at intermediate radii.

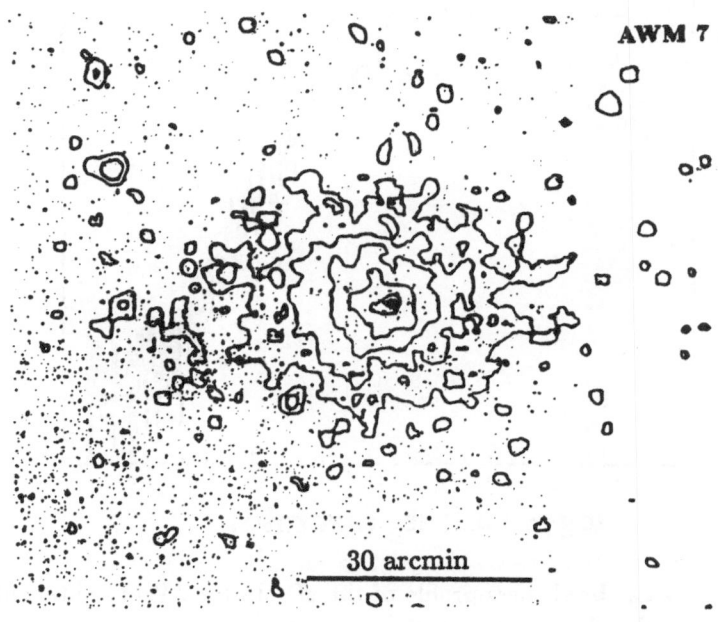

Fig. 16. X-ray image of the galaxy cluster AWM7 from the ROSAT All Sky Survey. The X-ray emission is almost centred on the central dominant cD galaxy NGC 1129. The x-ray emission can be traced out to a radius of $\sim$ 1 Mpc.

An azimutally averaged X-ray surface brightness profile for the cD-group AWM7 is shown in Fig. 17. An asymmetric component of the X-ray surface brightness located at the position of NGC 1129 has been cut out before the azimutal average was taken. The surface brightness profile was fit by the formula for an isothermal cluster model (Jones and Forman, 1984):

$$S = S_o \left(1 + \left(\frac{r}{r_c}\right)^2\right)^{-3\beta+1/2} \tag{1}$$

The scale height parameter determined from Fig. 17 is $\beta = 0.6$ and the core radius $r_c = 5.1$ arcmin (~ 150 kpc). The total gas mass out to a radius of 0.9 Mpc is about $2.5 \cdot 10^{13}$ M$_\odot$.

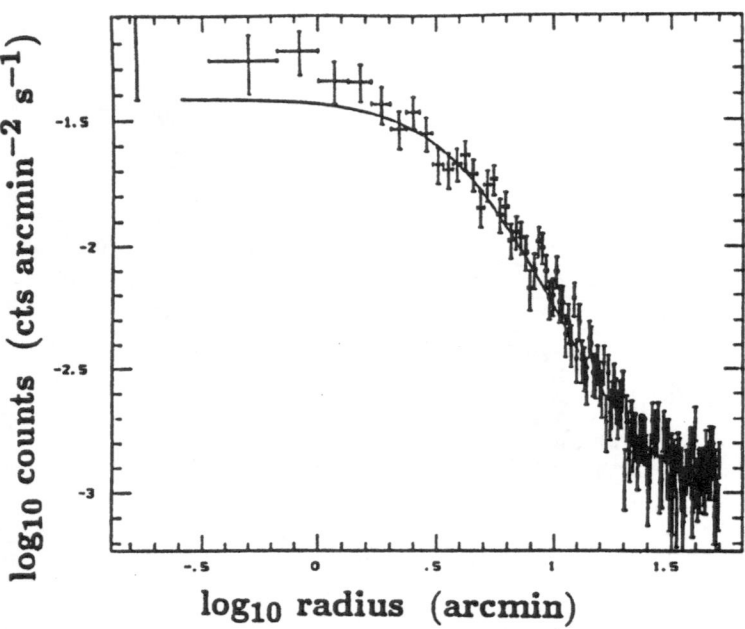

Fig. 17. X-ray surface brightness profile of the cD-cluster AWM7 obtained from the data shown in Fig. 11. An asymmetric peak centred on NGC 1129 near the centre of the cluster has been cut out from the image for the determination of the surface brightness profile.

5 RELEVANCE OF X-RAY OBSERVATIONS OF CLUSTERS OF GALAXIES FOR COSMOLOGICAL RESEARCH

The first results from observations of clusters of galaxies with ROSAT have shown that a number of very interesting cosmological research programs can be conducted with this X-ray observatory concerning the study of individual clusters as well as the study of large statistical samples. On the very general level of understanding the formation of our Universe the most important application of these observations are related to the study of the evolution of the large scale structure in the Universe and the cosmological parameters that govern the cosmic dynamics. Four major research programs will be conducted that are directed towards these applications: a careful determination of the present day X-ray luminosity function of clusters, a correlation of various physical and observable properties for a well studied sample of clusters, a study of cluster evolution towards high redshifts, and the analysis of the spatial correlation of X-ray selected clusters of galaxies.

For a large statistically complete sample of the order of 1000 clusters of galaxies the X-ray luminosity function should be determined which gives an approximate account of the size and mass spectrum of clusters of galaxies at the present epoch. In addition the statistics of more detailed cluster properties such as the spatial extent, the X-ray temperature, the morphology and state of evolution should be derived from more detailed observations of a smaller sample of clusters. These observations will then allow to statistically correlate X-ray luminosities to cluster masses and other properties.

These data are important for the comparison with theoretical predictions for the evolution of clusters from primordial density fluctuation fields for different cosmological scenarios (e.g. Frenk *et al.*, 1990). In particular the distribution of morphologies, as for example the fraction of observed clusters that show a merger or other signs of ongoing evolution at present, may provide the basis to discriminate between a closed or open Universe.

Another important clue to the understanding of large scale structure should come from the investigation of the evolution of clusters of galaxies. The above estimates show that such studies should be possible for small statistical samples (20 to 40 objects) out to redshifts of $z \sim 0.5$ where significant evolutionary effects are expected. A small number of clusters should even be discovered out to distances of $z = 1$. The comparison of these data with models will provide another serious constraint on the cosmological model. The magnitude of the evolutionary effect for accessible redshifts depends very sensitively on the form of the power spectrum of the primordial fluctuation field at wavelengths around 10 Mpc (Perrenod, 1980; Kaiser, 1986; Evrard, 1989; Böhringer, 1991).

Finally the large sample of X-ray selected clusters should be used to study the cosmography and the spatial correlation of the clusters. One of the most important first results of the ROSAT survey is an indication that this cluster sample will be different from the present optical samples and will probably also have quite different properties. The X-ray data will in particular permit the construction of a large cluster sample with clear selection criteria that are homogeneous over the sky except for the galactic plane. This is essential for a precise spatial correlation analysis.

The final goal is a study of the spatial distribution in three dimensions for which redshifts for the X-ray selected cluster sample are needed. Several optical follow-up programs are devoted to this aim. A statistically complete all sky study can probably be done out to a redshift of $z = 0.1$. Another program is aiming for a deeper sampling scale out to $z = 0.25$ for a few 1000 deg^2 in the southern sky and a very deep program is planed for the north ecliptic pole region.

These results will provide information on the power spectrum of the primordial fluctuation field on scales of up to 1000 Mpc. Considering the very dramatic implications the recently obtained results of the galaxy correlation had on the validity of different cosmological models (e.g. Saunders et al., 1989) one can expect further interesting surprises if the above mentioned projects are successful.

Besides this work on statistical samples of clusters it is also important in this context to understand the evolution and present structure of clusters of galaxies in detail. One important topic which was highlighted by these first results from ROSAT is the question of the dynamical state of these clusters. Cluster of galaxies that have attained a quasi equilibrium configuration (which can for example quite well be described by the King model) should be spherical symmetric and should have no signs of particular internal structure. Two clusters of galaxies, Coma and A 2256, which had been thought of as good examples for relaxed clusters previously, have revealed signature of internal structure in the first ROSAT images. Previously substructure in clusters has been studied mainly by optical and spectroscopic observations (e.g Geller, 1984; Dressler and Shectman, 1988; Zabludoff et al., 1990). But due to the limited statistics which can be achieved with the number of galaxies that can be studied spectroscopically, evolutionary effects as shown in the ROSAT images have never been discovered at such detail. The examples of A 2256 and Perseus of which the ROSAT images were shown here imply that these clusters are currently growing due to further infall of groups of galaxies or even mergers of clusters. These ROSAT results show that X-ray imaging is a very sensitive means to study the internal structure of clusters of galaxies.

Further important problems that require a detailed study of individual clusters are the cluster mass, the distribution of the gravitating mass in the cluster and the luminous to dark mass ratio within the cluster. ROSAT observations can provide detailed information on the gas density distribution in the cluster and give some constraints on the temperature profile of the gas. From these data the gravitational potential of the cluster can be reconstructed within certain limits. In the context of the cosmological studies metioned above it is important to have the correlation of the cluster mass and X-ray luminosity as well defined as possible which could be obtained from a small but well studied sample.

Another interesting application for the modelling of the cluster potential from X-ray images is the understanding of the lensing properties of clusters of galaxies that contain distorted, lensed images of distant galaxies.

The investigation of the gas contend of clusters of galaxies which can easily be deduced from the X-ray images is related to further interesting topics:

the origin of the intracluster gas which is certainly linked to the galaxy evolution in the cluster and the phanomenon of cooling flows. Also to investigate the Sunyaev-Zel'dovich effect in clusters of galaxies that should eventually result in an independent check on the Hubble constant, a knowledge of the distribution of the gaseous intracluster medium is essential.

All these topics constitute an extensive program of cluster of galaxy observations that should be carried out with ROSAT. One can only hope for a long lifetime of the satellite that a major part of these goals can be achieved.

Acknowledgement I like to thank the ROSAT team for the help in the data preparation and data reduction and for making these unique ROSAT observations possible. In particular I am grateful to R.A. Schwarz, H. Ebeling, W. Voges, A.C. Edge, U.G. Briel, G. Hartner, R.G. Cruddace, and S. Schindler for the help in the data analysis and for providing data from their work. I also like to thank H.T. MacGillivray, C.A. Collins, and D. Yentis for providing the optical data base and the help in the identification of the ROSAT cluster sources.

References

Abell, G.O., 1958,*Astrophys. J. Suppl.,* **3**,211.

Abell, G.O., Corwin, H.G., Olowin, R.P., 1989,*Astrophys. J. Suppl.,* **70**, 1.

Albert, C.E., White, R.A., Morgan, W.W., 1977, *Astrophys. J.,* **211**, 309.

Arnaud, K.A., Johnstone, R.M., Fabian, A.C., Crawford, C.S., Nulsen, P.E.J., Shafer, R.A., Mushotzky, R.F., 1987, *Mon. Not. R. astr. Soc.,* **227**, 241.

Bahcall, N.A. 1977, *Astrophys. J. (Letters),* **217**, 177.

Binggeli, B., Tammann, G.A., Sandage, A., 1987, *Astron. J.,* **94**, 251.

Böhringer, H., 1991, in: *Traces of Primordial Structure in the Universe* (in press), Böhringer, H. & Treumann, R.A. (eds.), MPE Report No. 227.

Branduardi-Raymont, G., Fabricant, D., Feigelson, E., Gorenstein, P., Grindley, I., Soltan, A., & Zamorani, G., 1981, *Astrophys. J.,* , **248**, 55.

Bridle, A.H. and Fomalont, E.B., 1976, *Astron. Astrophys.,* **52**, 107.

Bridle, A.H., Fomalont, E.B., Miley, G.K., and Valentijn, E.A., 1979, *Astron. Astrophys.,* **80**, 201.

Briel, U.G., Henry, J.P., Schwarz, R.A., Böhringer, H., Ebeling, H., Edge, A.C., Hartner, G.D., Schindler, S., & Voges, W., 1991, *Astron. Astrophys.,* **246**, L10.

Canizares, C.R., Clark, G.W., Jernigen, J.G., and Markert, T.H., 1983, **262**, 33.

Dressler, A. & Shectman, S.A., 1988, *Astron. J.,* **95**, 985.

Ebeling, H., 1991, in *Clusters and Superclusters of Galaxies,* Colles, M.M., Babul, A., Edge, A.C., Johnstone, R.M., and Raychaudhury (eds.), Institute of Astronomy, Cambridge, p27.

Edge, A.C., Stewart, G.C., Fabian, A.C., & Arnaud, K.A., 1990, *Mon. Not. R. astr. Soc.,* **245**, 559.

Evrard, A.E., 1989, *Astrophys. J.,* **341**, L71.

Fabian, A.C., Hu, E.M., Cowie, L.L., Grindley, J., 1981, *Astrophys. J.,* **248**, 47.

Fabricant, D., Lecar, M., and Gorenstein, P., 1980, *Astrophys. J.*, **241**, 552.

Fabricant, D., Kybicki, G., & Gorenstein, P., 1984, *Astrophys. J.*, **286**, 186.

Fabricant, D., Kent, S., & Kurtz, M., 1989, *Astrophys. J.*, bf 336, 77.

Frenk, C.S., White, S.D.M., Efstathiou, G., & Davis, M., 1990, *Astrophys. J.*, **351**, 10.

Geller, M.J., 1984, *Comments Ap.*, 10, 47.

Gioia, I.M., Henry, J.P., Maccacaro, T., Morris, S.L., Stocke, J.T., & Wolter, A., 1990a, *Astrophys. J.*, **356**, L35.

Gioia, I.M., Maccacaro, T., Schild, R.E., Wolter, A., Stocke, J.T., Morris, S.L., & Henry, J.P., 1990b, *Astrophys. J., Suppl.*, **72**, 567.

Heydon-Dubleton, N.H., Collins, C.A., & MacGillivray, H.T., 1989, *Mon. Not. R. astr. Soc.*, **238**, 379.

Huchra, J.P., Henry, J.P., Postman, M., & Geller, M.J., 1990, *Astrophys. J.*, **365**, 66.

Hughes, J.P., 1989, *Astrophys. J.*, **337**, 21.

Jones, C., Forman, W., 1984, *Astrophys. J.*, **276**, 38.

Kaiser, N., 1986, *Mon. Not. R. astr. Soc.*, **222**, 323.

Kent, S.M., and Sarget, W.L.W., 1983, *Astron. J.*, **88**, 697.

Koyama, K., Takano, S., and Tawara, Y., 1991, *Nature*, **350**, 135.

Kriss, G.A., Cioffi, D.S., & Canizares, C.R., 1983, *Astrophys. J.*, **272**, 439.

Lahav, O., Edge, A.C., Fabian, A.C., & Putney, A., 1989, *Mon. Not. R. astr. Soc.*, **238**, 881.

Perrenod, S.C., 1980, *Astrophys. J.*, **236**, 373.

Ponman, T., 1991, results presented at the 28th Yamada conference on *Frontiers in X-ray Astronomy*, Nagoya, April 8 to 12, 1991.

Raymond, J.C., & Smith, B.W., 1977, *Astrophys. J. Suppl.*, **35**, 419.

Saunders, W., Rowan-Robinson, M., Lawrence, A., Kaiser, N., Efstathiou, G., Ellis, R.S., & Frenk, C.S., 1989, *Nature*, **349**, 32.

Schwarz, R.A., Edge, A.C., Voges, W., Böhringer, H., Ebeling, H., and Briel, U.G., 1991, *Astron. Astrophys.*, (submitted).

Snyder, W.A., Kowalski, M.P., Cruddace, R.G., Fritz, G.G., 1990, *Astrophys. J.*, **365**, 460.

Stewart, G.C., Canizares, C.R., Fabian, A.C., Nulsen, P.E.J., 1984, *Astrophys. J.*, **278**, 536.

Sutherland, W.J., 1988, *Mon. Not. R. astr. Soc.*, **234**, 159.

Takano, S., Awaki, H., Koyama, K., Kunieda, H., Tawara, Y., Yamauchi, S., Makashima, K., and Ohashi, T., 1989, *Nature*, **340**, 289.

Trümper, J., *et al.*, 1991, *Astron. Astrophys.*, (in press).

Zabludoff, A.I., Huchra, J.P., & Geller, M.J., 1990, *Astrophys. J. Suppl.*, **74**, 1.

Statistics of Gravitational Lensing 1: Strong Lenses

Nick Kaiser

CIAR Cosmology Program, CITA, University of Toronto

1 Introduction

Gravitational lensing provides a direct probe of mass fluctuations in the Universe. In this first lecture I will consider 'strong lenses'; lenses with surface density greater than the critical value required to produce multiple images of distant sources. In the following lecture I will consider 'weak lensing' generated by structures with sub-critical surface density whose primary observable effect is to distort the images of background galaxies.

In this lecture I will first review some simple but illustrative models of gravitational lenses. To convert these models into predictions for lensing statistics requires two extra ingredients: the first is some statistical prescription for the lenses; e.g. from the galaxy or cluster mass function. The second is a description of the background objects being lensed; the quasar luminosity function (QLF). By combining these ingredients one can predict observables such as the frequency of resolvable multiple images (macro-lensing). In modelling these macro-lenses—we consider galaxies and clusters—we assume that the distribution of dark matter is locally smooth, and we find that the optical depth for lensing—by which we mean the fraction of observers who see multiple images of a given source—is very small. This is because only the central parts of galaxies or clusters are capable of producing multiple images and most of the mass resides outside these regions. However, if the dark matter is clumpy on a microscopic scale then a much larger optical depth is possible. Small clumps do not give rise to resolvable multiple images—for solar mass lenses, for instance, the splitting will be on the order of 10^{-6} arc-seconds—but this 'microlensing' can amplify the luminosity of quasars. This has several observable consequences. Here we will consider quasar-galaxy correlations caused by the brightening of quasars which lie behind galaxies, and which potentially provides a very direct measure of the density in compact objects around galaxies. Another interesting situation is if much of the dark mass is in compact objects, in which case the Universe as a whole may have a significant optical depth to microlensing.

My goal here is to emphasis the statistics of lenses and to illustrate how one can combine the various ingredients to confront observations. More elaborate lens models exist, and the data will no doubt improve, so the results obtained should not be considered definitive. They should however give reasonable rough estimates for many interesting cases, and they should illustrate the general dependence of lensing probabilities on such factors as source redshift and luminosity. Interesting aspects of strong lensing which are not considered here are time delays; giant arcs—though arc*lets* are considered in the second lecture; lensing of extended sources; and the temporal behaviour of microlensing events.

2 Theory

For simplicity, I will only consider lensing on an Einstein-de Sitter background and, since we will be considering deflections by systems with size $\ll H^{-1}$, in calculating the deflection of light rays we can use twice the Newtonian result for a particle moving at $v = c$. In comoving coordinates, the line element for the background cosmology is $ds^2 = a(\eta)^2(d\eta^2 - \delta_{ij}dw^i dw^j)$. Let the conformal time $\eta = 1$ at the present. The scale factor is $a = a_0\eta^2$, so the age of the universe is $t_0 = a_0/3$ and the Hubble parameter is $H = H_0\eta^{-3}$ with $H_0 = 2/a_0$. The lookback time is related to redshift by $(1 + z) = (1 - w)^{-2}$. Note that with this convention the unit comoving length is $a_0 = 6000$ Mpc/h, where $h \equiv H/(100 \text{ km/s/Mpc})$.

2.1 General Formalism

Let us consider a point source at the origin of our coordinates which emits radiation isotropically and let us set up a projector screen at some distance w_Q which is perpendicular to the z-axis; this is the 'observer plane'. In the absence of perturbations, the observer plane will be uniformly illuminated and a ray which leaves the quasar with angle $\vec{\theta}$ (i.e. with direction vector $\{\theta_1, \theta_2, \sqrt{1 - \theta^2}\} \simeq \{\theta_1, \theta_2, 1\}$) will pierce the observer plane at $\vec{x} = w_Q\vec{\theta}$. This defines our Lagrangian coordinates $\vec{x}$.

Let us now impose density fluctuations $\Delta(\mathbf{w}) \equiv (\rho(\mathbf{w}) - \bar{\rho})/\bar{\rho}$, with associated Newtonian (peculiar) potential ϕ given by Poisson's equation, which, in our coordinates, is

$$\nabla^2\phi(\mathbf{w}) = -6\eta^{-2}\Delta(\mathbf{w}), \qquad (2.1.1)$$

where ∇ denotes differentiation with respect to comoving coordinate: $\nabla_i f \equiv \partial f/\partial w_i \equiv f_{,i}$. These density perturbations will deflect the rays and the *displacement vector* $\vec{s}$ satisfies

$$\ddot{s}_i = -2\phi_{,i} \qquad (2.1.2)$$

and integrating this twice give the displacement for a ray with initial direction $\vec{\theta} = \vec{x}/w_Q$ at distance w from the quasar

$$s_i(\vec{x}, w) = -2 \int_0^w dw'(w - w')\phi_{,i}(\mathbf{w}' + \vec{s}(\vec{x}, w')), \qquad (2.1.3)$$

where $\mathbf{w} = \{w\theta_1, w\theta_1, w\}$ is the unperturbed light path, so the mapping from Lagrangian to Eulerian coordinates in the observer plane is:

$$\vec{r}(\vec{x}) = \vec{x} + \vec{s}(\vec{x}, w_Q). \qquad (2.1.4)$$

The generic behaviour of this kind of mapping is familiar to afficionados of the Zeldovich approximation or to anyone who has attended an open air swimming pool and contemplated the pattern of illumination on the bottom of the pool. The *deformation tensor* is

$$D_{ij} \equiv \frac{\partial r_i}{\partial x_j} = \delta_{ij} + \Phi_{ij}(\vec{x}) \qquad (2.1.5)$$

where the *shear tensor* is $\Phi_{ij} \equiv \partial s_i/\partial x_j$. If the shear becomes of order unity then caustic lines—the intersection of caustic surfaces with the observer plane—appear wherever $|D| = 0$. The lines in Lagrangian space where $|D|$ vanishes are called 'critical curves', and the caustics are the mapping of these curves into Eulerian space. As one moves across the observer plane and crosses a caustic line then two images of the source will appear (or disappear) together on the sky with infinite intensity (the amplification of an image is just $1/|D|$).

One natural way to proceed would be prescribe some statistical description of the $\Delta(\mathbf{w})$ field and then calculate the probability distribution for $|D|$ and therefore for the image brightnesses. However, things rapidly become complicated; equation 2.1.3 is an integral equation and we are usually interested in those special observers which fall inside the regions delineated by caustics rather than the full ensemble of observers. With one notable exception (see §2.2.6 below) this approach has not yielded analytic results. To try to gain more insight we will now consider some highly idealised special cases.

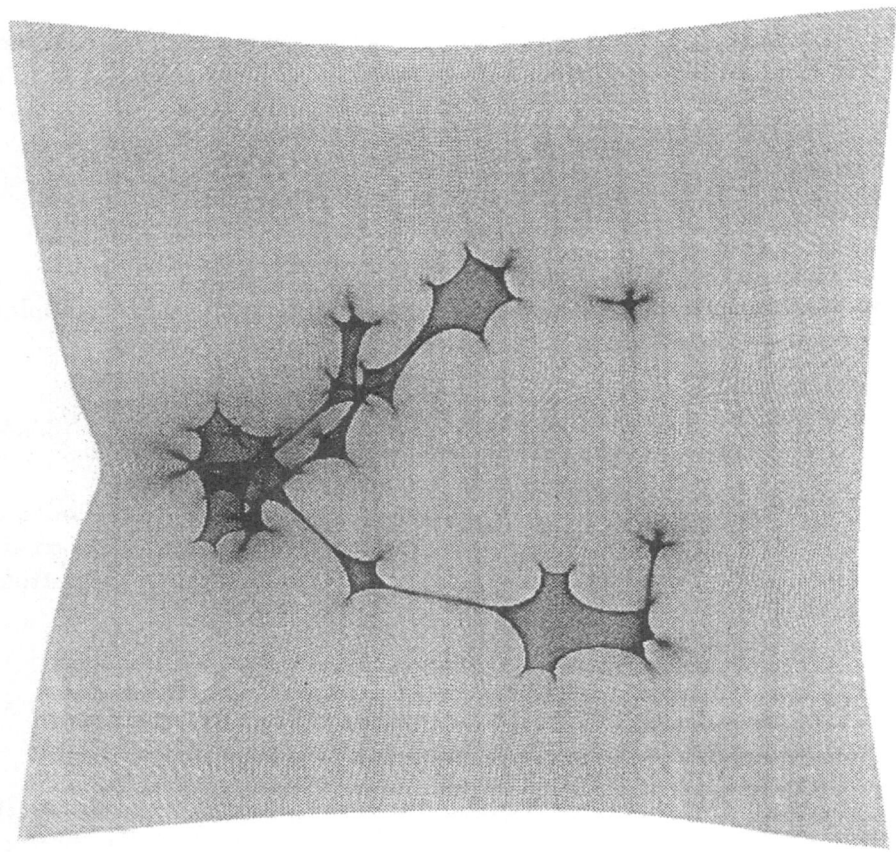

Fig. 1. Illumination pattern for a group of 20 point mass lenses. The lenses were laid down in a Poissonian manner within a circle of radius 0.4 times the figure width. The mean surface density is 0.3 time the critical value. The intensity of light (summed over the images) is proportional to the density of ink. The caustic structure is quite complex.

2.2 Simple Lens Models

We will now specialise to the situation where there is a single lens at distance w_L from the observer, and where the lens is circularly symmetric. The lenses we consider—point masses and isothermal spheres—have been studied in detail by Turner, Ostriker and Gott, (1984), and many of the results below can be found in that paper.

With spherical symmetry, the mapping is just

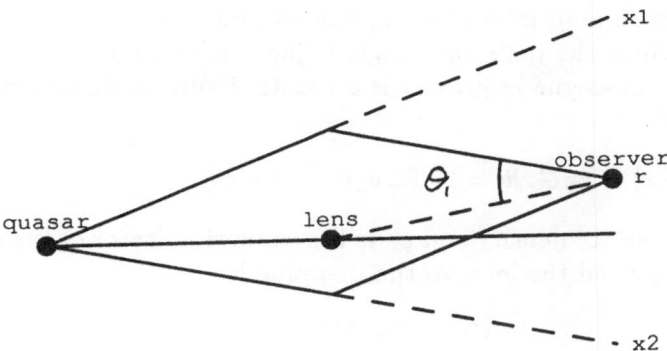

Fig. 2. Geometry for the circularly symmetric lens. An observer at Eulerian position r, relative to the axis defined by the source and the lens, sees multiple images
of the source—here the rays for two images are shown. In general, we wish to
solve for the number of images $N(r)$; the Lagrangian coordinates of these images
$x_1, x_2 \ldots x_N$; the observed angles $\theta_1, \theta_2 \ldots \theta_N$ of these images relative to the lens,
and their amplifications $A_1, A_2 \ldots A_N$, where $A = 1/|D(x)|$. We can then obtain a
probability distribution (over the ensemble of observers) for lensing configurations
$p(N, \theta_1, \theta_2 \ldots \theta_N, A_1, A_2 \ldots A_N)$.

$$r = x - w_{\mathrm{L}}\theta_{\mathrm{b}}, \tag{2.2.1}$$

where the bending angle is

$$\theta_{\mathrm{b}} = \frac{4GM(<b)}{b} \tag{2.2.2}$$

and where $M(<b)$ is the mass interior to the physical impact parameter
$b(x) = a(w_{\mathrm{L}})(1 - w_{\mathrm{L}}/w_{\mathrm{Q}})x$. The angles of the images relative to the lens
centre are

$$\theta = (1 - w_{\mathrm{L}}/w_{\mathrm{Q}})x/w_{\mathrm{L}}, \tag{2.2.3}$$

and the Jacobian of the mapping is

$$|D| = dr^2/dx^2 = (1 - w_{\mathrm{L}}\theta_{\mathrm{b}}/x)(1 - w_{\mathrm{L}}d\theta_{\mathrm{b}}/dx) \tag{2.2.4}$$

2.2.1 Constant Surface Density Lens

In problems concerning laboratory optics one often considers the perfect 'thin lens' in which the deflection angle is just proportional to the impact parameter. The analogue in gravity is a constant surface density disc. The bending angle is

$$\theta_b(x) = 4\pi G \Sigma b = 4\pi G \Sigma a_L (1 - w_L/w_Q)x, \qquad (2.2.1.1)$$

where Σ is the surface density and a_L is the expansion factor of the universe when the light passed the lens, so the mapping is

$$\vec{r} = \vec{x}(1 - \Sigma/\Sigma_c)) \qquad (2.2.1.2)$$

where

$$\Sigma_c(w_L, w_Q) \equiv \frac{1}{4\pi G a_L w_L(1 - w_L/w_Q)} = \frac{H_0}{8\pi G w_L(1 - w_L)^2(1 - w_L/w_Q)}. \qquad (2.2.1.3)$$

For any $\vec{r}$ there is only one $\vec{x}$ and therefore only one image. The amplification of the image is

$$A = 1/|D| = (1 - \Sigma/\Sigma_c)^{-2} \qquad (2.2.1.4)$$

which is infinite in the focal plane where $\Sigma_c(w_L, w_Q) = \Sigma$. The critical surface density Σ_c is, for lenses and sources at cosmological distances, of order $H/G \sim H^{-1}\bar{\rho}$, the Hubble length times the mean density.

Lenses of this kind are common in the optics lab but, beyond defining the critical surface density, they are of limited use in gravitational optics since Nature tends to produce centrally concentrated or irregular lenses.

2.2.2 Point Mass Lens

A more interesting model for a gravitational lens, particularly for microlensing, is the point mass lens. The bending angle is now $\theta_b = 4GM/b = 4GM w_Q/a_L(w_Q - w_L)x$ so the mapping is

$$r = x - x_E^2/x, \qquad (2.2.2.1)$$

where we have defined

$$x_E \equiv \sqrt{\frac{4GM w_L w_Q}{a_L(w_Q - w_L)}} \qquad (2.2.2.2)$$

which is the (Lagrangian) radius of the Einstein ring.

Equation 2.2.2.1 has two solutions

$$x(r) = \frac{r \pm \sqrt{r^2 + 4x_E^2}}{2} \qquad (2.2.2.3)$$

so the observer always sees two images. For an observer on the axis ($r = 0$) the separation of the images (from the lens) is given by $\theta_E = \sqrt{4GM(1 - w_L/w_Q)/a_L w_L}$, and as the observer moves from the axis the separation between the images increases.

The image separation is of little interest for microlensing, but the probability distribution for the net amplification is. If the images lie at x_1, x_2, the amplification for the sum of the images is

$$A = A_1 + A_2 = \frac{dx_1^2 - dx_2^2}{dr^2} = \frac{r^2 + 2x_E^2}{r\sqrt{r^2 + 4x_E^2}} \qquad (2.2.2.4)$$

the minus sign here arises because the determinant has opposite sign for the two images—the images are said to have opposite *parity*—and the last equality follows from equation 2.2.2.3. The amplification is infinite, $A \sim 1/r$, for observers along the axis. Another interesting feature of the point mass lens is that $A_1 - A_2 = 1$, so the difference between the image luminosities is constant and is equal to the true luminosity. The amplification of the weaker image varies as x_E^4/r^4 for large r.

Equation 2.2.2.4 gives the amplification factor as a function of r and therefore of $\theta = r(1 - w_L/w_Q)/w_L$, the angle between the direction to the lens and the (unperturbed) direction to the source. It is useful to define a differential cross section $\sigma(A)$ such that $\sigma(A)dA$ is the solid angle of the annulus about the unperturbed direction to the source where a lens would give amplification $A : A + dA$. Since, from 2.2.2.4, $r^2/x_E^2 = 2\sqrt{1 + 1/(A^2 - 1)}$,

$$\sigma(A) = \frac{\pi}{w_L^2} \frac{dr^2}{dA} = \frac{2\sigma_E}{(A^2 - 1)^{3/2}}, \qquad (2.2.2.5)$$

where $\sigma_E \equiv \pi\theta_E^2$ is the solid angle subtended by the Einstein ring. In §2.3 we will show how one can construct a useful approximation to the microlensing amplification probability distribution using this result.

2.2.3 Singular Isothermal Sphere Lens

A singular isothermal sphere (SIS) with rotation velocity v produces a bending of $\theta_b = 2\pi v^2$, so the mapping is

$$r = x - 2\pi v^2 w_L \text{sign}(x) \qquad (2.2.3.1)$$

which, if $r < x_E \equiv 2\pi v^2 w_L$, gives two images

$$x = r \pm x_E. \qquad (2.2.3.2)$$

The separation between the images is $4\pi v^2(1 - w_L/w_Q)$ and is independent of r, and the amplification factors are, from equation 2.2.4,

$$A = \frac{x_E \pm r}{r} \qquad (2.2.3.3)$$

As with the point mass, the amplification diverges with $A \sim 1/r$ for observers close to the axis, and the difference between the image brightnesses is again constant, though here it is twice the intrinsic image brightness.

The total angular cross-section for multiple imaging is

$$\sigma_E = \pi \theta_b{}^2 (1 - w_L/w_Q)^2. \qquad (2.2.3.4)$$

Multiplying this by the number density of lenses on the sky gives the fraction of observers who will see multiple images of a particular quasar, regardless of the image luminosities. To obtain the actual frequency of lensing events for realistic surveys, we need to multiply this by some 'amplification bias factor' which depends on the observer's selection criteria and the intrinsic QLF. This is discussed in §3, where we will use the differential cross-section $\sigma(A)dA = \sigma_E p(A)dA$, and where the pdf for the summed amplification is

$$p(A)dA = 8A^{-3}dA \qquad \text{for } A > 2; \qquad (2.2.3.5)$$

the pdf for the amplification for the brighter image is

$$p(A_1)dA_1 = 2(A_1 - 1)^{-3}dA_1 \qquad \text{for } A_1 > 2; \qquad (2.2.3.6)$$

and the joint pdf for A_1 and A_2 is

$$p(A_1, A_2) = p(A_1)\delta(A_2 - A_1 - 2). \qquad (2.2.3.7)$$

2.2.4 Spherical Lens + Weak Perturbation

In §2.1 we argued that in the general situation caustic surfaces form behind the lenses, but in the two examples considered above we found that the amplification factor diverges for a line of observers along the axis; the symmetry of a spherically symmetric lens having caused the caustic surface to degenerate to a line. We shall now show, following Chang and Refsdal (1984) and Nityananda and Ostriker (1984) and using the point mass lens as an example, how a small additional perturbation will break this symmetry, and that we then retrieve the generic caustic surface stucture.

A simple generalisation of the isolated point mass lens is to add a small external tidal field. For the low-τ random star field this should be quite a good approximation to describe the effect of neighbouring stars on a ray that passes quite close to one of the stars. If we scale the coordinates so $x_E = 1$, the mapping is

$$r_i(\vec{x}) = (1 - 1/x^2)x_i + Sx_i(\delta_{i0} - \delta_{i1}) \qquad (2.2.4.1)$$

where S is the small external shear. For low optical depth the typical shear is $S \sim \tau$. The Jacobian of this mapping is

$$|D| = |\partial r_i / \partial x_j| = 1 - \left(\frac{\cos 2\phi}{x^2} + S \right)^2 - \frac{\sin^2 2\phi}{x^4}. \qquad (2.2.4.2)$$

where $x_1 = x \cos \phi$, $x_2 = x \sin \phi$.

In the unperturbed case the critical line (where $|D| = 0$) is just the unit circle $x = 1$, and this maps to the origin in Eulerian coordinates. With the small perturbation the critical line is deformed to the slightly squashed ellipse

$$x(\phi) = 1 + d(\phi) = 1 + \frac{S \cos 2\phi}{2}, \qquad (2.2.4.3)$$

where the second equality is only valid to first order in S, and the mapping of this to Eulerian space is given parametrically by

$$\begin{bmatrix} r_1 \\ r_2 \end{bmatrix} = 2S \begin{bmatrix} \cos^3 \phi \\ \sin^3 \phi \end{bmatrix}. \qquad (2.2.4.4)$$

There are now four images inside the caustic and two outside. For example, the observer at $\vec{r} = 0$ sees images at $\phi = \pi/2$, $3\pi/2$ with $x = 1/\sqrt{1 - S}$ and at $\phi = 0$, π with $x = 1/\sqrt{1 + S}$.

In §2.3 we will need $\sigma(A; S)$ the differential cross-section for amplification for this lens. From (2.2.4.2) it is easy to see that for large impact parameters the effect of the shear perturbation is negligible, and one finds that the cross section is asymptotically identical to the unperturbed case (2.2.2.5) for $A \ll 1/S$. One can also show that $\sigma(A)$ is unchanged for very high amplifications as follows: The trick is to note that in the very highest amplification events the observer lives very close to a caustic and sees a pair of very bright images, the rays from which pass in a ribbon straddling the critical line in Lagrangian plane. In this plane, the probability for a ray to have $A > 1/\epsilon$, where $\epsilon \ll 1$ is just proportional to ϵ times the gradient of $\det(D)$ normal to the critical line, so $p(> A) \propto \epsilon \propto A^{-1}$. The probability in the Eulerian plane—i.e. over observers, which is what we want—is

$$p(> A) = \frac{1}{2A} p_L(> A) \qquad (2.2.4.5)$$

the factor two arising because both positive and negative parity images map to the same point in Eulerian space, and the factor $1/A$ just being the Jacobian of the map. This gives $\sigma(> A) \propto A^{-2} \rightarrow \sigma(A) \propto A^{-3}$, just as for an isolated point mass, and in fact this asymptotic scaling of $\sigma(A)$ is a general feature of images near fold catastrophes. Now while the structure of the caustics is very sensitive to external perturbations, the position of the critical line is not—for the small external shear the critical line becomes slightly elliptical—and the area of the ribbon where $A > 1/\epsilon$ is only weakly perturbed $p_L(A) \rightarrow p_L(A) + \delta p_L(A)$ with $\delta p_L(A) \sim S p_L(A)$ and the same is therefore true for $p(A)$. Thus, $p(A)$ is robust to small perturbations for both small ($A \ll 1/S$) and large ($A \gg 1/S$) amplifications but, from a perusal of figure 3, we would expect there to be a feature in between.

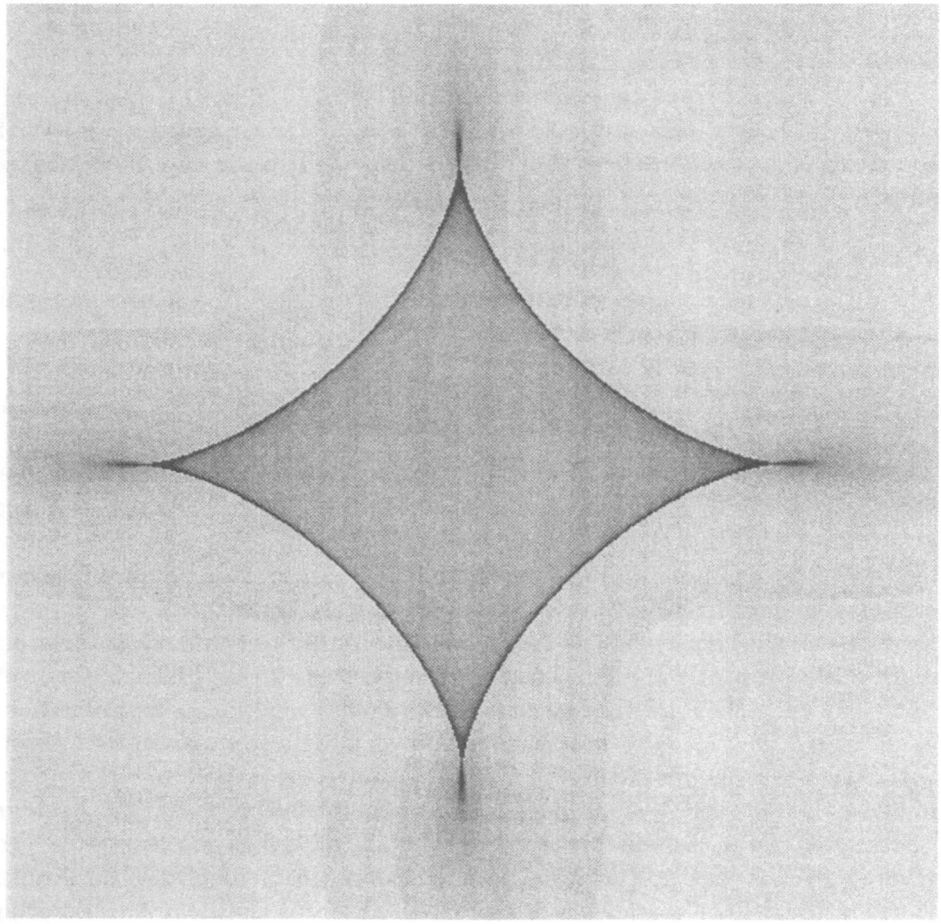

Fig. 3. Illumination pattern on the Eulerian (observer) plane for a point mass lens perturbed by a small external shear. The caustic lines are the intersection of the plane with generic 'fold' caustic surfaces. A similar pattern is obtained for an elliptical lens.

The shape of the feature around $A \sim 1/S$ was first calculated by Nityananda and Ostriker (1984). To calculate $\sigma(A)$ for the summed images—the *macroimage* amplifications—it is necessary to find the two or four images for a given Eulerian position; sum the inverse Jacobians; and then calculate the distribution over all positions. For high amplification we must be close to the critical line $x = 1$ so we set $x = 1 + d$. The mapping 2.2.4.1 then becomes

$$r_1 = (2d + S)\cos\phi$$
$$r_2 = (2d - S)\sin\phi \qquad (2.2.5.1)$$

If we now rescale variables $d' = d/S$, $r' = r/S$ and drop the primes we find that d satisfies the quartic equation

$$d^4 - \frac{2 + r^2}{4} d^2 + \frac{r_1^2 - r_2^2}{4} d + \frac{1 - r^2}{16} = 0, \qquad (2.2.4.6)$$

the real solutions of which give the two or four image positions. Having found the images one can then sum the amplifications to get the macroimage amplification factor.

$$A = \sum |4d - 2S(r_1^2/(2d-1)^2 - r_2^2/(2d-1)^2)|^{-1}. \qquad (2.2.4.7)$$

The differential cross-section is a universal function of SA, and we have

$$\sigma(A; S) = 2\sigma_E A^{-3} f(SA) \qquad (2.2.4.7)$$

All lenses produce the same illumination pattern, and it is only the size of the pattern is influenced by the strength of the perturbation. We can see from figure 4 that the departures from the cross-section for the unperturbed point mass case are considerable.

2.3 Amplification Distribution for Microlensing

We now wish to extend the calculation of sections 2.2.2 and 2.2.4 to treat a collection of point mass lenses. The physical situation we are most interested in is if we view the background of distant quasars through a galaxy composed wholly or in part of compact objects such as stars. Now in dealing with a real galaxy there is a very nice separation of scales between the overall size of the galaxy and the scale on which the mass distribution is locally clumpy, and one can consider the potential to be a smooth component generated by the overall mass distribution plus a random component generated by the locally poissonian distribution of stars. This presents us with a very precisely defined mathematical problem: If we view distant point sources through such a deflecting screen, what is $P(A)$, the probability distribution for (macro)image amplifications? Our ultimate goal is to convolve this pdf with the QLF to calculate the enhancment of quasars around foreground galaxies or *vice versa*.

This problem has a long history and has been approached from a number of directions and there are a number of results for $P(A)$ of various levels of sophistication and treating various limiting cases and regimes. Most progress has been made for the case of low optical depth (i.e. deflecting screens where the Einstein rings of the individual lenses rarely overlap) and, as we will describe, a useful result can be obtained by a superposition of individual point

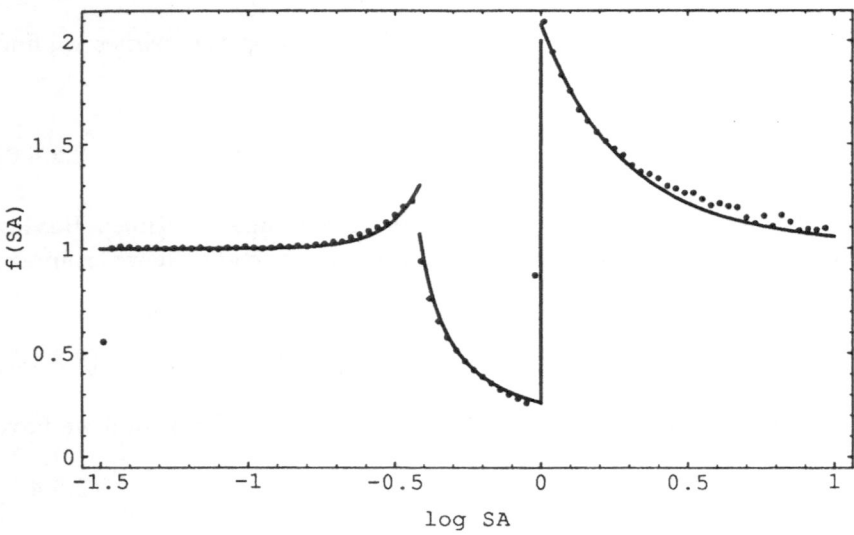

Fig. 4. Plot of $f(SA) = A^3\sigma(A; S)$ for the weakly perturbed point mass lens. This quantity is the deviation of the cross-section from the asymptotic result for an unperturbed point mass, and was calculated from 10^9 points scattered over the observer plane. For each observer we solve (2.2.4.6) for the image locations and then sum the amplifications as in (2.2.4.7). The solid line is a semi-analytic fit obtained by Kofman and Kaiser (1992).

mass lenses, once allowance has been made for the shear from neighbouring stars as for the general focusing of the light rays by the macroscopic mass distribution. We will then review the attempts that have been made to extend these results to treat lenses of finite optical depth.

2.3.1 Low Optical Depth, Thin Screen

In §2.2.2 we treated an isolated point mass lens. Here we want to calculate the pdf for amplification for a sheet of randomly placed point masses. Suppose, for simplicity, that we have a disk of stars. We want to shine light from a point source through this sheet and calculate the amplification probability distribution $P(A)$ for the illumination pattern on the observer plane. We can define an optical depth τ as the fraction of the sky covered by the Einstein rings – this is $n\sigma_E$ if n is the mean surface number density of stars in the sheet, and is also equal to the surface density of the sheets in units of Σ_{crit} – and we will consider the case where $\tau \ll 1$. Even with this assumption the final form for $P(A)$ is rather involved:

$$P(A) = \frac{2\tau f_1(\tau A) f_2((A-1)/\tau^2)}{(1-\tau)^2 (A^2-1)^{3/2}} \qquad (2.3.1.1)$$

and displays features at $A \sim 1/\tau$, $A \sim 1$ and $A - 1 \sim \tau^2$. The derivation of this result (Kofman and Kaiser, 1992) makes use of two approximation schemes which apply in two overlapping regimes: $(A-1) \gg \tau^2$ and $A \ll 1$; reassuringly these give the same result in the region where both are applicable.

The high amplification regime is obtained as follows: The typical shear perturbation in the vicinity of one star due to its neighbours is $S \sim \tau \ll 1$. A simple way to see this is to note that the typical distance to the nearest star is $x \sim n^{-1/2} \simeq x_E \tau^{-1/2}$ and the shear is $S = x_E^2/x^2$. Thus each star will generate an astroid caustic like that shown in figure 3, with the size of each astroid being determined by the particular value of the shear. The fraction of the observer plane covered by these astroids is $\sim \tau^2$ and is small. Consider observers who live close to or inside one of these astroids. We can calculate $P(A)$ for these observers as some kind of convolution of the cross section $\sigma(A; S)$ given by (2.2.4.7) with the probability distribution for the shear (the derivation of which is reviewed in the next section). An additional wrinkle is that one must allow for the net convergence of the light rays by the macroscopic surface density of the disk; this enhances the surface density of astroids on the observer plane by a factor $\bar{A} = (1-\tau)^{-2}$. The upshot of this is

$$P(A; \tau) dA = \frac{2\tau f_1(\tau A)}{A^3 (1-\tau)^2} \qquad (2.3.1.2)$$

where

$$f_1(y) = y \int \frac{dy' y' f(y')}{(y^2 + y'^2)^{3/2}}. \qquad (2.3.1.3)$$

We have expressed $P(A)$ here as a suitably scaled version of the isolated point mass cross-section with a correction term $f_1(\tau A)$ which, like the correction term $f(y)$ for an individual perturbed lens from which it is derived is asymptotically unity for $A \ll 1/\tau$ and $A \gg 1/\tau$, but which has a feature at $A \simeq 1/\tau$. This is plotted in figure 5. Note that while each individual star shows large departures from the isolated point mass behaviour, when we average over the rather broad random distribution of shears the effect, while still technically of order unity, tends to get washed out.

The factor $(1-\tau)^{-2}$ in (2.3.1.2) arises from the focussing of the caustics due to the macroscopic surface density. One might question the logic of retaining this correction since we are working under the assumption that $\tau \ll 1$. There are really two justifications in keeping this term: One is that we are not ultimately interested in $P(A)$ – which is the probability that a particular quasar seen through the deflecting screen has been amplified by A – rather we want to know how how the surface density of quasars on the

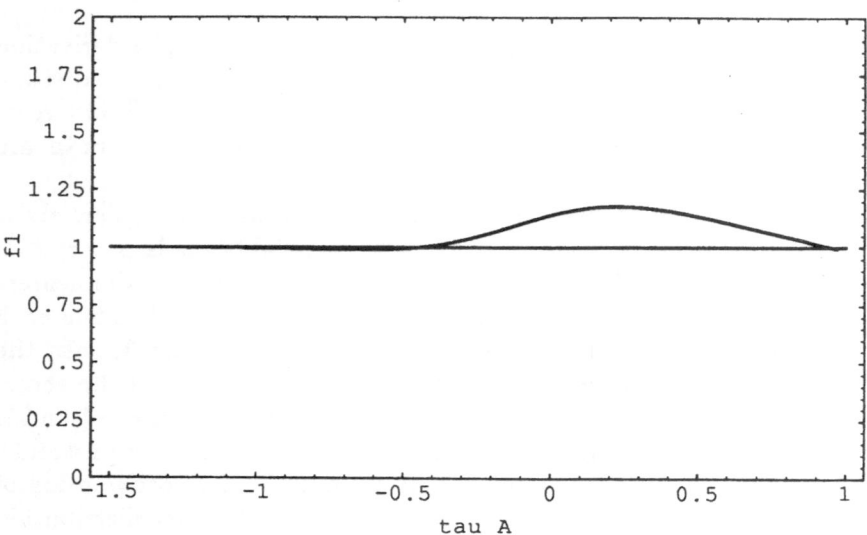

Fig. 5. Comparison of $f(\tau A)$, the deviation from the simple $p \propto A^{-3}$ scaling for a single star and for an ensemble of stars with shear distribution generated from the neighbouring stars.

sky is modified. To make this calculation (see §3.2 below) we must first allow for the net dilution of the surface number density of sources caused by the net focussing – recall that lensing does not change the surface brightness of extended sources; sources are made brighter by enhancing their size, but at the same time this reduces their number density – and this just cancels the factor $(1-\tau)^2$ appearing in (2.3.1.2). It would really make more sense to calculate the observably relevant quantity directly, but in passing through the intermediary of $P(A)$ we are conforming to common practice, and then consistency requires that we keep this term in. A more pragmatic justification for this factor is that it seems to work when we compare with numerical experiments.

Equation (2.3.1.2) is only valid for amplifications corresponding to observers who lie particularly close to one astroid and this, it turns out, requires $(A-1) \gg \tau^2$. Another way to see that this formula requires modification at these low amplifications is to note that the total probability would otherwise exceed unity.[1]

---

[1] While the total probability $\int dA\, P(A)$ diverges at low A with $P(A)$ given by (2.3.1.2), the excess amplification $\int dA\,(A-1)P(A)$ is convergent. In fact the excess amplification is dominated by $A \sim 1$, and, for low τ, is just equal to 2τ

The necessary modification to (2.3.1.2) at $\delta A \equiv A-1 \sim \tau^2$ is obtained in a very different manner. Here we are concerned with observers who do not live particularly near to an astroid. These observers will see one primary image of a distant source which is barely deflected from its unperturbed position on the sky, and in addition will see a number of faint secondary images, one close to each lens created by rays which suffer large deflections. As we will show in the next section, the amplification for the primary image (which we will denote by A_0) is obtained by summing the shears from the stars around the primary ray in a tensorial manner to give $S = \sum s_i$ and using $A = |1 - S^2|^{-1} \simeq 1 + |S^2|$. In §2.2.2 we found that the amplification for the fainter image is $A_i = x_E{}^4/x^4 = s_i^2$ and the net excess amplification is

$$\delta A = (A_0 - 1) + \sum A_i = \left| \sum s_i \right|^2 + \sum |s_i|^2. \qquad (2.3.1.4)$$

Since the stars have a Poissonian distribution, the calculation of $P(\delta A)$ is well defined. Schneider (1986) has obtained an expression for $P(\delta A)$ under the assumtion that the two terms here are independent. Unfortunately, while that gives a cut-off at about $\delta A \sim \tau^2$, the asymptotic behaviour for $\delta A \gg \tau^2$ fails to match the normalisation of (2.3.1.2). This is not altogether too surprising since for $\delta A \gg \tau^2$, both of the sums in (2.3.1.4) are dominated by the nearest star and the two terms are then perfectly correlated. Allowing for the correlations between the primary and secondary images is rather complicated, and while it is not possible to present the result for $P(\delta A)$ in a simple closed form, one can reduce the result to a single integral for which Kofman and Kaiser (1992) have obtained a useful analytic fit.

The final result is (3.2.1.1) with the fitting formulae for f_1 and f_2 given by Kofman and Kaiser which incorporates these features. As it turns out the feature at large amplification, while technically of order unity and therefore in need of calculation, is in fact rather weak, f_1 having roughly a 15% enhancement at the its peak. One should really include the systematic shear from the galaxy, and this requires some assumptions about the mass profile of the galaxy, but since in a realistic survey one will be averaging over galaxies with a range of systematic shear, it is unlikely that this will make a large difference to the result. It should also be admitted that the detailed form for the modification at low amplifications is also largely of academic interest since we will ultimately convolve $P(A)$ with the QLF, and if this is locally smooth then a useful practical approximation is to take

$$P(A) = \frac{2\tau}{(A^2 - 1)^{3/2}(1 - \tau)^2} \qquad (2.3.1.5)$$

but with a cut-off at $\delta A \sim \tau^2$ to render the total probability equal to unity. More precisely, the cut-off should be placed at $A_{\min} = \cosh x$, with $x =$

---

which is the mean excess amplification one obtains with a uniform sheet with the same surface density. This seems very reasonable.

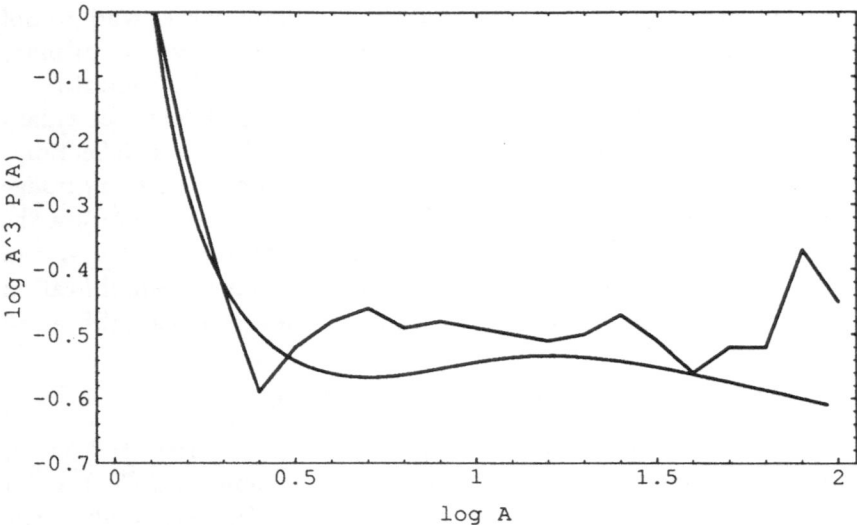

Fig. 6. A comparison of the probability distribution for amplification with that obtained numerically by Rauch *et al.* 1991.

$\coth^{-1}(1-(1-\tau)^2/2\tau)$. While this might seem a rather crude approximation to (2.3.1.1) it turns out that the impact on the final result is rather mild.

These results are only valid for low optical depths $\tau \ll 1$, but, as we shall see, this is the case for several interesting applications. It turns out that the results are also valid for a 3-D distribution of lenses (rather than a thin sheet as we have assume here) since one can map the 3-D problem into the 2-D one. Another nice feature of the low optical depth case is that $P(A)$ is independent of the mass function of the lenses. Unfortunately, neither of these properties hold for finite optical depths.

2.2.6 Amplification Distribution for Finite Optical Depth

The results obtained above apply only for low τ. While this is the relevant regime for some applications, it would be nice to know how $P(A)$ behaves as τ becomes appreciable. One particularly interesting application would be to microlensing of an image which is macroscopically spit since the rays for such an image are essentially guaranteed to pass through a region of surface density comparable to the critical value.

The only exact analytic result that has been obtained is the result by Schneider for $P(A)$ in the high-A limit. The assumption made by Schneider is that very high magnification events are dominated by a pair of very bright images formed close to a fold catastrophe. In that case one can calculate the pdf for amplifications for a single *image* (rather than for the summed intensities, which is much more difficult), and then double the luminosity. This is one of the rare situations where the general formalism of §2.1 is useful.

The calculate $P(A)$ over rays, the procedure is to calculate the pdf for the shear for randomly chosen ray, and then transform this to give $P(A)$. The shear pdf for a random star field was obtained by Nityananda and Ostriker (1984) as follows. We can write the shear at an arbitrary point on the deflector plane (which we take to be the origin) as the sum of contributions from all the stars (with distance b and direction ϕ) to obtain

$$\Phi_{ij} = \begin{bmatrix} S_r & S_i \\ S_i & -S_r \end{bmatrix} \qquad (2.2.6.1)$$

where $S_r = \sum \cos 2\phi/b^2$ and $S_i = \sum \cos 2\phi/b^2$, so if we think of the shear as being a complex number $S = S_r + iS_i$, then the net shear S is just the sum of the contributions from individual stars $S = \sum s$. Now the pdf for the individual star shears is independent of ϕ by symmetry, and as the stars are Poisson distributed has radial profile $P(s) \propto 1/s^2$. Since the stars are statistically independent, the final pdf for S is obtained by taking the Fourier transform of the 1-star pdf and raising this to the number of stars and then inverse transforming. The result was first expressed in terms of Bessel functions by Nityananda and Ostriker and a more convenient form was found by Schneider (1987):

$$P(S)dS = \frac{\tau S dS}{(\tau^2 + S^2)^{3/2}}. \qquad (2.2.6.2)$$

The amplification is, from equations 2.2.6.1, 2.1.5,

$$A = |D|^{-1} = |1 - S^2|^{-1}. \qquad (2.2.6.3)$$

For high amplification, $S = 1 + \delta S$ with $|\delta S| = 1/2A$, so $dA/d\delta S = 2A^2$ and the pdf for A on the Lagrangian (deflector) plane is

$$P_L(A) = P(S = 1)(dA/d\delta S)^{-1} = \frac{\tau}{2A^2(1 + \tau^2)^{3/2}}. \qquad (2.2.6.4)$$

Just as we did in the previous section, we can obtain the probability distribution over the Eulerian plane by multiplying by $\overline{A}/A$, and using $A_{\text{tot}} = 2A$ we find for the final macroimage amplification pdf

$$P(A_{\text{tot}})dA_{\text{tot}} = \frac{2\tau A_{\text{tot}}^{-3} dA_{\text{tot}}}{(1 - \tau)^2(1 + \tau^2)^{3/2}}. \qquad (2.2.6.5)$$

As expected, this produces the generic A^{-3} tail expected from simple fold catastrophes and for low optical depth the constant of proportionality agrees with equation (2.3.1.5), but the amplitude of the high-A tail becomes suppressed as τ increases.

This is a very nice result, but is limited in applicability to sufficiently high amplifications such that the macroimage is dominated by two sources. As we saw in the previous section, for low-τ, one actually has to go to extremely large amplifications $A \gg 1/\tau$ for this to be a good approximation. In the limit of high τ it is not clear that the situation will be much better since the typical macroimage will then be composed of a very large number of microimages, so the assumption of two dominant images may again not be very good.

A rather different approach to the finite-τ $P(A)$ was taken by Peacock (1986) who in essence treated the problem rather like one would treat the problem of absorption by a series of screens which are individually of low optical depth. Another important conceptual point emphasised by Peacock is that we are not really trying to calculate how the QLF is modified relative to the 'true' QLF, rather the operationally useful question is to ask, how would the QLF change if one were to take some homogeneous FRW universe and then make the mass locally clumpy? In the formalism we have adopted this is accomplished by taking the source term for the gravitational potential fluctuations to be a Poissonian distribution of point masses superimposed on a compensating negative density perturbation.

Numerical studies can be used to explore $P(A)$ in the finite-τ regime. Recently, Rauch et al., (1991) have given results for $P(A)$ obtained numerically (both from ray-tracing and image finding). There are some noticeable differences between their results and the low-τ $P(A)$ obtained in §2.3.1. In particular, there seems to be a sharper dip and rise in $A^3 P(A)$ around $A = 3$. They ascribe this to the effect of double lenses, noting that the double lenses have a rather large cross-section at $A \simeq 3$, and we would concur with this view. If correct, this means that the finite-τ $P(A)$ will, unlike the low-τ result, depend on the details of the mass spectrum of lenses. As already noted, the other rather nice property of low-τ lensing – that the results are equally applicable to a 3-D distribution of lenses as to a thin 2-D sheet – does not carry over to the finite-τ case. It is interesting in this regard that Rauch et al. have noted differences in the form of $P(A)$ for 2-D and 3-D lens distributions.

3 Applications

3.1 Multiple Imaging of Quasars

We first review the estimate of the optical depth from galaxies obtained by Fukugita and Turner (1991) using the observed luminosity function for galaxies and the Tully-Fisher and Faber-Jackson relations to convert this to a distribution function in bending angle. While this can account quite well for the small $\lesssim 2''$ splittings, it is hard to account for the larger splittings as lensing by single galaxies; these require multiple galaxy systems—indeed, in several cases a cluster can be seen—and in the latter part of this section we shall compare the frequency of large splittings with that expected based on the observed X-ray temperature distribution function for clusters.

If we denote the differential number density of sources at lookback time w_L and bending angle θ_b by $dn(w_L, \theta_b)/d\theta_b$, then the optical depth is

$$\frac{d^2\tau}{d\ln\theta_o dw_L} = \frac{\pi\theta_o{}^3 a_0^3 w_L^2}{4} \frac{w_Q}{2(w_Q - w_L)} \frac{dn}{d\theta_b}\left(w_L, \frac{\theta_o w_Q}{2(w_Q - w_L)}\right). \quad (3.1.1)$$

To calculate the fraction of multiply imaged quasars in a magnitude limited sample one must multiply this by a 'magnification bias factor'. This depends on the selection criteria and quasar luminosity function (QLF), and is difficult to calculate precisely. However, if the lenses all have similar mass density profiles (as we shall assume here) then (3.3.1) can still be used to predict the shape of the distribution of splitting angles and the relative contributions from galaxies and clusters.

3.1.1 Galaxies

Following Fukugita and Turner (1991), we assume that galaxies come in 3 varieties: E's, S0's and S's, and that each of these has the same shape of luminosity function: $dn = F\phi_*(L/L_*)^\alpha \exp(-L/L_*)dL/L_*$, with $\alpha = -1.1$, and with $\phi_* = 1.56F \times 10^{-2}h^3/\text{Mpc}^3$ with $F = (0.12, 0.19, 0.69)$ for the respective galaxy types. To convert dn/dL to $dn/d\theta_b$ we will use the Tully-Fisher and Faber-Jackson relations and assume that the galaxies have flat rotation curve halos (this should be a good approximation for spiral galaxies, but is rather more speculative for the early type galaxies). Writing $L/L_* = (V/V_*)^\beta = (\theta/\theta_*)^{\beta/2}$, with $\beta = (4.0, 4.0, 2.62)$, $V_* = (389, 356, 189)$ km/s, and $\theta_* = 2\pi V_*^2/c^2 = (2.17'', 1.82'', 0.51'')$. We can immediately infer from this that the optical depth ($\propto \theta_*{}^2\phi_*$) will be divided in the proportions $(0.41, 0.47, 0.13)$, and is therefore dominated by the early type galaxies.

With these assumptions, the optical depth to multiple imaging for objects at w_Q by any one variety of galaxies is distributed over observed splitting angle and lens distance as

$$\frac{d^2\tau}{d\ln\theta_o dw_L} = \frac{\pi\theta_o{}^3\beta F\phi_* a_0^3 w_L^2}{8} z^{\alpha+1} \exp(-z) \quad (3.1.1.2)$$

where $z = (\theta_o w_Q / 2\theta_* (w_Q - w_L))^{\beta/2}$ and so

$$\frac{d\tau}{d\ln\theta_o} = \pi\theta_*^2 F\phi_* a_0^3 w_Q^3 g(\theta_o/2\theta_*) \qquad (3.1.1.3)$$

where

$$g(y) = y^3\,\Gamma(\alpha+1-2/\beta, y^{\beta/2}) - 2y^4\,\Gamma(\alpha+1-4/\beta, y^{\beta/2}) + y^5\,\Gamma(\alpha+1-6/\beta, y^{\beta/2})$$
$$(3.1.1.4)$$

The four bell shaped curves in figure 7 are the contribution to the optical depth from the three classes of galaxies and their sum. Also shown in this figure are the observed splittings from Turner, (1989). The indication seems to be that the observed splittings are larger than one would expect from galaxies alone. It is easy to see what is required: a population with potential wells (and therefore splitting angles) about 3 times larger than those for the galaxies, but with number density about 10 times smaller so that the optical depth ($\tau \propto n\theta_b^2$) is roughly flat. These objects would have virial temperatures of about 1.5 keV, and it is interesting to ask how this abundance matches the temperature function (TF) determined at somewhat higher temperatures from X-ray observations.

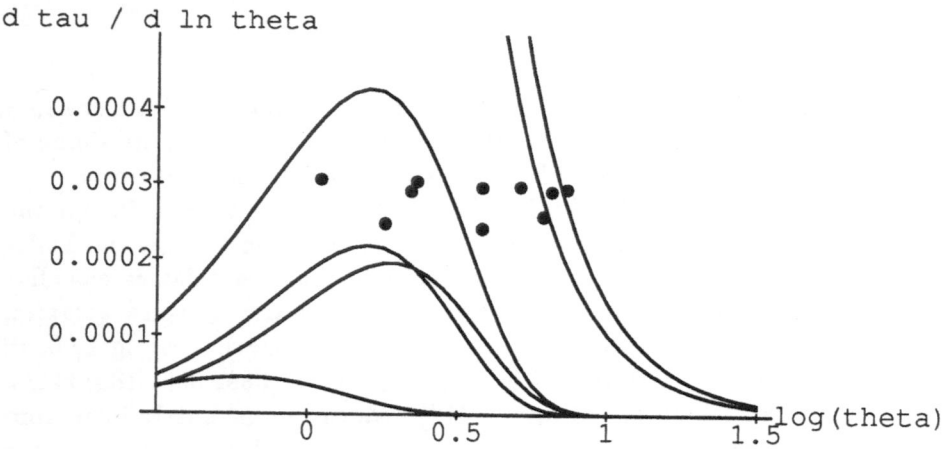

Fig. 7. Optical depth for multiple imaging of quasars at $z = 2$ from galaxies and from clusters.

3.1.2 Clusters

Here, we take as input the present epoch TF for X-ray clusters from Edge
et al. (1990): The cumulative TF is well fitted by a power law $dn(T)/dT = AT^{-5}$, with $A \simeq 8 \times 10^{-4}h^3\text{Mpc}^{-3}\text{keV}^{-4}$ for $3 < T < 10$ keV. There seems
to be a rather sharp drop for $T > 10$ keV, and there is a hint that the TF
flattens at lower temperatures. We want to use the TF to predict $d\tau/d\ln\theta_o$,
much as we did for the galaxies. There is something of a problem here in
that the X-rays only probe directly radii $\gtrsim 200$ kpc, while even for the most
massive clusters the surface mass density at this radius is somewhat below
the critical value required for lensing (for example, a 10 keV SIS cluster
produces a 45" splitting, in which case the impact parameter for the light
rays is about $100h^{-1}$kpc).

To make a firm prediction for $d\tau/d\ln\theta_o$ we would need to know the
mass profile, but this is very uncertain. On the one hand, the X-rays show
a core radius, but stacked against this there is now a broadening consensus
that the galaxy distribution in clusters is much peakier (Beers and Tonry,
1986; Merrifield and Kent, 1989). From the theoretical side there are indications from cosmological N-body simulations (Carlberg and Dubinski, 1991;
Quinn, Salomon and Zurek, 1991) with very large numbers of particles that
the mass profiles of clumps are very peaky, and the indication from arcs and
distortion of faint background galaxies by clusters (Hammer, 1991; Tyson,
1991) is that the dark matter in these clusters has quite a small core radius.
There is no necessary conflict between the X-ray observations and the other
indications since it is quite possible to make models in which the mass distribution is singular, but in which the X-ray surface brightness is very flat
within some radius. While the uncertainties preclude a firm prediction of
the optical depth for lensing vs. angle, it seems interesting to compare the
observed frequencies with a simple SIS cluster population model, if only as
a benchmark.

If we model the clusters as a non-evolving population of SIS's then the
optical depth distribution is

$$\frac{d^2\tau}{d\ln\theta_o dw_L} = \frac{\pi\theta_o^{3-\gamma}a_z^{\gamma-1}Aa_0^3w_L^2}{4}\left(\frac{w_Q}{2(w_Q - w_L)}\right)^{1-\gamma}, \qquad (3.1.3.1)$$

and integrating this, for $\gamma = 5$, gives

$$\frac{d\tau}{d\ln\theta_o} = \frac{4\pi\theta_o^{-2}a_z^4Aa_0^3w_Q^3}{105} \simeq 1.44 \times 10^{-2}(w_Q/0.42)^3(\theta_o/1")^{-2}, \quad (3.1.2.2)$$

which is plotted as the higher of the upper two curves in figure 1.

An alternative is to assume that the cluster population is evolving in a
self-similar manner, so that, at epoch z,

$$N(> T; z) = (1 + z)^{6/(n+3)}N_0(> (1 + z)^{(1-n)/(n+3)}T), \qquad (3.1.2.3)$$

and so

$$\frac{dn(T;z)}{dT} = (1+z)^{(7-n)/(n+3)}\frac{dn((1+z)^{(1-n)/(n+3)}T;0)}{dT}. \qquad (3.1.2.4)$$

These equations state that the clusters in the past are a scaled replica of the present epoch population (e.g. Kaiser, 1986). For example, for a 'CDM-like' spectrum $n = -1$, the population at $z = 1$ has a characteristic number density 8 times higher than at the present, but any characteristic temperature in the distribution (e.g. the 'break' at 10 keV) will shift to half the present value. For pure power law $dn(T;0)/dT = AT^{-\gamma}$,

$$\frac{dn(T;z)}{dT} = (1+z)^{q(\gamma,n)}AT^{-\gamma} \qquad (3.1.2.5)$$

where $q(\gamma,n) \equiv (7-n-\gamma(1-n))/(n+3)$. For $\gamma = 5$ the number of clusters at fixed T varies with redshift as $(1+z)^{(2+4n)/(n+3)}$, so in this case there will be greater or fewer clusters in the past depending on whether n is greater than or less than -1/2. The optical depth $d^2\tau/d\ln\theta_o dw_L$ is given by (3.1.2.1), but with the inclusion of a factor $(1-w_L)^{-2q(\gamma,n)}$, and $d\tau/d\ln\theta_o$ has the same form as equation 6, but, for $n \simeq -1$ and $w_Q = 0.42$ (corresponding to redshift $z = 2$), reduced in amplitude by about 30%. This is also plotted in figure 1.

It appears that a simple extrapolation of the observed temperature function (down by about a factor 2 from 3 keV) seems to account reasonably well for the 'missing lenses'. This model also requires an extrapolation in radius from the scales $\simeq 200h^{-1}$kpc probed by the X-ray observations (though by only a rather modest amount for the most massive clusters). If extrapolated to still lower temperatures, the model would predict too many lenses at smaller separation, since $\tau \propto \theta_o^{-2}$, but this divergence can be cut off if the dark matter has a core radius of a few tens of kpc (or at least if the mass density profile flattens within this radius). The predicted behaviour is very sensitive to the assumed density profile interior to the X-ray radius. For example, with an SIS, $\rho \propto Tr^{-2}$, so the surface density $\Sigma \propto Tr^{-1}$, and therefore the cross-section for lensing, which is the area where Σ exceeds the critical value, scales as $\sigma \propto T^2$ and hence the optical depth scales as $d\tau/d\ln\theta_o \propto \theta_o^{-2}$. With the seemingly minor modification of taking $\rho \propto r^{-3/2}$, the cross-section now scales as $\sigma \propto T^4$ and in that case $d\tau/d\ln\theta_o$ is constant. In that case, the effect due to clusters would be very small indeed. All one can reasonably conclude is that the frequency of large ($\sim 6''$) splittings seen seems to fit quite well with a continuous distribution of clusters with profiles close to $\rho \propto r^{-2}$ down to a few tens of kpc, and with dn/dT which is a simple extrapolation from the slightly higher temperature clusters counted by Edge et al.

3.1.3 Magnification Bias

As mentioned, to convert these raw optical depth distributions into predictions for the absolute number of multiply imaged quasars one must allow for magnification bias. The basic idea here is that since if images are brightened by lensing and if one has a sample of quasars which are magnitude limited then multiply imaged quasars will be overrepresented. This bias will be most strong for the brightest quasars where the QLF becomes steep; clearly if there were a sharp cut-off in the QLF and one set a magnitude limit above this then only lenses quasars would get into the sample and the amplification bias would be infinite. A very simple estimate of this bias shows that the effect may rise to be of order ten or so for very bright quasars, though one rapidly runs out of quasars. Calculating this bias in detail requires intimate knowledge of the selection criteria (redshift, magnitude etc.) for the particular survey considered. In terms of the models presented here, one could, for instance, use the model of §3.1.1 for the number density of lenses as a function of splitting angle combined with the joint pdf for the image brighnesses for the SIS lens (2.2.3.7) and a model for the observed QLF to obtain the predicted number density of lenses as a function of splitting angle and luminosity. One this plane, one could then decide what area would have been observable (small splittings, for example, would not have been detectable in a survey with poor seeing or in which the quasar would have been rejected as non-stellar due to the contaminating light of the lensing galaxy).

There are a couple of simple calculations which may give some results of rather general interest. For the case of very small (sub-arc-second) splittings (e.g. Crampton *et al.*, 1991) it might be realistic to assume that the quasars are selected in the first instance on the magnitude of the combined images, with the instances of lensing being found on further inspection with very high resolution photometry. For larger splittings it may be more relevant to assume that the initial detection is controlled by the magnitude of the brighter image, and that instances of lensing are then those cases where the second image is not too much fainter than the brighter image. The calculation of the magnification bias for these two situations is left as an exercise.

3.2 Quasar-Galaxy Associations

Macrolensing of quasars is, we have see, a very rare phenomenon. This is because such a small fraction of the universe resides in the central parts of galaxies and clusters of galaxies where Σ exceeds $\Sigma_{\rm crit}$. Most of the mass in galaxies resides at larger radius. This mass may potentially be revealed through its lensing effect as it brightens the background quasars, and this would result in a statistical cross-correlation between galaxies and much more distant background quasars.

The first observational indication of such an effect is the result of Webster *et al.*, 1988, (hereafter WHHW), who found that the surface density of galaxies around quasars is enhanced by a factor $\simeq 3$ at separations $4'' < \theta < 6''$. This statistic has the same expectation value as (but is easier to measure than) the enhancement of quasar surface density around a randomly chosen foreground galaxy. One possible interpretation is that this effect is by microlensing due to compact objects associated with the foreground galaxy amplifying the background QSO's. If so, this statistic is a direct measure of the mean projected surface density at $\theta \simeq 5''$ times a factor which depends on the details of the apparent luminosity distribution function of the background quasars and the probability distribution for amplifications.

While the idea of measuring the mass profile around galaxies in this way is attractive, it turns out that the strength of the observed effect is very hard to understand, at least in terms of conventional models for the mass distribution and the intrinsic QLF. There are two lines of argument that have been used to show this. The first, which is independent of the mass distribution in the 'lensing' galaxies, is that there is a maximum enhancement which may be calculated from the QLF for any given magnitude limit, and which, for the WHHW set up, is actually less, by about a factor 2, than the claimed effect (Narayan, 1989; Kovner, 1989). The way this comes about is that lensing affects the images in two ways: it can stretch the images and make them bigger and therefore brighter (since lensing conserves surface brightness), but at the same time, it dilutes their number density on the sky by the same amount. For bright magnitudes the QLF is steep and the net result is an enhancement in the number of galaxies at a given apparent magnitude. At faint magnitudes the QLF is very flat, and the net effect is a diminution. As one increases the strength of a lens, and therefore the amplification, there is then a maximum enhancement.

The second line of argument is to include ones knowledge of the mass distribution around galaxies from their internal velocity dispersions, and various workers (Hogan, Narayan and White, 1989; Schneider, 1989; Peacock, 1986) have concluded that the expected effect is rather small (though there is considerable variety in the estimates due to differences in the assumed properties of the lensing galaxies and the background QLF). In the first part of this section we repeat this calculation, but using more accurate estimates of the mass profiles and QLF. The result is that the expected effect is about a factor 20 smaller than what is seen. This is very interesting; perhaps these observations are telling us there is something drastically wrong with our understanding of mass clustering, or perhaps with the QLF. To explore this, we consider various possible loopholes in the argument, but find that most possible modifications to the model do not help resolve the problem.

The solid angle subtended by the Einstein ring for a point-mass lens of mass M at distance w_L and a source at w_Q is (from §2.2.2)

$$\sigma_E = \frac{4\pi GM(1 - w_L/w_Q)}{a_L w_L}. \tag{3.2.1}$$

The optical depth (for Einstein rings) for an object of physical surface density Σ is then

$$\tau = 4\pi G \Sigma a_L w_L (1 - w_L/w_Q) = \Sigma/\Sigma_c. \tag{3.2.2}$$

For a flat rotation curve halo with rotation velocity V, $G\Sigma = V^2/4b = V^2/(4a_L w_L \theta)$, so for such an object

$$\tau = \pi(V^2/c^2)(1 - w_L/w_Q)\theta^{-1}, \tag{3.2.3}$$

so provided that the galaxy is much closer than the quasar the expected effect is independent of the distance to the galaxy, which is convenient. All this assumes that the galaxy halo is composed entirely of compact objects.

For a galaxy with a rotation speed of 200 km/s then, the optical depth at $\theta = 5$" is about 6% (this is the value for a low redshift galaxy with $w_L \ll w_Q$), so the assumption of low optical depth should be very good. Let us imagine for the moment that we consider only quasars from a small range of redshift corresponding to a single distance w_Q, and that the distribution of apparent luminosities of these quasars is $\phi(l)$. Using the results from §2.3.1 we find that the luminosity function for quasars seen through the galaxy is

$$\phi'(l)dl = (1 - \tau)^2 \int dA \, P(A)(\phi(l/A)dl/A). \tag{3.2.4}$$

The $(1 - \tau)^2$ factor here is the dilution of the number density of sources on the sky due to the macroscopic focussing – note that with standard candles, i.e. a delta-function for $\Phi(m)$ we would actually see a suppression of the number density of quasars behind the galaxy – which, as advertised, just cancels the corresponding factor in $P(A)$. Combining (3.2.3), (3.2.4) and (2.3.1.5) then gives us the perturbation to the luminosity function for a galaxy of given circular velocity at distance w_L and quasars at distance $w_Q > w_L$.

To obtain the angular quasar-galaxy cross-correlation function for a realistic survey requires that we integrate this over the redshifts and magnitude for the quasars and over the galaxy distance, paying attention to the fact that the distribution of circular velocities will be a function of distance. This is straightforward but rather messy. A great simplification is possible is we assume that the quasars are all much more distant than the galaxies $w_Q \gg w_L$, in which case the strength of the lens is independent of both its distance and that of the quasar[2] and one then finds that the fractional

---

[2] This might seem to be at odds with the result that the strength of a lens is weakened if it is close to the observer. For a constant surface density lens, the strength is indeed proportional to $1/w_L$, but for a flat rotation curve halo, if we look at a fixed angle then for the nearby galaxies we are peering through a higher surface density and this just cancels the $1/w_L$ factor.

enhancement is just

$$w_{qg}(\theta) = \frac{\phi'(l) - \phi(l)}{\phi(l)} = \langle \tau(\theta) \rangle f(m) \qquad (3.2.5)$$

where $\langle \tau \rangle$ is the average optical depth for a galaxy which gets into the magnitude limited sample and $f(m)$, which we write as a function of the magnitude of the quasars $m = 2.5 \log_{10} l$, is a magnification bias factor which may readily be calculated from (3.2.4). What is observationally more relevant is the angular cross-correlation for a magnitude limited quasar survey. This is also easily determined from (3.2.4) and we can express the result as $w_{qg}(\theta) = \langle \tau(\theta) \rangle F(m)$.

We see from (3.2.4) that the appropriate observational datum here is the apparent magnitude distribution for the quasars, subject to the selection criteria relevant to the particular survey. The WHHW sample covered a wide range of redshifts – only very low z quasars being rejected – and so a good approximation to $\phi(m)$ for this sample should be the apparent luminosity distribution obtained by Boyle, Shanks and Peterson (1988). This is quite accurately represented by a 2-power-law form $\phi(l) \propto l^{-3}$, for high luminosities and $\phi(l) \propto l^{-1.7}$ for low, with the break at an apparent luminosity corresponding to $m = 19.5$.

All that remains is to calculate $\langle \tau(\theta) \rangle$, where the averaging is done over the galaxies that enter a magnitude limited survey. Using the combination of galaxy luminosity functions and luminosity-velocity relations as in the previous section, but now to predict the expected optical depth for microlensing $5''$ from a galaxy randomly selected from a magnitude limited catalogue, one finds

$$\langle V^2 \rangle = (223\,\mathrm{km/s})^2, \qquad (3.2.6)$$

which, in equation 3.2.3, gives a mean optical depth of 7%.

Now WHHW were working at $M = 18.8 - 19$ (Narayan, 1989), which is $0.5 - 0.7$ magnitudes brighter than the break, so from figure 8 we can see that $F(m) \simeq 3$, and the expected net enhancement is only about 20%, with again a diminution by $(1 - w_L/w_Q)$ for the more distant galaxies, much less than the factor 3 observed. The estimate found here is somewhat smaller than that found by Schneider (1989). This is due in part to his steeper assumed slope for the QLF, and in part due to his larger assumed halo velocity dispersion.

The predicted effect is depressingly small, and it seems very unlikely that the effect found by WHHW is due to lensing by mass in galaxies. Various authors have speculated on the possibility that there is extra matter clustered around galaxies which is not accounted for in our simple flat-rotation curve model. Unfortunately this does not work; there are very strong constraints on how much mass there is around galaxies from the cosmic virial theorem (Davis and Peebles, 1983; Kaiser and Tribble, 1991) and the maximum mass allowed by this constraint would negligibly enhance the estimate

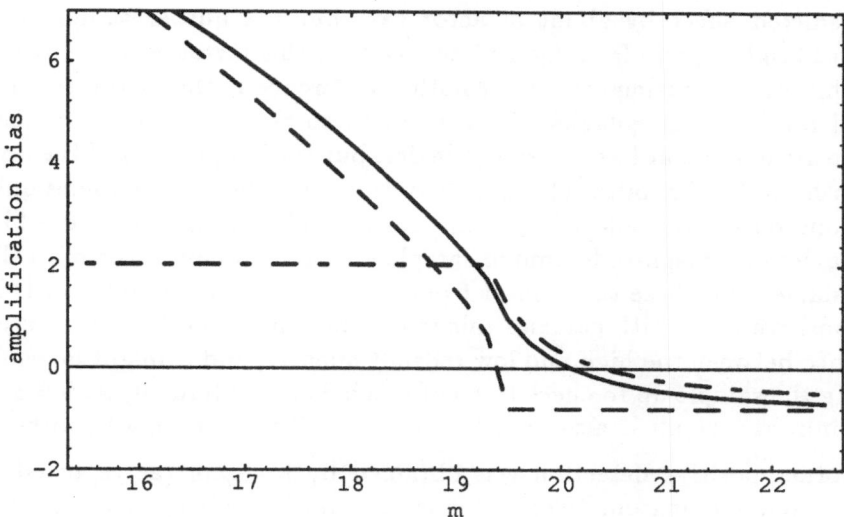

Fig. 8. The solid line is the amplification bias factor $F(m)$ computed from the Boyle *et al.* quasar luminosity function. The dashed line is the differential form of this function $f(m)$. Also shown (dot-dash) is the quantity analogous to $F(m)$, but now giving the enhancement arising from a smooth mass distribution. This bears out a point emphasised by Narayan that a smooth mass distribution is about as efficient at causing quasar galaxy associations – except, it would seem for very bright quasars where microlensing is more efficient.

we have made. In fact, it is much more likely that we have overestimated the effect; we have neglected the factor in (3.2.3) which reduces the optical depth below what we have used; we have assumed that the galaxies are just like typical galaxies today whereas we know from the combination of counts and redshifts for faint galaxies that even at 20th mag we are seeing a population with a higher space number density than their present day counterparts (and which are therefore most likely less massive); there must be some absorption due to dust even at the impact parameters of about 15 kpc/h which are relevant here and since the predicted effect from lensing is so weak it is not implausible that absorption would reduce or even reverse the sign of the effect.

Since the original WHHW paper there have been a number of studies of a similar kind. A positive detection has been found by Drinkwater *et al.* (1991). They used a rather different statistic to the angular correlation function (which is, as we have seen, most readily predicted from the theory), but one can easily model their statistic (the distribution of nearest neighbour distances) and the result of this exercise is that one requires a cross-correlation of about the same amplitude as the original WHHW result but at about twice the angular scale. The larger angular scale reduces the predicted effect by about a factor two, but the quasar sample used here was brighter, and from figure 8 we see that this increases the effect by a roughly compensating amount. Another difference is that Drinkwater *et al.* used much fainter galaxies (R mag ~ 22), so not only is it more likely that we are looking at less massive galaxies, but the suppression of the optical depth due to the finite ratio of galaxy to quasar distance becomes quite large (one reason this effect is perhaps surprisingly strong is that the high V^2 galaxies in a magnitude limited sample are predominantly the most distant galaxies and these suffer most from the $1 - w_L/w_Q$ suppression. On a more positive note, with galaxies this faint, one might well be able to see a difference between the high and low redshift quasars, and it might be worth dividing the sample up to check to see if this is found. All in all, we feel that the Drinkwater result is also hard to reconcile with the lensing hypothesis.

Another positive detection was obtained by Margain (as reported by Narayan and Wallington, 1992). This is at a much smaller angle scale and so is + more reasonable, though still rather stronger than one would expect.

Yee (1992) on the other hand, finds a null result $w_{qg} \simeq 0 \pm 0.3$. The magnitude selection criteria were broadly similar to those of Drinkwater *et al.*. Yee's galaxies were slightly brighter but the quasars were slightly fainter. An important difference between these studies is that Yee excluded low redshift quasars. This is important for a number of reasons: i) As mentioned, when the galaxies are faint then one will expect to see a bigger effect for higher z quasars, all other things being equal. ii) The luminosity function tends to steepen if one restricts the range of redshifts used, and this increases the amplification bias factor. iii) If the quasars are at high z then one can be more confident that any correlations (if found) are not due to physical associations.

The situation is rather depressing. We have seen that while there is potentially a handle on dark mass around galaxies from these studies, the expected effect is rather small and may well be contaminated by other effects such as absorption in galaxies. We have also seen that it is not easy to discriminate between microlensing and amplification by a smooth component from this kind of observation. The positive detections on scales $\gtrsim 5''$ are implausibly large (though they may reflect real physical associations). Yee's result, while still a null detection, is the most encouraging since it suggests that with an increase in the data one might start to see the effect of lensing. Unfortunately, the bright quasars are a finite commodity, and the prospect for an increase in the data base is limited.

3.3 Cosmological Density of Compact Objects

Another interesting application of microlensing is if a substantial fraction of the total mass of the Universe is in compact objects. If the density in lenses is Ω_L then the differential optical depth for microlensing of sources at w_Q is

$$\frac{d\tau}{dw_L} = \frac{6\Omega_L(w_Q - w_L)w_L}{(1 - w_L)^2 w_Q}. \qquad (3.3.1)$$

Integrating this out to the quasar redshift gives

$$\tau(w_Q) = 6\Omega_L((1 - 2/w_Q)\log(1 - w_Q) - 2). \qquad (3.3.2)$$

For sufficiently low Ω_L and/or w_Q, the universe will be optically thin, and so we can apply the same methods used in the previous section to calculate the distortion of the QLF. For example, for $z = 2$ ($w_L \simeq 0.42$) and $\Omega_L = 1$ this formula gives $\tau \simeq 0.29$, which is probably about the limit to which the low-τ formulae can be trusted.

The optical depth for the redshifts of typical quasars is quite modest, even if $\Omega_L = 1$, but one can get an interesting effect for the most distant quasars. Provided the masses of the lenses are sufficiently large then the luminosity function for the lensed quasars will be asymptotically

$$dn/d\ln L \propto L^{-2} \qquad (3.3.3)$$

for high L, and if the intrinsic luminosity function is steeper than this then the lensed quasars will eventually come to dominate the counts.

Now according to Boyle (1991), the QLF for high z-galaxies is almost one power of L steeper than this at high L. If this is correct, then one may provide a constraint on the net density in compact objects (Webster 1991). This is illustrated in figure 9, where we have plotted the Boyle's model for the QLF for $z \simeq 2$ quasars, and also the QLF one would observe if the intrinsic QLF looked like Boyle's model. In this calculation we have assumed $\Omega = 0.3$ in compact objects, so $\tau \simeq 0.1$, and have used the simple form for $P(A)$ given in (3.2.1.5). It is interesting to note that even with this rather high density in compact objects and high quasars redshifts, the modification to the QLF is, for the most part, rather minor, but as expected the QLF after lensing is somewhat shallower at the bright end. It is not clear to me that this QLF is firmly excluded by Boyle's data – there are precious few quasars in the brighter bins – but it is starting to look embarrassing, and a larger density in lenses would probably be excluded. Following Webster, I have also shown that happens if the intrinsic QLF is truncated (I set the cut-off at $M = -27$) and now the post-lensing QLF is nicely compatible with that observed.

The idea that one might be able to put some constraint on the mass in compact objects from this type of observation is an exciting one. As we have seen, it would be hard to reconcile the steep slope found by Boyle with

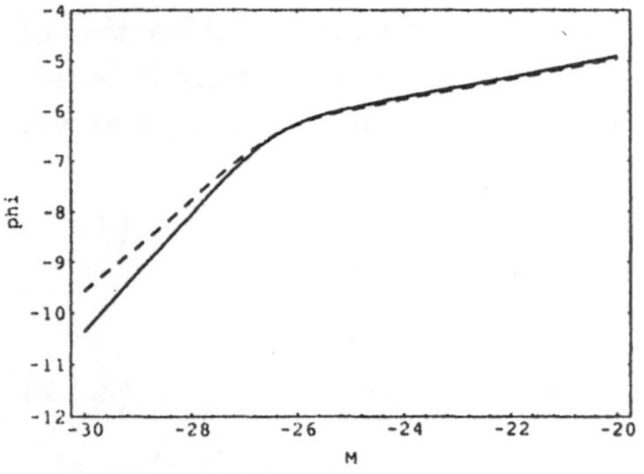

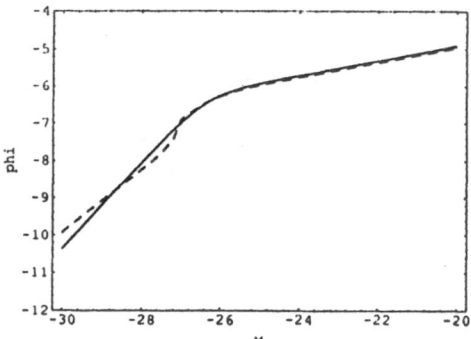

Fig. 9a. The solid line here is Boyle's model for the luminosity function for $z \simeq 2$ quasars. The dashed line is the QLF one would see using this model for the intrinsic luminosity function and assuming $\Omega = 0.3$ in compact objects.

Fig. 9b. The solid line is as before, and now the dashed line is the QLF one would see if the intrinsic luminosity function is truncated brightward of $M = -27$.

say a closure density in lenses. The problem at the present is that the test relies on the last couple of points in Boyle's analysis, and there is not great deal of unanimity on the part of the observers. Hartwick and Shade (1991), for instance, find a much shallower QLF than Boyle, and if they are right then there is really no interesting constraint from this test.

References

Beers, T. and Tonry, J., 1986. Ap.J., **308**. 510.

Boyle, B.J., 1991. In proceedings of Texas-ESO/CERN meeting on relativistic astrophysics.

Boyle, B.J., Shanks, T., and Peterson, B.A., 1988. MN, **235**, 935.

Carlberg, R. and Dubinski, J., 1991. Preprint.

Chang, K. and Refsdal, S., 1984. Astr. Astrophys., **132**, 168.

Crampton, D., McClure, R.D., and Fletcher, J.M., 1991. In "The Space + Distribution of Quasars", ed D. Crampton, Astronomical Soc. of the Pacific conference series.

Davis, M., and Peebles, P.J.E., 1983., Ap.J., **267**, 465.

Drinkwater, M.J., Webster R.L. and Thomas, P., 1991. In "The Space Distribution of Quasars", ed D. Crampton, Astronomical Soc. of the Pacific conference series.

Edge, A., Stewart, S., Fabian, A. and Arnaud, K., 1990. MN, **245**, 559.

Fukugita, M. and Turner, E.L., 1991. MNRAS in press.

Hammer, F., 1991. Preprint.

Hartwick, F.D.A. and Schade, D., 1990. Ann. Rev. Astron. & Astrophys., **28**, 437.

Hogan, C.J., Narayan, R., and White, S.D.M., 1989. Nature, **339**, 106.

Kaiser, N., 1986. MN, **219**, 785.

Kaiser and Tribble, 1991. In "The Space Distribution of Quasars", ed D. Crampton, Astronomical Soc. of the Pacific conference series.

Kofman, L. and Kaiser, N., 1992. In preparation

Kovner, I., 1989. Ap.J., **341**, L1.

Merrifield, M.R., and Kent, S.M., 1989. AJ., **98**, 351.

Narayan, R., 1989. Ap.J., **339**, L53.

Narayan, R. and Wallington, S., 1992. Preprint.

Nityananda, R. and Ostriker, J.P., 1984. J. Astr. Ap. **5**, 235.

Peacock, J.A., 1986. MN, **223**, 113.

Quinn, P., Salomom, P. and Zurek, W., 1991. In Beyond the 1st Three Minutes. eds. Holt, Bennett an Trimble, AIP Conference series.

Rauch, K., Mao, S., Wambsganss, J. and Paczynski, B., 1991, preprint.

Schneider, P., 1989. Astron. Astrophys., **221**, 221.

Schneider, P., 1987. Ap.J., **319**, 9.

Schneider, P. 1986. Astron. Astrophys., **179**, 80.

Turner, E.L., 1989. 14th Texas Symposium on Relativistic Astrophysics.

Turner, E.L., Ostriker, J.P. and Gott, J.R., (1984) Ap.J., **284**, 1

Tyson, A., 1991. In "Beyond the 1st Three Minutes" eds Holt, Bennett and Trimble, AIP Conference series.

Webster, R.L., Hewett, P.C., Harding, M.E., and Wegner, G.A., Nature, **336**, 358 (WHHW).

Webster, R., 1991. In "The Space Distribution of Quasars", ed D. Crampton, Astronomical Soc. of the Pacific conference series.

Yee, H., Filippenko, A. and Tang, D., 1992. To appear in proceedings of + 1991 Hamburg conference on Gravitational lenses.

Statistics of Gravitational Lensing 2:
Weak Lenses

Nick Kaiser

CIAR Cosmology Program, CITA, University of Toronto

1 Introduction

In the previous lecture we considered strong lenses and their effect on distant quasars. Here we will be concerned with lensing by structures with surface density less than the critical value for multiple imaging, and their influence on images of distant galaxies. The optical depth for strong lensing of quasars is very uncertain since one cannot rule out the possibility that the dark matter is 'microscopically' clumpy. For extended images like galaxies, the effect of microlensing is washed out and since we know that the optical depth for strong macrolensing is tiny we can be sure that nearly all galaxies are at most weakly lensed.

Weak lensing maps the 'true' sky surface brightness pattern – by which we mean the fictitious sky brightness one would see in the absence of intervening mass fluctuations – into the observed sky: $\mathcal{I}_{\mathrm{obs}}(\vec{\theta}) = \mathcal{I}_{\mathrm{true}}(\vec{\theta} + \vec{\delta\theta}(\vec{\theta}))$, where $\vec{\delta\theta}(\vec{\theta})$ is the transverse deflection of the light rays caused by the mass fluctuations. This deflection field is not directly observable since we do not know the 'true' surface brightness pattern, but its gradient, the shear tensor $\Phi_{ij} \equiv \partial\delta\theta_i/\partial\theta_j$, is.

One effect of a positive foreground overdensity will be to enhance the apparent luminosity of the background galaxies. We explored this somewhat in the previous lecture when we considered the quasar-galaxy correlations that would result if the mass around the foreground galaxies is smooth. There we found a positive correlation for high quasars luminosities and an anti-correlation for luminosities below the break in the QLF. The reason for this is that there are two competing effects operating: On a patch of the sky where the lensing has made the images larger (and therefore brighter and more likely to enter one's magnitude limited catalogue) their surface number density will also have been suppressed. The enhancement due to the first term depends on the slope of the number-magnitude relation. It turns out that for faint galaxies the number counts go approximately like $n(> l) \propto l^{-1}$, and for this slope the two terms cancel.

A more interesting observable effect of weak lensing is a stretching of the images of distant galaxies which is encoded in the trace-free components of the shear, and would cause intrinsically circular objects to appear slightly squashed. Real galaxies have substantial random intrinsic ellipticities, so observing a single galaxy will not reveal a small gravitational distortion. The effect of extended mass fluctuations, however, will produce stretching acting coherently on a large number of background galaxies, and this can, in principle, be measured.

The distortions can be exploited in three ways. First there is cross-correlation analysis where one searches for a systematic distortion of background galaxies seen around prominent foreground objects – typically bright galaxies or clusters of galaxies. This approach yields the spatial cross-correlation of mass with the foreground objects – $\xi_{g\rho}$ and $\xi_{c\rho}$ for galaxies and clusters respectively. Second there is the possibility of measuring the angular auto-correlation function of the background galaxy ellipticities; this can tell us about the mass auto-correlation function $\xi_{\rho\rho}$. Thirdly, with very massive foreground clusters and very distant background galaxies it is possible to measure the effect significantly for an individual object (rather than for a stacked 'composite' as in the cross-correlation analysis) and one then wants to be able to invert the lensing equation to determine the projected mass density.

The organisation of the lecture reflects this division. I shall first establish some results of general applicability and then consider these three applications in turn.

2 Some Useful Results

As in the previous lecture we will consider lensing on a Einstein-de Sitter background and we will use the same notation for coordinates etc.. For the necessary background for more general cosmological models see the seminal papers by Gunn (1967a,b). The effect we are concerned with can best be visualised if one imagines a narrow circular bundle of rays emerging from one's telescope. If one propagates these rays back in time then any spatial density fluctuations will deflect these rays in a random manner and the cross-section of the beam will be distorted. If the wavelength of the fluctuations is large compared to the width of the beam, then the distortion is just a constant shear, and the beam cross-section will, to first order, be an ellipse. This is the ellipticity of a galaxy which would appear circular, and is therefore just minus the apparent ellipticity of an intrinsically circular galaxy image. If we examine all the galaxy images on some small patch of sky, then the mean ellipticity provides an estimate of the gravitationally induced shear, plus of course noise arising from the fact that one is looking at a finite number of intrinsically randomly oriented galaxies.

Let us consider a small patch of sky around the z-axis and define angular coordinates $\vec{\theta}$ such that the initial direction vector of a ray is $\hat{\mathbf{w}} = (\theta_1, \theta_2, 1)$. If we consider an image plane at some distance w, then, in an unperturbed universe, a ray with direction $\vec{\theta}$ will pierce this plane at $\vec{r} = w\vec{\theta}$. In reality, the ray will intercept the plane at $\vec{r} + \vec{s}$, so the observed surface brightness will be

$$\mathcal{I}_{\text{obs}}(\vec{\theta}) = \mathcal{I}_{\text{true}}(\vec{\theta} + \vec{\delta\theta}(\vec{\theta})), \tag{2.1}$$

where $\vec{\delta\theta} = \vec{s}/w$. We will work in the weak field limit—this should be an excellent approximation—so the displacement vector is $\delta\theta_i = s_i/w$, where the displacement of the ray in comoving coordinates is obtained from the geodesic equation $\ddot{s}_i = -2\phi_{,i}$. From (2.1.3) of lecture 1 we find

$$\delta\theta_i = \frac{-2}{w} \int_0^w dw'(w - w')\phi_{,i}(\mathbf{w}' + \mathbf{s}, 1 - w'), \tag{2.2}$$

which is an integral equation for the deflection since the integration is performed along the perturbed ray trajectory. For sufficiently weak fields the perturbed path will be very close to the unperturbed path $\mathbf{w} = \{w\theta_1, w\theta_2, w\}$ and the shear tensor is then

$$\Phi_{ij}(\vec{\theta}, w) \equiv \partial\delta\vec{\theta}_i/\partial\theta_j = \frac{-2}{w} \int_0^w dw' w'(w - w')\phi_{,ij}(\mathbf{w}', 1 - w'), \tag{2.3}$$

and is now a linear function of ϕ and therefore of the mass density also. The big question of course is how weak does the potential fluctuation have to be for this to be valid. In (2.3) we are obtaining the relative deflection for neighbouring rays using their unperturbed separation. That is valid provided the shear itself is small compared to unity – i.e. that the lensing be weak – and this should be a very good approximation (for further justification see §4). There would also appear to be a much stronger restriction, since we are Taylor expanding the potential field, that the deflection of the bundle as a whole should be much less than the scale on which the potential is varying. This would appear to be a rather poor approximation for our universe, but is really of no consequence if we assume that the density fluctuations in regions separated by cosmological distances are statistically uncorrelated since all we are concerned with is the statistical properties of the tidal field ϕ_{ij} and these may safely be determined using the unperturbed rays provided again only that the shear is small. Note that we have not assumed that the density fluctuations giving rise to $\phi(\mathbf{w})$ are of small amplitude, only that their shear is small, or, equivalently, that the fluctuations in projected surface density be much less than Σ_{crit}.

Equation (2.3) gives the shear for a particular image plane at distance w. Really, we will measure the average shear $\Phi_{ij}(\vec{\theta}) = \int dw\, W(w)\Phi_{ij}(\vec{\theta}, w)$, where $W(w)$ is some normalised weighting function (just proportional to dn/dw if we weight the galaxies equally), and this is

$$\Phi_{ij}(\vec{\theta}) = 2 \int_0^1 dw\, g(w)\phi_{,ij}(\mathbf{w}, 1 - w), \tag{2.4}$$

where

$$g(w) \equiv w \int_w^1 dw'\, W(w')(1 - w/w') \tag{2.5}$$

measures the propensity for a mass fluctuation at distance w to distort background galaxies with distance distribution $W(w)$.

We see from (2.4) that the shear is just a radial projection of the transverse components of the tidal field. The trace of Φ_{ij} describes the overall compression or dilatation of the images. Here we are more interested in the changes in the shapes of the images. These are encoded in the two quantities $\epsilon_1(\vec{\theta}) \equiv \epsilon_+(\vec{\theta}) \equiv \Phi_{11} - \Phi_{22}$ and $\epsilon_2(\vec{\theta}) \equiv \epsilon_\times(\vec{\theta}) \equiv 2\Phi_{12}$. The aliases ϵ_+ and $\epsilon_\times$ are meant to be visually suggestive: A stretching of the images along the x-axis (y-axis) corresponds to a positive (negative) ϵ_+, and a stretch along the upper right (upper left) diagonal gives a positive (negative) $\epsilon_\times$. In general, a stretch which would make a circular image have ellipticity ϵ with semimajor axis at angle ϕ is $\{\epsilon_+, \epsilon_\times\} = \{\epsilon\cos 2\phi, \epsilon\sin 2\phi\}$.

Now these ellipticity coefficients are related in a very simple (and linear) way to quadrupole moments of the galaxy images $Q_{ij} \equiv \int d^2\theta(\theta_i - \langle\theta_i\rangle)(\theta_j - \langle\theta_j\rangle)\mathcal{I}_{obs}(\vec{\theta})$ (e.g. Tyson et al. 1984). The easiest way to see this is to rotate into the frame in which Φ is diagonal, so the distortion is simply a scale transformation $\theta_1 \to \theta_1' = (1 + \Phi_{11})\theta_1$, $\theta_2 \to \theta_2' = (1 + \Phi_{22})\theta_2$ and we have $Q_{11}^{obs} = (1 + \Phi_{11})^3(1 + \Phi_{22})Q_{11}^{true}$ and $Q_{22}^{obs} = (1 + \Phi_{11})(1 + \Phi_{22})^3 Q_{22}^{true}$, so if Φ is small and we define

$$\epsilon_1 \equiv (Q_{11} - Q_{22})/(Q_{11} + Q_{22})$$
$$\epsilon_2 \equiv 2Q_{12}/(Q_{11} + Q_{22}) \tag{2.6}$$

then we find

$$\epsilon_i^{obs} = \epsilon_i^{true} + \epsilon_i^{grav}. \tag{2.7}$$

Thus we see that this simple combination of quadrupole moments (2.6) provides an unbaised linear estimator of the gravitationally induced ellipticity – the intrinsic ellipticity being considered here a source of random noise – and this in turn is a linear function of the density fluctuations we wish to measure.[1] If one makes a scatter plot of the locations on the ϵ-plane for all the galaxies on a patch of sky then, in the absence of lensing, the distribution should be symmetric about the origin. Any component of the shear which is coherent over the patch will cause a shift of the centroid of the distribution to ϵ^{grav}. With N galaxies one should in principle be able to detect a shift $N^{-1/2}\bar{\epsilon}$ where $\bar{\epsilon}$ is some measure of the width of the intrinsic distribution.

It is also interesting to compare the gravitationally induced ellipticity to the density fluctuations $\Delta(\mathbf{w})$ along the line of sight. The fluctuation in the projected density, averaged over some distance distribution $W(w)$, is

$$\Delta_p(\vec{\theta}) = \int_0^1 dw\, W(w)\Delta(\mathbf{w}, 1 - w). \tag{2.8}$$

Note the similarity between (2.4) and (2.8). The latter is the projection of the spatial density contrast $\Delta(\mathbf{w})$, which is just the Laplacian of the potential $\phi_{,ii}$, while the shear is the projection of $\phi_{,ij}$. For a perturbation of amplitude Δ and length scale λ, the projected density will be $\Delta_p \sim (\lambda/\bar{w})\Delta$, where $\bar{w}$ is a measure of the depth of the survey, while the shear and hence the ellipticity is $\epsilon \sim (\lambda/\bar{w}^3)\Delta \sim \bar{w}^2 \Delta_p$. Thus we anticipate that the ellipticity-galaxy cross-correlation function and the ellipticity autocorrelation function will have a similar dependence on scale as w_{gg}, but with correlation strength smaller by $\bar{w}^2$ and $\bar{w}^4$ respectively.

3 Cross-Correlation Analysis

The problem here is to calculate the systematic distortion of a faint background galaxy found at an angle θ from some suitably selected foreground object. Let us consider first only foreground lenses at a some distance w_l and background galaxies at w_g; we will later have to perform a double integration over lens and background galaxy distances to calculate the final effect. Now in any instance, the shear from (2.3) will receive contributions both from the mass physically associated with the foreground object and from random clutter along the line of sight. The latter will average to zero and we have

$$\langle \Phi_{ij}(\vec{\theta}) \rangle = 2w_l(1 - w_l/w_g) \int dz \langle \phi_{,ij}(w_l\theta_1, w_l\theta_2, w_l + z) \rangle \tag{3.1}$$

---

[1] The simple definition of ϵ_i in terms of quadrupole moments as we have presented it is not very useful in practice since there is a divergent contribution to Q_{ij} from photon counting fluctuations. One can generalise the analysis to treat quadrupole moments measured with a weighting function to define an effective aperture and this preserves the linear relation (2.7).

where the expectation values are understood to be subject to the condition that there is a foreground object at $\mathbf{w} = (0, 0, w_l)$. Provided our selection criteria do not prefer objects of a particular orientation $\langle\phi\rangle$ must have cylindrical symmetry[2] about the line of sight: $\langle\phi\rangle = \phi(b, z)$, and we find from (2.2) that

$$\langle\delta\theta_i(\vec{\theta})\rangle = (1 - w_l/w_g)(\theta_i/\theta)D(\theta) \tag{3.2}$$

where $D(\theta) = -2 \int dz \; (\partial\phi(b, z)/\partial b)_{b=w_l\theta}$ is the bending angle, and from this we obtain

$$\langle\Phi_{ij}\rangle \equiv \frac{\partial\langle\delta\theta_i\rangle}{\partial\theta_j} = (1 - w_l/w_g)\left[\frac{D(\theta)}{\theta}\left(\delta_{ij} - \frac{\theta_i\theta_j}{\theta^2}\right) + \frac{dD}{d\theta}\frac{\theta_i\theta_j}{\theta^2}\right]. \tag{3.3}$$

If we let $(\theta_1, \theta_2) = (\theta\cos\phi, \theta\sin\phi)$ we have $\langle\epsilon_1\rangle = \epsilon_T \cos 2\phi$ and $\langle\epsilon_2\rangle = \epsilon_T \sin 2\phi$ which is simply a stretching perpendicular to the line joining the objects with amplitude

$$\epsilon_T(\theta) = (1 - w_l/w_g)(D/\theta - dD/d\theta). \tag{3.5}$$

This then is the fundamental link between the observable ϵ_T and the bending angle $D = 4GM(b)/(a_lb)$, where $M(b)$ is the average excess projected mass interior to comoving impact parameter b. This in turn is simply related to the galaxy-mass cross-correlation function since $M(b) \propto \int_0^b db \; b \int_{-\infty}^{\infty} dz \; \xi_{g\rho}(\sqrt{b^2 + z^2})$.

We will model the galaxy lenses as a population of flat rotation curve halos, for which $D = 2\pi v^2/c^2$, and so

$$\epsilon_T = 2\pi(1 - w_l/w_g)\theta^{-1}v^2/c^2. \tag{3.6}$$

To predict the effect one must calculate an appropriate (i.e. pairwise) average value for ϵ_T. The analysis is greatly simplified if the foreground galaxies are very much closer than the background galaxies, in which case the expected effect is just $\langle\epsilon_T\rangle = 2\pi\theta^{-1}\langle v^2\rangle/c^2$, where $\langle v^2\rangle$ is the average rotation velocity for a galaxy selected at random from a magnitude limited sample. This is very similar to the analysis in §3.2 of lecture 1 in which we calculated the optical depth for lensing of very distant quasars by a flat rotation curve galaxies and, comparing (3.6) with (3.2.3) of lecture 1, we see that ϵ_T is just twice the optical depth. This may be estimated from the observed luminosity function of galaxies combined with the Tully-Fisher and Faber-Jackson relations and, following Fukugita and Turner (1991), in §3.2 of lecture 1 we found $\langle v^2\rangle = (220 \text{ km/s})^2$, so $\langle\epsilon_T\rangle = (\theta/0.7'')^{-1}$.

Tyson et al. (1984) (hereafter TVJM) noted that while the effect is rather small, with the large number of pairs of foreground and background galaxies

---

[2] In fact in modelling galaxy halos we will assume spherical symmetry, but it is worth bearing in mind that objects finding algorithms have a tendency to find objects with short axes perpendicular to the line of sight.

obtainable from their deep photometry one should be able to detect the effect at a reasonable level of significance. They classified $\simeq 10^5$ galaxies from a number of deep ($m \simeq 24$) photographic plates, and split these into 'foreground' and 'background' samples by luminosity. This gave $\simeq 27,000$ foreground/background pairs with separation $\theta < 63''$, from which they estimated $\langle \epsilon_T(\theta) \rangle$ in logarithmically spaced bins[3] in θ. Surprisingly, they found a very small effect. The quantity they plot is $\mathcal{D} \equiv \theta \langle \epsilon_T(\theta) \rangle$. For the flat-rotation curve galaxy model this is $\mathcal{D} \simeq 0.7$, whereas (measuring the points from their graph[4]) I find $\mathcal{D} \simeq (3.6 \pm 3) \times 10^{-2}$ which is compatible with the noise and about a factor 20 lower than the theoretical expectation.

From the lack of any significant effect TVJM were led to consider models with truncated halos, and obtained what seems to be a rather conservative limit on the outer radius of a few tens of kpc, and this gave a correspondingly low bound on the density parameter of $\Omega < 0.03$. This is rather surprising, given that the cosmic virial theorem (Davis and Peebles 1983) and the observed pairwise velocities of galaxies in the CfA survey indicate the the statistical mass enhancement around a typical galaxy rises roughly as $\delta M \propto r^{1.2}$, in which case $\mathcal{D}$ would be expected to *increase* with scale, albeit rather slowly.

There are however two factors which decrease the predicted effect quite substantially. The first is the effect of the finite distance ratio w_l/w_g in (3.6). In the TVJM analysis the 'foreground' and 'background' samples were quite well separated in magnitude and they calculated that the overlap between the two samples was very small; the probability of confusing a background and foreground galaxy was $\simeq 15\%$. Even with little or no overlap however, the suppression due to the finite lens-galaxy distance ratio can be quite strong. The reason is that in a magnitude limited sample, the most luminous and therefore most massive objects are preferentially at the greatest distance, and these suffer most from the $(1 - w_l/w_g)$ suppression. One can make a quantitative estimate of the suppression if one assumes that $L \propto v^\beta$, in which case one finds for 'foreground' objects in a small magnitude slice around m_l and 'background' galaxies around m_g that $\langle \epsilon_T \rangle = 2\pi \alpha(m_l, m_g)\langle v^2 \rangle/c^2 \theta^{-1}$. The suppression factor is

$$\alpha(m_l, m_g) = \frac{\int\limits_0^\infty dw_g n_g \int\limits_0^{w_g} dw_l n_l w_l^{4/\beta}(1 - w_l/w_g)}{\int\limits_0^\infty dw_g n_g \int\limits_0^\infty dw_l n_l w_l^{4/\beta}}, \tag{3.7}$$

---

[3] It is interesting that the strength of the distortion for a flat rotation curve lens falls off inversely with separation θ, but the number of pairs increases as θ^2, so the signal to noise ratio for logarithmic bins is independent of θ.

[4] At very small angles TVJM found a marginally significant detection. This is thought to be an artefact of the image identification analysis (Tyson, private communication) and I have ignored this first point.

where n_l and n_g are just the differential distance distributions dn/dw for the foreground and background samples. To calculate the final average suppression factor is then simply a matter of performing an appropriate double integral over m_l and m_g. This calculation was made by Kovner and Milgrom (1987) and they estimated a suppression of a factor 3. Now since that estimate was made, considerable progress has been made in measuring redshifts for faint galaxies and I have recalculated the suppression using direct estimates of dn/dw obtained from Broadhurst *et al.*(1988), Colless *et al.*(1989) and Lilly *et al.* (1991). This exercise gives a suppression very similar to Kovner and Milgrom's estimate even though their assumption of constant space density for L_* galaxies is now seen to be quite inappropriate. Kovner and Milgrom concluded that with this correction the limit on v^2 for L_* galaxies is well above that measured by other means. I disagree with this; compared with our estimate of $\langle v^2 \rangle$ there is still a considerable discrepancy.

Another factor which might have suppressed the signal considerably is atmospheric seeing and imperfect focusing which would act to circularise the background galaxy images[5]. Indeed, since the number density of these galaxies is considerably higher than that of L_* galaxies today it is not unreasonable to imagine that what one is seeing are dwarf galaxies undergoing a burst of star formation, in which case they could be quite small. Now one indication that this effect is not enormous comes from the observations of by Lilly, Cowie and Gardner (1991) who find that their considerably fainter galaxies are at least marginally resolved. Another indication comes from the study of Tyson, Valdes and Wenk (1990) who have found significant distortion of galaxies behind clusters, again using fainter galaxies than those considered here. It is worth noting that both these studies used CCD photometry at CFHT rather than photography at Kitt Peak.

Another obvious factor which may help resolve the discrepancy is if our estimate of $\langle v^2 \rangle$ is too large. We may have been overgenerous in assigning early type galaxies extended halos, and even at $J \sim 21.5$, the limit for the 'foreground' galaxies, we may be dealing with a somewhat more numerous (and therefore presumably less massive) population that the present day bright galaxies from which our estimate of $\langle v^2 \rangle$ is made. If so, then this is bad news for people looking for quasar galaxy associations since these two probes measure the same quantity and, as we found in §3.2 of the previous lecture, the positive detections of quasar galaxy associations (if they are indeed attributed to lensing) would require a $\langle v^2 \rangle$ about a factor 20 *larger* than our estimate.

---

[5] Note that while seeing can considerably reduce the ellipticities from weak lensing this does not reduce the observability of the effect since the dominant source of noise – the random intrinsic ellipticities of the galaxies – is reduced by the same amount. It is an effect that must be carefully calibrated however.

To summarise, the failure of TVJM to measure a significant effect remains somewhat puzzling. Even allowing for the factor 3 reduction from finite distance ratio considerations, the observed result (which is compatible with the noise) is a factor 6 lower than that expected from properties of nearby galaxies (and, as already remarked, the cosmic virial theorem analysis would suggest that the untruncated flat rotation curve model is, if anything, conservative). The two likely culprits are poor seeing and the possibility that the lensing galaxies are less massive than galaxies studied locally. It would seem worthwhile to repeat this observation under better conditions in order to eliminate the former effect – though to obtain the required number of galaxy pairs using CCD photometry would be arduous to say the least. Another promising avenue is to do the same analysis using rich clusters as the foreground lenses and in this way probe $\xi_{c\rho}(r)$.

4 Auto-Correlation Analysis

We saw at the end of §2 that a single perturbation of scale λ and amplitude $\delta\rho/\rho$ produces an ellipticity $\epsilon \sim \lambda\overline{w}\delta\rho/\rho$. For $N \sim \overline{w}/\lambda$ such objects along the line of sight adding randomly one would expect a mean square ellipticity $\langle\epsilon^2\rangle \sim \lambda\overline{w}^3\langle(\delta\rho/\rho)^2\rangle$. This can be compared to the mean square projected density $w(\theta) \sim (\lambda/\overline{w})(\delta\rho/\rho)$, so we have $\langle\epsilon^2\rangle \sim \overline{w}^4 w(\theta)$ where $\overline{w}$ characterises the depth of the survey. The angular correlation function for the ellipticity can be calculated quantitatively (Blandford et al., (1991) hereafter BSBV; Miralda-Escude, (1991)) and expressed in terms of the power-spectrum of density fluctuations $P(k)$. A slightly different approach is to relate the 3-D spectrum to the angular power spectrum of the ellipticity – which is as easy to estimate as the angular correlation function – and this has some advantages; in any case, the angular power spectrum provides a fairly straightforward route to the angular ellipticity correlation function.

4.1 Power Spectrum Analysis

The angular power spectrum of e.g. the projected density (2.6) is defined by

$$\langle\Delta_p(\vec{\kappa})\Delta_p^*(\vec{\kappa}')\rangle = (2\pi)^2\delta(\vec{\kappa} - \vec{\kappa}')P_\Delta(\kappa) \tag{4.1.1}$$

where $\Delta_p(\vec{\kappa}) \equiv \int d^2\theta\,\Delta_p(\vec{\theta})\exp(i\vec{\kappa}\cdot\vec{\theta})$. The δ-function in (4.1.1) simply indicates that $\Delta_p(\vec{\theta})$ is a homogeneous and isotropic random field.

Limber's equation (Limber, 1954) relates the angular correlation of a projected quantity to the 3-dimensional correlation function. The analogue in Fourier space is

$$P_\Delta(\kappa) = \int_0^1 dw\frac{W^2(w)}{w^2}P(\kappa/w, w). \tag{4.1.2}$$

Similarly, we find for the two-point function of the shear $\Phi_{ij}(\vec{\kappa})$

$$\langle \Phi_{ij}(\vec{\kappa})\Phi_{lm}^*(\vec{\kappa}')\rangle = (2\pi)^2 \delta(\vec{\kappa} - \vec{\kappa}')\hat{\kappa}_i\hat{\kappa}_j\hat{\kappa}_l\hat{\kappa}_m P_\epsilon(\kappa), \qquad (4.1.3)$$

where $\hat{\kappa}_i \equiv \kappa_i/\kappa$, and therefore

$$\langle \epsilon_i(\vec{\kappa})\epsilon_j^*(\vec{\kappa}')\rangle = (2\pi)^2 \delta(\vec{\kappa} - \vec{\kappa}')\chi_i(\vec{\kappa})\chi_j(\vec{\kappa})P_\epsilon(\kappa), \qquad (4.1.4)$$

where the 'ellipticity power spectrum' is

$$P_\epsilon(\kappa) = 144 \int_0^1 dw \frac{g^2(w)}{w^2(1-w)^4} P(\kappa/w, w), \qquad (4.1.5)$$

where we have defined $\{\chi_1(\vec{\kappa}), \chi_2(\vec{\kappa})\} \equiv \{\hat{\kappa}_1^2 - \hat{\kappa}_2^2, 2\hat{\kappa}_1\hat{\kappa}_2\}$. Equations (4.1.4, 4.1.5) provide the basic link here between the observable $\epsilon_i(\vec{\kappa})$ and $P(k)$. Note that if we ignore the time dependence of the 3-D power spectrum then the angular spectra $P_\Delta(\kappa)$ and $P_\epsilon(\kappa)$ are simple linear convolutions of $P(k)$ in log-frequency (with kernels determined from the redshift distribution of the galaxies).

The angular dependence of the power spectra of the various ellipticity components through the $\chi_i(\vec{\kappa})$ factors in (4.1.4), can be understood in a qualitative manner. Consider the ϵ_1 component which describes stretching or compression along the coordinate axes. One would expect this to be excited most strongly by Fourier modes with wave vectors parallel to one or other axis, and not to be excited at all by modes along the diagonal direction, and since $\chi_1(\vec{\kappa}) = \cos^2 2\phi$, where ϕ is the angle of $\hat{\kappa}$ from the θ_1 axis, this expectation is fulfilled. This special form for the power spectrum results from the fact that the shear Φ_{ij} is not a general tensor; rather it is derived from a scalar $\Delta(\mathbf{w})$, and it is this which causes the characteristic dependence on direction in κ-space. This is not just a mathematical curiosity, but can be used, along with the general dependence on κ to help discriminate between gravitational stretching and artefacts.

If we Fourier transform $\langle \epsilon_i(\vec{\kappa})\epsilon_j^*(\vec{\kappa})\rangle$ we obtain the angular two-point functions $\langle \epsilon_i(\vec{\theta}')\epsilon_j(\vec{\theta}' + \vec{\theta})\rangle$, essentially as obtained by BSBV and Miralda-Escude though generalised slightly to allow for an arbitrary distance distribution for the galaxies and arbitrary evolution of the power spectrum. Like the power spectra, the ellipticity correlation functions have a hexadecapolar ($\cos 4\phi$) dependence on azimuthal angle ϕ, though here the amplitude of the angular modulation is model dependent. From a practical standpoint there is little to choose between power spectrum and real-space correlation analysis; one can readily construct estimators for both, and these have very similar precision. From the theoretical point of view there seem to be some advantages of the power spectrum analysis – though there is no harm in doing both – in that the angular power spectrum seems to be more simply related to the 3-D power spectrum, and the characteristic anisotropy of the ellipticity correlation is seen most clearly in the Fourier domain.

4.2 Relation to Other Probes of LSS

If we assume for simplicity that the galaxies all lie at a single distance w_S so $W(w) = \delta(w - w_S)$, and assuming linear theory evolution one finds from (4.1.4), (4.1.5) that the mean square ellipticity at a point is

$$\langle \epsilon^2 \rangle = \langle \epsilon_1^2 + \epsilon_2^2 \rangle = \int \frac{d^2\kappa}{(2\pi)^2} P_\epsilon(\kappa) = \frac{24\pi w_S^3}{5} \int \frac{d^3k}{(2\pi)^3} k^{-1} P(k). \qquad (4.2.1)$$

This agrees with equation 81 of BSBV for the mean square polarization at a point (i.e. $C_{\text{pp}}(\theta)$ evaluated at zero lag). Now this formula is very similar to the expression for the variance in peculiar velocities; the velocity for a mode $\Delta(\mathbf{k})$ is $\mathbf{v}(\mathbf{k}) = i\hat{\mathbf{k}}(H\lambda/2\pi)\Delta(\mathbf{k}) = 2i(\mathbf{k}/k^2)\Delta(\mathbf{k})$, so we have

$$\langle v^2 \rangle = 4 \int \frac{d^3k}{(2\pi)^3} k^{-2} P(k). \qquad (4.2.2)$$

It is also similar to the result of Sachs and Wolfe (1967) for the large angle microwave background anisotropy. Here, the fractional temperature fluctuation is just one third of the potential at the point of last scattering, so we have

$$\langle (\Delta T/T)^2 \rangle = 4 \int \frac{d^3k}{(2\pi)^3} k^{-4} P(k). \qquad (4.2.3)$$

We can use these equations, plus observational limits on the amplitudes of peculiar velocities and microwave anisotropies, to constrain the possible strength of gravitational distortions. On scales approaching the Hubble radius, the tightest constraint is from the microwave anisotropy limits which limit the rms distortion on these scales to be more than a few times 10^{-4} at most.

In the case of peculiar velocities we actually have a detection: The Local Group moves at $\simeq 600$ km/s in the microwave frame. Part of this motion derives from the attraction of the Local Supercluster, but the bulk of this is attributable to the pull of other neighbouring mass concentrations on somewhat larger scales. If our situation is representative then we have

$$\langle \epsilon^2 \rangle = 7 \times 10^{-3} w_S^3 \left(\frac{\lambda}{100 h^{-1} \text{Mpc}} \right)^{-1} \frac{\langle v^2 \rangle}{(700 \text{km/s})^2} \qquad (4.2.5)$$

which, for sources at $z = 1$ ($w_S = 0.293$), is 1.8×10^{-4}, or a rms stretch of $\simeq 1.3\%$.

4.3 Models

Figures 2a, 2b show the angular power spectra computed for two models for $P(k)$. The quantities plotted are actually the dimensionless quantities $\sigma^2(\kappa) \equiv \kappa^2 P(\kappa)/(2\pi)$, which give the contribution to the variance per log interval of κ. Figures 3a, 3b show the angular correlation functions. The first model is standard CDM with $h = 0.5$ and $b = 1.4$ (i.e. the linear theory density variance in a 8 Mpc/h top-hat sphere is $1/1.4^2$), and the second is a phenomenological model suggested by the IRAS power spectrum and has $P(k) \propto k(1 + (k/k_1)^4)^{-1/2}$.

In both cases, we see that we get an effect on the order of 10^{-4} at around $\theta = 1^0$, in accord with the rough estimate above. It is also apparent that, for either model, the shapes of the power spectra of ellipticity, projected density, and the cross-spectrum are very similar. The CDM model appears to have a well defined 'coherence angle'; i.e. $C_1(\theta)$ etc. become flat at zero lag, and this is also seen in the similar plots in BSBV and Miralda-Escude. However, this may be an artefact of using linear theory at small scales where non-linear clustering will have certainly set in, so these curves should really only be used to predict the large angle distortions.

At small scales a more interesting model is to assume that there are galaxies, with the well known $\xi \propto r^{-1.8}$ two-point function, each of which is surrounded be an extended dark halo. The ellipticity power spectrum for a model incorporating these features is shown in figure 3. At the highest spatial frequencies, greater than the inverse of the typical distance to a neighbour galaxy – this is about 300 kpc/h with the canonical number of $10^{-2}h^3\mathrm{Mpc}^{-3}$ for the space density of typical bright galaxies – the power spectrum of the galaxies is flat, so the power spectrum of the mass is $P(k) \propto k^{-2}$ and $\sigma_\epsilon^2(\kappa)$ is independent of frequency. At somewhat larger scales we start to see the effect of correlated neighbours; the galaxy power spectrum changes slope to $k^{-1.2}$ and that of the mass to $k^{-3.2}$, so σ_ϵ^2 starts to rise as $k^{-1.2}$. At some point we must reach the effective edge of a halo and at lower frequencies we should go over to the results obtained above assuming mass traces light. While it is uncertain quite how one matches these regimes together, from the amplitude of the fluctuations in the two bracketing regimes it appears that the peak power occurs at a frequency corresponding to $\simeq 1/(5 \mathrm{\ Mpc}/h)$ with maximum value $\sigma_\epsilon^2(\kappa) \simeq 10^{-3}$.

Let us estimate of what amount of data would be needed to make a detection. Consider a survey of side $\Theta = 1$ degree. At 26th magnitude, this would contain $\simeq 10^5$ galaxies (Lilly, Cowie and Gardner, 1991), and the precision on σ_ϵ^2 would be $\simeq 4 \times 10^{-6}(\kappa/200)$ per log interval of wavenumber (Kaiser, 1992), whereas the expected signal is $\simeq 3 \times 10^{-4}(\kappa/200)^{0.8}$, so this gives a formal signal to noise of about 100; a very strong signal indeed. Note that the limiting precision and the predictions from large scale structure have very similar κ-dependence, so one should be able to detect fluctuations simultaneously on a wide range of scales. At higher spatial frequencies we

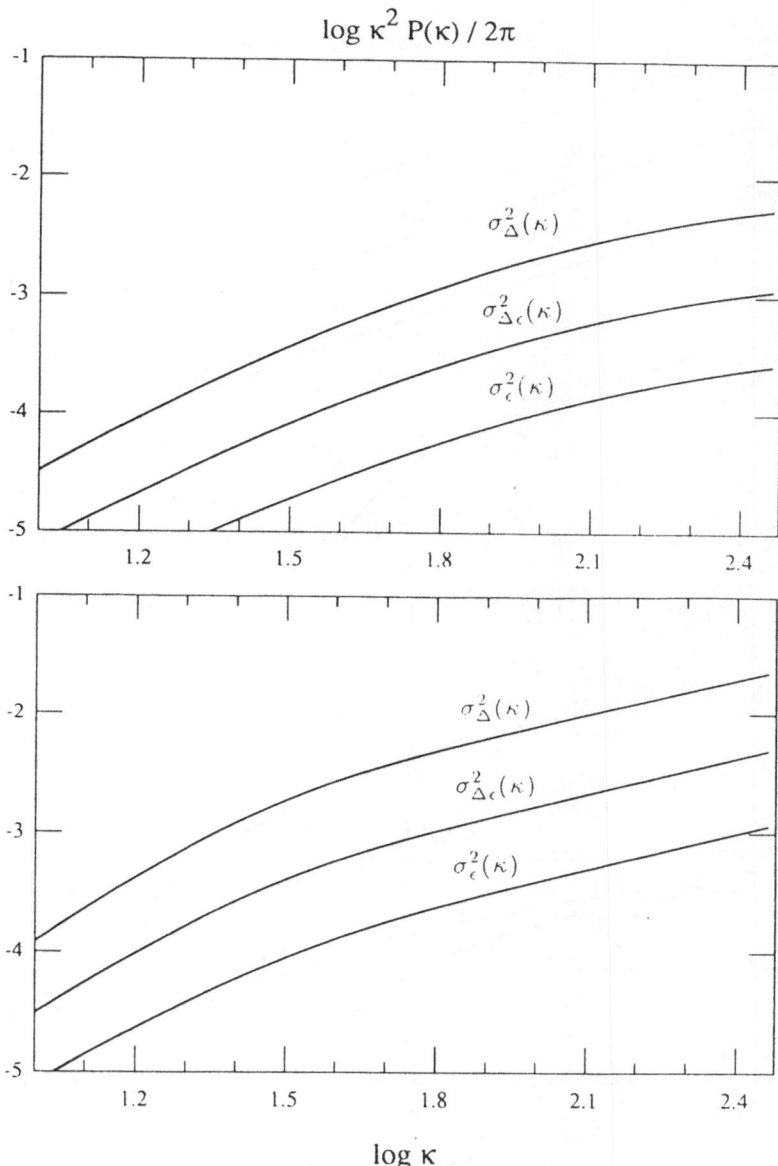

Fig. 1. Predicted angular power spectra for CDM model (upper panel) and the phenomenological model (lower panel). The quantities plotted give the contribution to the total variance per e-folding in angular frequency. The curves show the projected density power spectrum, the ellipticity power spectrum and the cross-spectrum (analogous to the cross-correlation function). I have assumed that the galaxy density fluctuations are measured for a sample of galaxies with roughly half the depth of the faint galaxies used for measuring the ellipticities.

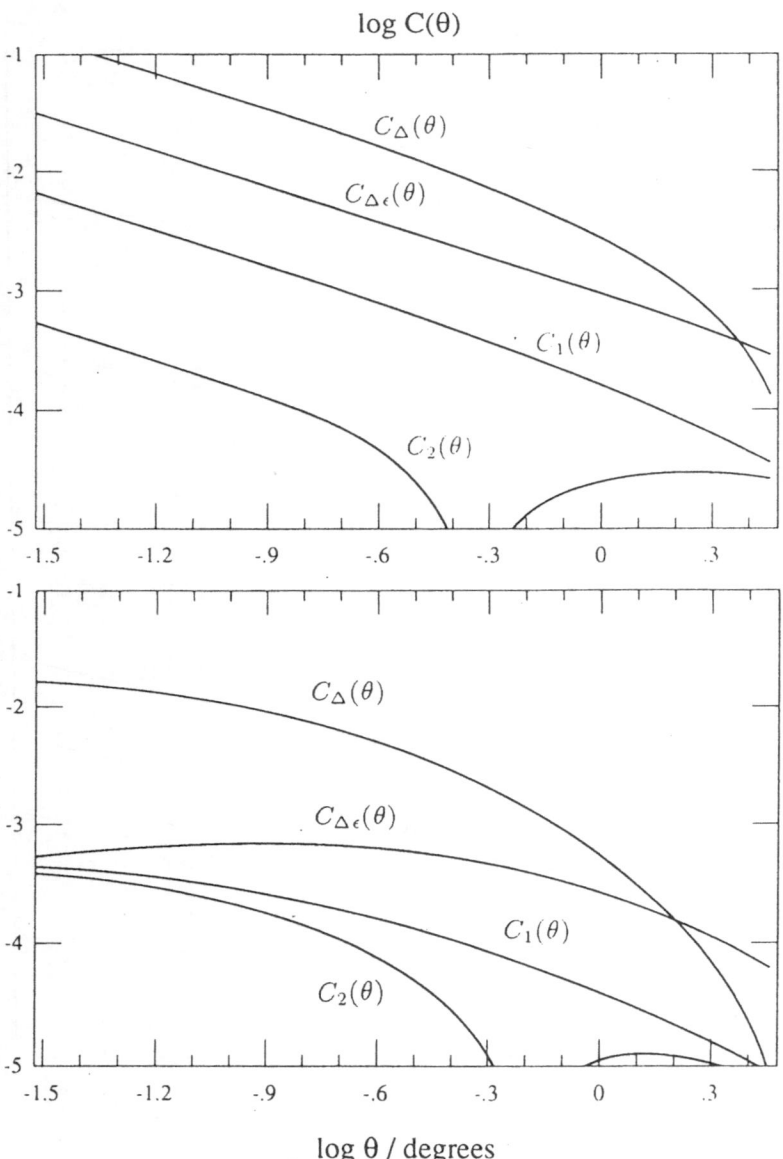

Fig. 2. Angular correlation functions obtained using the same models as in figure 1. Note that while the projected power spectra in figure 1 all have a very similar shape, the bend in the 3-D power spectrum results in weird bumps and wiggles in the angular correlation functions – particularly the ellipticity auto-correlation functions. It would be much easier to infer the behaviour of $P(k)$ from figure 1 than from figure 2.

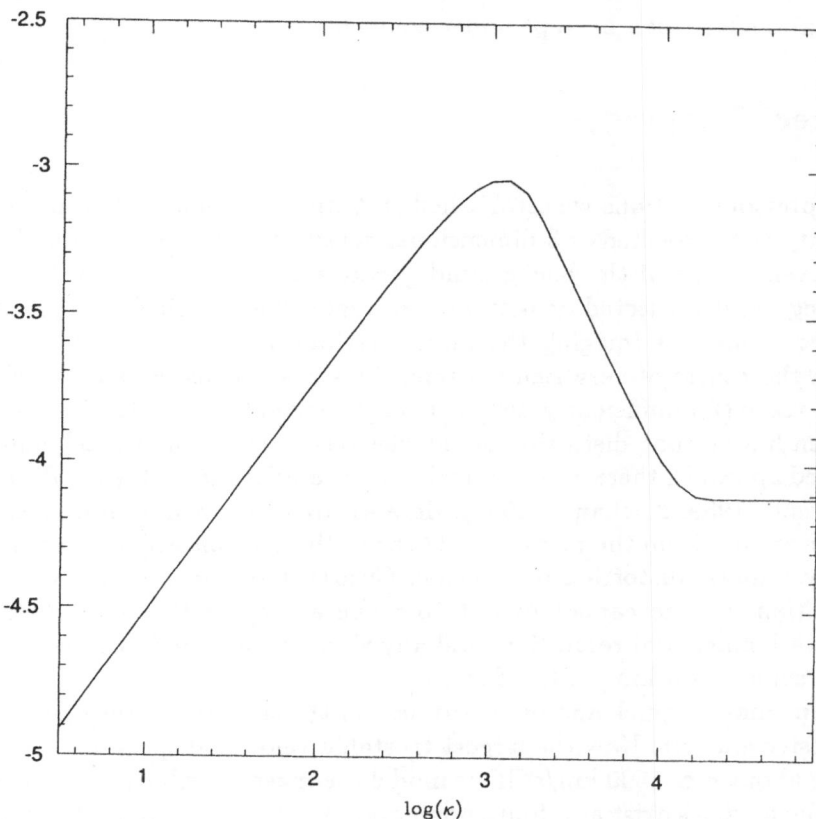

Fig. 3. Angular ellipticity power spectrum $\sigma_\epsilon^2(\kappa)$ for a model in which galaxies are distributed in space with the observed power-law two-point function, and have flat rotation curve haloes extending to $\simeq 5\,h^{-1}\mathrm{Mpc}$ (from Squires and Kaiser, 1992). The outer cut-off is chosen so that the total mass density equals the critical value. This model gives a plausible extrapolation of the model shown in the lower panel of figure 1 to higher frequencies. While the details of how real galaxy haloes are truncated is not known, and this affects the behaviour around the peak, the model should be fairly reliable for both lower and higher frequencies and shows that the total variance is $\simeq 10^{-3}$ at most (i.e. the rms shear or ellipticity is a few percent).

know the predicted power must turn over, so the autocorrelation technique would not be very powerful here. The normalisation for these predictions is really based on the assumption that the rms velocity from large-scale structure is about 700 km/s and that the mean distance to the galaxies is $\overline{w} \simeq 0.3$, but even if the mass is somewhat less strongly clustered, or if the distance is actually somewhat less there should no real obstacle from 'root-

N' statistics at least to detecting the effect. The real question is whether systematic effects can be kept under control.

5 Direct Imaging

In the previous sections we established statistical relations between the the ellipticity and projected or 3-dimensional density fields. Here we will discuss how, given a map of the background galaxy ellipticities, one can estimate the foreground projected density or *vice versa*. The particular application we have in mind is imaging the mass distribution in clusters of galaxies. Now in the centre of very rich clusters, the surface density can exceed the critical value (for sufficiently distant background galaxies at least), and one will then have strong distortion of galaxies resulting in 'giant arcs', and the preferred approach there is to constrain parameterised models (e.g. Hammer and Rigaut, 1989; Kochanek, 1990). Here we are concerned with weaker distortions arising from the mass outside the critical region which gives rise to weak systematic distortion or 'arclets'. Clearly, any one galaxy gives little information, so one cannot expect to make a map of the mass distribution with high spatial resolution, and any viable technique must necessarily involve some smoothing of the data.

Let us make a quick and dirty estimate of the strength of the distortions for massive clusters. Now the largest trustable velocity dispersions for clusters are about $\sigma \simeq 1500$ km/s. If we model the mass distribution as $\rho \propto r^{-2}$ and if the galaxies exist as a finite population within this potential well then $v_{\rm rot} = \sqrt{3}\sigma$. From (3.6) we found that for a $\rho \propto r^{-2}$ lens the strength is a diminishing function of lens distance, so let us assume we have chosen a reasonable nearby cluster and we then expect from (3.6) a tangential distortion $\epsilon_T = A\theta^{-1}$ with $A \simeq 97''$. The minimum uncertainty estimator for the strength of the lens is $\hat{A} = \sum_{\rm galaxies} \epsilon_T \theta^{-1} / \sum_{\rm galaxies} \theta^{-2}$ and the uncertainty in this estimator is $\langle (\hat{A} - A)^2 \rangle = \langle \epsilon_g^2 \rangle / (4\pi n \ln \theta_{\rm max}/\theta_{\rm min})$ where $\langle \epsilon_g^2 \rangle$ is the mean square intrinsic ellipticity for a galaxy and n is the density of background galaxies on the sky. If we take $n = 10^5$ per square degree and assume we probe one e-folding in radius then the noise is $\langle (\hat{A} - A)^2 \rangle^{1/2} = 3.2'' \langle \epsilon_g^2 \rangle^{1/2}$ so if the rms ellipticity for a galaxy is $\simeq 0.3$ then the cluster we have described would be detected at about the 100-σ level.

Tyson, Valdes and Wenk (TVW: 1990) have shown that, for some clusters at least, there does appear to be a significant measurable signal. Their approach is to construct a smoothed scalar measure of the stretching; the excess tangential alignment. The resulting maps of this quantity are, to my mind, very impressive and show that there is indeed a very strong signal present. Unfortunately, the scalar quantity calculated is not related in in a particularly simple way to the projected density. However, in angular frequency space it is relatively straightforward to construct the scalar surface density (or an unbiased estimate thereof) from the ellipticities (Kaiser, 1992). Moreover, since we know we will need to smooth the data, and this is easily done in Fourier space too, we kill two birds with one stone.

If we consider a single lens at some distance w_l then from (2.3) we have

$$\Phi_{ij}(\vec{\theta}) = g(w_l) \int dz \phi_{,ij}(w_l\theta_x, w_l\theta_y, z) = g(w_l)w_l^{-2}\frac{\partial^2\Psi(\vec{\theta})}{\partial\theta_i\partial\theta_j}, \quad (5.1)$$

where $\Psi(\vec{\theta}) = \int dz\, \phi(w_l\theta_1, w_l\theta_2, z)$ is the projected potential which satisfies the two-dimensional Poisson's equation

$$\frac{\partial^2\Psi}{\partial\theta_1^2} + \frac{\partial^2\Psi}{\partial\theta_2^2} = -4\pi Gw_l^2 \Sigma a_l. \quad (5.2)$$

This is easily obtained by projecting the 3-D version of Poisson's equation.

Fourier transforming (5.1), (5.2), we get

$$\Phi_{ij}(\vec{\kappa}) = -g(w_l)w_l^{-2}\kappa_i\kappa_j\Psi(\vec{\kappa}) \quad (5.3)$$

and

$$4\pi Gw_l^2 a_l \Sigma(\vec{\kappa}) = \kappa^2\Psi(\vec{\kappa}) \quad (5.4)$$

so

$$\Phi_{ij}(\vec{\kappa}) = 4\pi Ga_l g(w_l)\hat{\kappa}_i\hat{\kappa}_j \Sigma(\vec{\kappa}) \quad (5.5)$$

and therefore

$$\epsilon_i(\vec{\kappa}) = 4\pi Ga_l g(w_l)\chi_i(\vec{\kappa})\Sigma(\vec{\kappa}) \quad (5.6)$$

Given a map[6] of $\epsilon_i(\vec{\theta})$ one can Fourier transform, divide by $\chi(\vec{\kappa})$ and inverse transform to obtain the projected mass density $\Sigma(\vec{\theta})$. However, since $\chi_1(\vec{\kappa})$ and $\chi_2(\vec{\kappa})$ both vanish for particular directions it is preferable to use the weighted linear combination:

$$\Sigma(\vec{\kappa}) = \frac{\chi_1(\vec{\kappa})\epsilon_1(\vec{\kappa}) + \chi_2(\vec{\kappa})\epsilon_2(\vec{\kappa})}{4\pi Ga_l g(w_l)} \quad (5.7)$$

as the estimate of $\Sigma(\vec{\kappa})$. As already remarked, any method for calculating the projected density must invoke some smoothing of the data, and this may conveniently be performed as one transforms back from $\Sigma(\vec{\kappa})$ to $\Sigma(\vec{\theta})$.

An alternative use for this relation is to use the projected surface density for a suitable magnitude limited galaxy catalogue to predict the expected ellipticity pattern. Now the foregoing analysis assumed that the density enhancement was well localised around w_l. The analysis can be generalised and one finds an approximate relation between $\epsilon(\vec{\kappa})$ and the projected density of galaxies $\Delta_p(\vec{\kappa})$. The estimator for the ellipticity is

---

[6] The ellipticities are of course only defined at the locations of the galaxies. To construct an ellipticity field one might divide the sky up into a grid of cells and determine an average, or better still a median, ellipticity for each cell.

$$\tilde{\epsilon}_i(\vec{\kappa}) = (P_{\Delta\epsilon}/P_{\Delta})\chi_i(\vec{\kappa})\Delta(\vec{\kappa}). \tag{5.8}$$

where P_{Δ} is given by (4.1.2),

$$P_{\Delta\epsilon}(\kappa) = \int_0^1 dw \frac{W_{\mathrm{fg}}(w)g(w)}{w^2(1-w)^2} P(\kappa/w, w), \tag{5.9}$$

and W_{fg} is dn/dw for the foreground galaxies. The variance in (5.8) is

$$\frac{\langle(\tilde{\epsilon}_i(\vec{\kappa}) - \epsilon_i(\vec{\kappa}))^2\rangle}{\langle\epsilon_i(\vec{\kappa})^2\rangle} = 1 - \frac{P_{\Delta\epsilon}(\kappa)^2}{P_{\Delta}(\kappa)P_{\epsilon}(\kappa)} \tag{5.10}$$

and this is very small if the 'foreground' galaxies have a typical distance roughly half that of the background galaxies.

I have implicitly assumed above that the goal of all this is to map, or determine the power-spectrum of, the dark matter. In this picture the faint galaxies merely provide a convenient backdrop – the 'cosmic wallpaper' – to reveal the lenses. An alternative and equally legitimate goal is to use these studies as a probe of the nature of the cosmic wallpaper itself. For instance, TVW have claimed from their analysis of weak lensing by A 1689 and CL 1409+52 that they can put a lower limit of $z = 0.9$ on the mean redshift for the background population. While I think there is something in this line of argument, there are clearly a number of problems that need to be overcome before one can get quantitative estimates of the distance to the background galaxies in this way. One factor is the reduction of the ellipticity by seeing which needs to be calibrated and which requires a good understanding of the distribution of sizes of the galaxies (though it should be said that incorporating this effect will only push up the estimated distance). Another problem is that one really needs to have a good understanding of the lens properties. The conventional approach is to compare the strength of the lens with that expected based on the line of sight velocity dispersion. This seems to be comparing apples with oranges since if the clusters are selected primarily on the basis of surface luminosity density then one expects that the richest clusters found – and particularly the subset which show a strong measurable effect – will be structures which are elongated along the line of sight and it is not clear how to predict the surface density from the velocity dispersion; one might well do better just using the surface luminosity density and adopting a sensible mass-to-light ratio.

I think it is highly questionable whether it will be possible to obtain sufficiently precise estimates of the lens strength (from velocity dispersions or photometry of the cluster) to really pin down the distance to the background galaxies. What is more promising is the possibility of measuring the relative distances to the background galaxies as a function of luminosity. For instance, one might measure the distortion for a 'bright' and 'faint' subset of the background galaxies – this could be done for a single cluster, or better still using a collection of clusters at similar distances – and the ratio, from (5.7), (2.5) gives a direct measure of the relative values of $\int_{w_l} dw\,(1 - w_l/w)dn/dw$. If the redshift distribution for the brighter subset were known then this would tell us the distance to the fainter subset and if one had redshifts for both subsets one could hope to constrain cosmological parameters.

6 Concluding Remarks

We have reviewed techniques for measuring mass fluctuations on a wide range of scales from galaxy haloes through rich clusters to large-scale structure. While the failure to detect systematic distortions caused by galaxy haloes remains somewhat puzzling, the spectacular results of TVW for rich clusters show that here at least one has a practical tool for detecting dark matter. What I find particularly exciting about this approach is the possibility of probing the mass distribution much farther out in the cluster halo, and this should provide powerful constraints on theories for cluster and galaxy formation. Another exciting prospect is the possibility of measuring the power spectrum of mass fluctuations $P(k)$ from the auto-correlation function of ellipticities for randomly chosen fields. While this is largely unexplored territory, it should be kept in mind that the normalisation for the predictions here rests on the assumption that the gravity of large-scale structure drives streaming motions with rms amplitude $\simeq 700$ km/s. If this is correct then distortions must be present at an observable level. The observations necessary to reach the required precision will be arduous, and there will no doubt be many hurdles to overcome, but the scientific payoff from such a programme is enormous.

Rererences

Blandford, R.D., Saust, A.B., Brainerd, T., and Villumsen, J.V., 1991. (BSBV) MNRAS, **251**, 600.

Broadhurst, T., Ellis, R. and Shanks, T., 1988. MNRAS, **235**, 827.

Colless, M., Ellis, R., Taylor, K. and Hook, R., 1989. MNRAS, **244**, 408.

Davis, M. & Peebles, P.J.E., 1983. Ap.J., **267**, 465.

Fukugita, M. and Turner, E.L., 1991. MNRAS in press.

Gunn, J.E., 1967a. Ap.J., **147**, 61.

Gunn, J.E., 1967b. Ap.J., **150**, 737.

Hammer, F. and Rigaut, F., 1989. Astr. Astrophys., **226**, 45.

Kaiser, N., 1992. Ap. J. in press.

Kochanek, C., 1990. MNRAS, **247**, 135.

Kovner, I and Milgrom, M, 1987. Ap.J., **321**, L113.

Lilly, S.J., Cowie, L.L., and Gardner, J.P., 1991. Ap.J., **369**, 79.

Limber, D.N., 1954. Ap.J., **119**, 655.

Miralda-Escude, J., 1991. Ap. J., **380**, 1.

Peebles, P.J.E., 1980. "The Large Scale Structure of The Universe", Princeton: Princeton University Press.

Sachs, R.K., and and Wolfe, A.M., 1967, Ap.J., **147**, 73.

Squires, G. and Kaiser, N., 1992. In preparation

Turner, E., 1989. In 14th Texas Symposium on Relativistic Astrophysics.

Tyson, J., Valdes, F., and Wenk, R., 1990. Ap.J., **349**, L1. (TVW)

Tyson, J., 1991. In proceedings of "The 1st 3 Minutes", Ed. S. Holt.

Tyson, J., Valdes, F., Jarvis, J., and Mills, A., 1984. Ap. J., **281**, L59 (TVJM).

Valdes, F., Tyson, J.A., and Jarvis, J.F., 1983. Ap.J., **271**, 431.

Lecture Notes in Physics

For information about Vols. 1–374
please contact your bookseller or Springer-Verlag

New Series m: Monographs